D0025018

NUCLEAR AMERICA

This bibliography was conceived and compiled from the periodicals database of the American Bibliographical Center by editors at ABC-Clio Information Services.

Lance Klass, project coordinator

Susan Kinnell
Robert de V. Brunkow
Jeffery B. Serena

Pamela R. Byrne
Gail Schlachter

ABC-Clio Information Services

Santa Barbara, California
Oxford, England

Library of Congress Cataloging in Publication Data
Main entry under title:

Nuclear America, a historical bibliography.

Includes index.
1. Atomic weapons–Bibliography. 2. Atomic weapons and disarmament–Bibliography. 3. Atomic energy–Bibliography. I. ABC-Clio Information Services.
Z6724.A9N8 1983 [U264] 016.3558'25119 83-12227
ISBN 0-87436-360-8

ABC-Clio Information Services
2040 Alameda Padre Serra, Box 4397
Santa Barbara, California 93103

Clio Press Ltd.
55 St. Thomas Street
Oxford 0X1 1JG, England

Cover design and graphics by Lance Klass and Susan Kinnell
Printed and bound in the United States of America

ABC-CLIO RESEARCH GUIDES

The ABC-Clio Research Guides are a new generation of annotated bibliographies that provide comprehensive control of the recent journal literature on high-interest topics in history and related social sciences. These publications are created by editor/historians and other subject specialists who examine every article entry in ABC-Clio Information Services' vast history database and select abstracts of all citations published during the past decade that relate to the particular topic of study.

Each entry selected from this database—the largest history database in the world—has been reedited to ensure consistency in treatment and completeness of coverage. The extensive subject profile index (ABC-SPIndex) accompanying each volume has also been reassessed, specifically in terms of the particular subject presented, to allow precise and rapid access to the entries.

The titles in this series are prepared to save researchers, students, and librarians the considerable time and expense usually associated with accessing materials manually or through online searching. ABC-Clio's Research Guides offer unmatched access to significant scholarly articles on the topics of most current interest to historians and social scientists.

ABC-CLIO RESEARCH GUIDES

Gail Schlachter, Editor
Pamela R. Byrne, Executive Editor

1.
World War II from an American Perspective
1982 LC 82-22823 ISBN 0-87436-035-8

2.
The Jewish Experience in America
1982 LC 82-24480 ISBN 0-87436-034-x

3.
Nuclear America
1983 LC 83-12227 ISBN 0-87436-360-8

4.
The Great Depression
1983 LC 83-12234 ISBN 0-87436-361-6

5.
Corporate America
1983 LC 83-11232 ISBN 0-87436-362-4

6.
Crime and Punishment in America
1983 LC 83-12248 ISBN 0-87436-363-2

7.
The Democratic and Republican Parties
1983 LC 83-12230 ISBN 0-87436-364-0

8.
The American Electorate
1983 LC 83-12229 ISBN 0-87436-372-1

CONTENTS

INTRODUCTION . . . xi

1. THE ROAD TO HIROSHIMA . . . 1
2. THE DEVELOPMENT OF NUCLEAR WEAPONRY . . . 13
3. THE BALANCE OF TERROR . . . 33
4. ATTEMPTS AT NUCLEAR ARMS CONTROL . . . 77
5. NUCLEAR REACTORS AND PUBLIC REACTION . . . 115

SUBJECT INDEX . . . 137

AUTHOR INDEX . . . 180

LIST OF ABBREVIATIONS . . . 184

INTRODUCTION

BACKGROUND

The first controlled nuclear chain reaction took place in a secret laboratory built under the stands of Stagg Field at the University of Chicago in 1942. That signal event created its own reaction, as the scientists who had labored to unravel the mysteries of the atom came to appreciate both the awesome power they had helped unleash and their responsibility for controlling its applications.

In those early years it was argued that atomic energy offered a safe, inexpensive, and virtually inexhaustible source of power which could usher in a golden age of global prosperity. In the military sphere, it was believed that the terrible power of atomic weaponry would make another world war unthinkable and would thus bring about an era of international peace. Yet the development of the Cold War and the encystment of East and West into nuclear-armed power blocs led not to world peace but to a fragile balance of terror that continues to threaten mankind with annihilation. The development of peacetime applications of nuclear energy, moreover, has brought problems of radioactive contamination, nuclear waste disposal, and industrial malfunction, as the accident at Three Mile Island showed so clearly.

These issues have given rise to a vast body of scholarship devoted to recording, analyzing, and understanding the controversies surrounding nuclear energy. As a result, there has developed a critical need for bibliographical control of the literature. *Nuclear America: A Historical Bibliography* is designed to be an indispensable research tool for all who are interested in the promise and the perils of nuclear power. It encapsulates a decade of research as published in the journal literature worldwide and gives focus to this complex, and terribly important, area of study.

SCOPE

The 824 abstracts of articles presented in this volume were drawn from ABC-Clio Information Services' data base—the largest history data base in the world—which includes abstracts of scholarship in more than 2,000 journals in 42 different languages, published in 90 countries. All of the entries in this volume concern the United States directly; none concerns nuclear energy or weapons development in other countries except as related to the United States or to American interests.

In order to create this unique bibliographical volume, the editors did an exhaustive search through every one of the many thousands of abstracts of articles written from 1973 to 1982, selecting every entry that related to the subject. Thus the present volume represents the broad range of scholarship on the topic of nuclear energy in America and far exceeds what one could expect to find through an online search of the data base or through even an exhaustive manual search of the subject index for the ABC-Clio Information Services history data base as a whole.

ORGANIZATION

The abstracts are organized into five chapters. The variance in size of these chapters represents not any predisposition of the editors but rather the amount of scholarship on each topic that was published in the journal literature during the decade of scholarship covered by this volume.

The first chapter, "The Road to Hiroshima," includes studies relating to the development and use of the atomic bombs dropped on Hiroshima and Nagasaki in 1945, and the costs and consequences of those attacks. It covers the early years of atomic research, the Manhattan Project, the decision to use the bomb against Japan, the actual Enola Gay mission which dropped the bomb, the effects of the nuclear attacks on Hiroshima and Nagasaki and on the course of the war, and the aftermath, both human and political, of this first offensive use of atomic weapons in warfare.

Chapter 2, "The Development of Nuclear Weaponry," encompasses studies of the development of atomic weapons in the United States after 1945 and their effects on American society. This chapter covers the continued development, refinement, and testing of atomic bombs, the development of the hydrogen bomb and the neutron bomb, and the institutionalization of atomic weapons research. Chapter 2 also includes studies of the concurrent development of nuclear carrying systems and related tactical weaponry and the use of nuclear energy to power the Rickover generation of American naval ships. The societal issues of nuclear weapons development are also addressed, from the days of the Oppenheimer case and the Rosenberg atomic secrets trial, to the furor surrounding the threat of radiation from atomic blasts, fallout shelters, the ABM and MX controversies, the ecological and human effects of surface and subsurface bomb tests, and the growing realization by the American people of the responsibilities and threats nascent in nuclear weapons development.

The third chapter, "The Balance of Terror," includes studies that focus on the growth and institutionalization of the nuclear arms race within the context of the Cold War. It focuses on strategic studies of nuclear deterrence, confrontation politics, brinkmanship, the question of nuclear parity, the strategic deployment of nuclear weapons and delivery systems, and the expanding nuclear arsenals in the United States and the USSR. The selection of materials in this chapter focuses not on generalized Cold War studies or those dealing with international tension, but on scholarship that highlights and clarifies the central role of doomsday weapons and the threat of their use in the polarization of international politics. Studies of the Cuban Missile Crisis are not included unless they deal specifically with the threat of nuclear war or with the incisive change that occurred in the general level of nuclear tension between the US and USSR that resulted from the crisis. Studies of nuclear proliferation are included only when they deal with an implicit threat of nuclear weapons development, as in the case of India, or with American military or diplomatic involvement.

The fourth chapter, "Attempts at Nuclear Arms Control," includes materials which discuss international efforts to retreat from the brink of nuclear cataclysm. The chapter covers such topics as the Helsinki accords, the Strategic Arms Limitation Treaties, the drive for mutual balanced force reductions where related to nuclear weapons or the threat of their use, efforts toward unilateral nuclear disarmament, domestic protest movements against nuclear arms, attempts by the American government to limit or control the proliferation of nuclear weapons, and other tentative or significant steps toward arms reduction. Studies of the European antinuclear movements are not included unless they specifically deal with American policy, involvement, or activities.

The final chapter, "Nuclear Reactors and Public Reaction," includes studies of the development of atomic power as a viable energy alternative, the issue of atomic reactor proliferation when related primarily to the availability of nuclear energy, the disposal of nuclear wastes, fusion energy research, the threat of nuclear terrorism, protests against nuclear reactors and nuclear energy development, the question of safeguards, and the accident at Three Mile Island.

These chapters are followed by ABC-SPIndex—one of the most advanced and comprehensive indexing systems yet developed, and allows fast, analytical, pinpoint access by the user. ABC-SPIndex links together the key subject terms and historical period of each abstract to form a composite index entry that provides a complete subject profile of the journal article. Each set of index terms is then rotated so that the complete profile appears in the index under each of the subject terms. Thus the number of access points is increased severalfold over conventional hierarchical indexes, and irrelevant material can be eliminated early in the search process, often without recourse to the abstract or article itself. The explanatory note at the beginning of the subject index provides more information about ABC-SPIndex.

Great care has been taken to eliminate inconsistencies that might have appeared in the subject index as a result of combining a decade of data base material on this special subject. The subject profile index thus reflects a highly labor-intensive effort—the product of many hours of editing and reediting—and allows easy access to the materials included in this volume. Additional cross-references have been added to ensure fast and accurate searching.

1

THE ROAD TO HIROSHIMA

1. Abelson, Philip H. A SPORT PLAYED BY GRADUATE STUDENTS. *Bull. of the Atomic Scientists 1974 30(5): 48-52.* Reminisces about working in the University of California Lawrence Radiation Laboratory, 1935-40, in nuclear physics, primarily in transuranic elements.

2. Adamson, June. FROM BULLETIN TO BROADSIDE: A HISTORY OF BY-AUTHORITY JOURNALISM IN OAK RIDGE, TENNESSEE. *Tennessee Hist. Q. 1979 38(4): 479-493.* The Oak Ridge *Journal* was developed under the authority of the Army Corps of Engineers to serve the atomic energy project at Oak Ridge, Tennessee. Because the project was top secret, the paper carried mainly news of stores, public health clinics, churches, and innocuous local events. Covers 1943-49. Based on personal interviews, letters, and the *Journal;* 31 notes.
W. D. Piersen

3. Almaráz, Felix Díaz, Jr. THE LITTLE THEATRE IN THE ATOMIC AGE: AMATEUR DRAMATICS IN LOS ALAMOS, NEW MEXICO 1943-1946. *J. of the West 1978 17(2): 72-82.* The development of the little theatre in Los Alamos was a response to the social and cultural needs of the public employees assigned to this isolated community. A company of amateur performers organized the Little Theatre Group on 8 November 1943. Some initial difficulties were encountered in scheduling rehearsal space and performance dates, but audience response was enthusiastic and the group's treasury reflected its success. Under the guidance of dedicated leaders, the amateur actors experimented with a variety of drama projects. Minutes of the Board of Directors of the Los Alamos Little Theatre and secondary sources; illus., 4 photos, 29 notes.
B. S. Porter

4. Alvarez, Luis W. BERKELEY: A LAB LIKE NO OTHER. *Sci. and Public Affairs 1974 30(4): 18-23.* In 1934 the author, a graduate student in physics at the University of Chicago, visited the Ernest Lawrence Radiation Laboratory in Berkeley, California.

5. Anderson, Herbert L. EARLY DAYS OF THE CHAIN REACTION. *Sci. and Public Affairs 1973 29(4): 8-12.* Reminisces about the first controlled nuclear chain reaction, in a laboratory at the University of Chicago in 1942 where the first nuclear pile was built.

6. Anderson, Herbert L. [FERMI, SZILARD, AND ATOMIC ENERGY]. THE LEGACY OF FERMI AND SZILARD. *Bull. of the Atomic Scientists 1974 30(7): 56-62.* Discusses Leo Szilard's and Enrico Fermi's research on nuclear fission during 1933-39, particularly at Columbia University.
FERMI, SZILARD AND TRINITY. *Bull. of the Atomic Scientists 1974 30(8): 40-47.* Reminiscences on working with Fermi and Szilard at Columbia, at the University of Chicago collaborating on the first controlled chain reaction, and at Los Alamos, New Mexico, 1940-45.

7. Badash, Lawrence. RADIUM, RADIOACTIVITY, AND THE POPULARITY OF SCIENTIFIC DISCOVERY. *Pro. of the Am. Phil. Soc. 1978 122(3): 145-154.* Investigates the causes of the popularity in Europe and the United States of radium and radioactivity, 1898-1935. People wrote poems, staged parties, named race horses, and invented games in their honor. Numerous articles in newspapers and magazines attempted to explain the phenomena, or at least to make wild claims for their powers. Other scientific discoveries, equally important, failed to attain such notoriety. The appeal of radium and radioactivity was rooted in the culture of the past, when magical claims and unseen powers had never really lived up to public expectations, a task the newcomers seemed destined to finish. 65 notes. V. L. Human

8. Bainbridge, Kenneth T. A FOUL AND AWESOME DISPLAY. *Bull. of the Atomic Scientists 1975 31(5): 40-46.* Recounts participation in the Manhattan Project construction of the first nuclear bomb at Los Alamos, New Mexico, and his impressions of its detonation in 1945.

9. Baldwin, Paul H. THE ENDING OF WORLD WAR II. *Aerospace Hist. 1976 23(3): 136-139.* The author graduated in physics from Carnegie Institute of Technology in 1937 and applied to the Mellon Institute of Industrial Research for work as an atomic physicist. He felt that Hitler would be the one to use the atomic bomb in World War II. During the war the author was a Radar Observer with the 547th Night Fighter Squadron in the Philippines. Describes the training and utilization of night fighter squadrons to intercept enemy fighters. The night fighter operation and the A-bomb would have helped end the war sooner. President Truman made the correct decision in using the bomb. 3 photos.
C. W. Ohrvall

10. Barnett, Lincoln. HOW ALBERT EINSTEIN, A GENTLE, MODEST GENIUS BORN 100 YEARS AGO, FOUND SANCTUARY AND INSPIRATION IN PRINCETON. *Smithsonian 1979 9(11): 68-79.* Einstein decided to accept an appointment to the newly established Institute for Advanced Study in Princeton, New Jersey, in 1933 after observing with horror the Nazi assault on the intellect. He renounced his German citizenship while he was a visiting professor at Caltech, shortly after Hitler became chancellor of Germany. In 1939 he wrote President Roosevelt a prophetic letter warning about the atomic bomb. It was a prologue to the Manhattan Project and Hiroshima. Illus.
E. P. Stickney

11. Bernstein, Barton J. HIROSHIMA RECONSIDERED—THIRTY YEARS LATER. *Foreign Service J. 1975 52(8): 8-13, 32-33.* Reviews the decision to use atomic warfare against Japan in 1945. S

12. Bernstein, Barton J. THE PERILS AND POLITICS OF SURRENDER: ENDING THE WAR WITH JAPAN AND AVOIDING THE THIRD ATOMIC BOMB. *Pacific Hist. Rev. 1977 46(1): 1-27.* The ambiguous American response to Japan's 10 August 1945 surrender offer strengthened the militarists in Japan and nearly prolonged the war. President Truman and Secretary of State James F. Byrnes were reluctant to retain the Emperor. Byrnes and Truman were concerned about domestic political effects and feared a popular backlash if the surrender terms were not harsh enough. They were therefore willing to consider using a third atomic bomb or mounting a costly invasion of Japan. Secretary of War Henry L. Stimson and Admiral William Leahy urged acceptance of Japan's surrender terms in order to end the war quickly, keep Russia out of the peace settlement, and avoid world-wide horror at the use of a third atomic bomb. Based on documents in numerous manuscript collections; 92 notes.

W. K. Hobson

13. Bernstein, Barton J. THE QUEST FOR SECURITY: AMERICAN FOREIGN POLICY AND INTERNATIONAL CONTROL OF ATOMIC ENERGY, 1942-1946. *J. of Am. Hist. 1974 60(4): 1003-1044.* Overriding the objections of some scientists, the administration of Franklin D. Roosevelt made the decision *not* to share atomic secrets with the USSR. Henry Stimson considered promising such secrets in return for an opening of Soviet society, and had F.D.R. lived he might quite possibly have used "atomic diplomacy" to ease America's postwar role. President Truman later saw the United States as a trustee for the awesome weapon. Although at times he seemed disposed to listen to Stimson and Acheson who suggested a more open and direct approach to the Soviet Union, he was averse to sharing secrets that might end America's nuclear monopoly. The Acheson-Lilienthal report was so amended by Bernard Baruch (with provisions for inspections and giving up veto power in the UN) as to make it obviously unacceptable to the Soviet military chief. The climate of political opinion in 1946 made it difficult for the United States to offer any proposal which might have been acceptable to the Soviet Union. The controversy over atomic energy was both cause and consequence of the Cold War. 81 notes.

K. B. West

14. Bernstein, Barton J. ROOSEVELT, TRUMAN, AND THE ATOMIC BOMB: A REINTERPRETATION. *Pol. Sci. Q. 1975 90(1): 23-69.* Reinterprets the Roosevelt and Truman policies on the construction and use of atomic weapons. Concludes that FDR's decisions to treat the bomb as a legitimate weapon and to exclude the Soviets from a nuclear partnership prepared the way for the Truman administration's atomic bombing of Japan, the toughening of policy at Potsdam, and the postwar practice of atomic diplomacy. J

15. Bernstein, Barton J. SHATTERER OF WORLDS: HIROSHIMA AND NAGASAKI. *Bull. of the Atomic Scientists 1975 31(10): 12-22.* US policymakers in the Roosevelt and Truman administrations had little doubt about the desirability of using atomic weapons both for ending the war in the Pacific and for intimidating the USSR.

16. Bernstein, Barton J. THE UNEASY ALLIANCE: ROOSEVELT, CHURCHILL, AND THE ATOMIC BOMB, 1940-1945. *Western Pol. Q. 1976 29(2): 202-230.* Reinterprets the wartime Anglo-American relationship on

atomic energy, defined primarily by Roosevelt and Churchill. Atomic energy represented, potentially, the cornerstone of a postwar Anglo-American entente in which only these "two policemen" (the United States and Great Britain) would have the atomic bomb. Roosevelt was a shrewd administrator in this important area, as revealed by his wartime foreign policy: his understanding of power, his attitudes toward the Soviet Union, his view of the United Nations, and his expectations about the postwar world. He was not naïve or innocent, but astute, about power in international affairs. He was not a Wilsonian internationalist but a firm believer in big-power politics. Based on the recently declassified American and British archives. J/S

17. Boller, Paul F., Jr. HIROSHIMA AND THE AMERICAN LEFT: AUGUST 1945. *Int. Social Sci. Rev. 1982 57(1): 13-28.* Many Leftists insist today that Japan was thoroughly beaten in August 1945 and that the United States dropped atomic bombs mainly to intimidate the Soviet Union. Careful examination of American Leftist opinion in 1945, however, reveals that the strongest defenders of the atomic bombing of Hiroshima and Nagasaki were the groups friendliest to Stalinist Russia and that the bitterest critics were anti-Stalinist liberals and radicals. The former also supported and the latter condemned a harsh unconditional-surrender policy toward Japan. J

18. Daniels, Gordon. BEFORE HIROSHIMA: THE BOMBING OF JAPAN, 1944-5. *Hist. Today [Great Britain] 1982 32(Jan): 14-18.* Describes the bombing of Nagoya, Osaka, Kobe, Okinawa, and Iwojima by American bombers, which resulted in the death of 200,000 and the razing of cities despite Japan's civil defense preparations, which began in 1937 with the Air Defense Law and with preparation training by the Greater Japan Air Defense Association, beginning in 1939.

19. Eggleston, Noel C. ROLE PLAYING: THE ATOMIC BOMB AND THE END OF WORLD WAR II. *Teaching Hist. A J. of Methods 1978 3(2): 52-58.* Describes in detail the development of an historical role playing exercise for classroom use revolving around the question, "Should the United States drop an atomic bomb on Japan?" Discusses setting in 1945, character participants, discussion questions, and the historical works used as well as the specific goals, results, and benefits of this exercise. Examines the value of role playing technique for history instructors. 8 notes.

20. Ekebjär, Göran. PROBLEM RÖRANDE ATOMBOMBINSATSEN MOT JAPAN 1945 [The problem concerning the atomic bombing of Japan in 1945]. *Kungliga Krigsvetenskaps Akademiens Handlinger och Tidskrift [Sweden] 1965 169: 469-486.* Describes the events leading up to the atomic bombing of Hiroshima and Nagasaki in 1945. The bombs were dropped in the belief that they would prevent a longer war and thus save many lives. The Americans, however, did not consider alternative military strategies; and Japan, weakened by years of war and a shortage of raw materials, would have capitulated within five months at the most. Secondary sources; biblio. U. G. Jeyes/S

21. Feld, Bernard T. EINSTEIN AND THE POLITICS OF NUCLEAR WEAPONS. *Bull. of the Atomic Scientists 1979 35(3): 5-16.* Chronicles Albert Einstein's work on the creation of nuclear arms, 1939-42, which he believed to

be a necessary deterrent because of his Zionism and his fear of Nazi Germany, and his efforts to control and discourage the further use of the weaponry, which was rooted in his basic pacifism, 1943-55.

22. Fitch, Val L. THE VIEW FROM THE BOTTOM. *Bull. of the Atomic Scientists 1975 31(2): 43-46.* An enlisted soldier present at the first test of the atomic bomb at Los Alamos, New Mexico, 1945, describes the preparations and his impressions of that portentous explosion.

23. Freedman, Lawrence. THE STRATEGY OF HIROSHIMA. *J. of Strategic Studies [Great Britain] 1978 1(1): 76-97.* Follows the Roosevelt Administration's decisionmaking on the operational use of the first atomic bombs. Cities were attacked because of the lack of significant military targets and to increase the bomb's shock value over conventional strategic bombing, and not for experimental value or to intimidate the Soviet Union. The bombing of Nagasaki was the logical extension of the decision to use the bomb in the first place. Primary sources; 45 notes. A. M. Osur

24. Frisch, Otto R. "SOMEBODY TURNED THE SUN ON WITH A SWITCH." *Sci. and Public Affairs 1974 30(4): 12-18.* The author, a Danish physicist, discusses his career in nuclear physics and the construction of the first atomic bomb at Alamogordo, New Mexico, in 1945.

25. Gowing, Margaret. REFLECTIONS ON ATOMIC ENERGY HISTORY. *Bull. of the Atomic Scientists 1979 35(3): 51-54.* Secrecy surrounding nuclear arms testing and experimentation caused ill feelings between scientific communities in Great Britain and France and those in the United States, since progress was closely guarded even against allies, 1939-45.

26. Halász, Nicholas and Halász, Robert. LEO SZILÁRD, THE RELUCTANT FATHER OF THE ATOM BOMB. *New Hungarian Q. [Hungary] 1974 15(55): 163-173.* Leo Szilárd (1898-1964), a Hungarian émigré in the United States, was one of three Hungarian American scientists who induced Albert Einstein to sign the famous letter to President Franklin D. Roosevelt urging establishment of an American atomic weapons program during World War II; later he became a crusader for world peace and disarmament.

27. Hammond, Thomas T. "ATOMIC DIPLOMACY" REVISITED. *Orbis 1976 19(4): 1403-1428.* The theses of Gar Alperovitz in his *Atomic Diplomacy—Hiroshima and Potsdam* (New York: Simon and Schuster, 1965), "are either implausible, exaggerated or unsupported by the evidence" and do not stand up under careful analysis. 86 notes. A. N. Garland

28. Hewlett, Richard G. BEGINNINGS OF DEVELOPMENT IN NUCLEAR TECHNOLOGY. *Technology and Culture 1976 17(3): 465-478.* In both the Manhattan Project and the early Atomic Energy Commission, tension arose between the scientists, who wished to keep charge of the projects they had begun, and the engineers, who were better able to bring them to fruition. The "engineer is better qualified both technically and psychologically than the scientist to determine when the change of command should occur." Based on published works by the author; 23 notes. C. O. Smith

29. Horder, Mervyn. IF NO ATOM BOMB? *Blackwood's Mag. [Great Britain] 1975 318(1918): 110-115.* Discusses the military strategy and circumstances of the release of two atom bombs over Japan in 1945.

30. Kamen, Martin D. THE BIRTHPLACE OF BIG SCIENCE. *Bull. of the Atomic Scientists 1974 30(9): 42-46.* Reminisces about work and research at the Ernest O. Lawrence Radiation Laboratory at the University of California, Berkeley, 1936-39.

31. Kaspi, André. CONTROVERSE: FALLAIT-IL BOMBARDER HIROSHIMA? [Debate: should the bomb have been dropped on Hiroshima?]. *Histoire [France] 1981 (32): 87-91.* Examines the significance of the use of the atomic bomb in the closing days of World War II after President Harry S. Truman returned from the Potsdam Conference. Its use was as a weapon of diplomacy, marking the beginning of the Cold War.

32. Keegan, John. NUCLEAR REVELATIONS: A REVIEW ESSAY. *Int. Security 1981 6(3): 195-206.* Reviews Michael Mandelbaum's *The Nuclear Revolution: International Politics before and after Hiroshima* (1981), and analyzes its major theme that "change" in the Marxian sense was over and the world passed into another age with Hiroshima.

33. Leopold, Richard W. HIROSHIMA THIRTY YEARS LATER. *Rev. in Am. Hist. 1976 4(3): 464-470.* Review article prompted by Margaret Gowing's two-volume study, *Independence and Deterrence: Britain and Atomic Energy, 1945-1952* (New York: St. Martin's Pr., 1974) and Martin J. Sherwin's *A World Destroyed: The Atomic Bomb and the Grand Alliance* (New York: Alfred A. Knopf, 1975) which discuss the impact of science on government, warfare, and foreign policymaking, 1938-52.

34. Lifton, Robert Jay. THE CONCEPT OF THE SURVIVOR. Dimsdale, Joel E., ed. *Survivors, Victims, and Perpetrators: Essays on the Nazi Holocaust* (Washington: Hemisphere Publ., 1980): 113-126. Describes common psychological responses among survivors of the atomic bomb blast at Hiroshima, the World War II concentration camps, the Vietnam War, and the Buffalo Creek flood disaster (1972).

35. Lilley, Stephen Ray. A MINUTEMAN FOR YEARS: CLARENCE CANNON AND THE SPIRIT OF VOLUNTEERISM. *Missouri Hist. Rev. 1980 75(1): 33-50.* Representative Clarence Cannon of Missouri achieved some fame during World War II when, as chairman of the House Appropriations Committee, he fought to secure funds for the Manhattan Project that was revolutionizing warfare. A believer in activism, Cannon fought for 20 years to gain a military commission during the Spanish-American War, the US punitive expedition against Mexico, and during World War I. Cannon's motives for making this long campaign are explored. Based on books, newspapers, interviews, the Cannon papers at the University of Missouri, Columbia; illus., 44 notes.

W. F. Zornow

36. Manley, J. H. ASSEMBLING THE WARTIME LABS. *Bull. of the Atomic Scientists 1974 30(5): 42-47.* Reminisces about wartime nuclear physics, work in the Massachusetts Institute of Technology Radiation Laboratory, and the Manhattan Project.

37. Mark, Eduard. "TODAY HAS BEEN A HISTORICAL ONE": HARRY S. TRUMAN'S DIARY OF THE POTSDAM CONFERENCE. *Diplomatic Hist. 1980 4(3): 317-326.* Contains the text of a hitherto-unpublished diary kept by President Harry S. Truman (1884-1972) while at the Potsdam Conference in 1945. Included in it are the president's comments on communism, the atomic bomb, Soviet dictator Joseph Stalin, British Prime Minister Winston Churchill, and other topics and personalities. Text of a primary document; 23 notes. T. L. Powers

38. Markov, M. A. REFLECTIONS OF A SOVIET SCIENTIST ON EINSTEIN. *Bull. of the Atomic Scientists 1979 35(3): 27-34.* Quoting extensively from the works of Albert Einstein, chronicles his attitudes on physics, the usefulness of science, and pacifism, 1919-55.

39. McDaniel, Boyce. A PHYSICIST AT LOS ALAMOS. *Bull. of the Atomic Scientists 1974 30(10): 39-43.* Describes graduate studies at Cornell University in physics and the construction of the first plutonium bomb, at Los Alamos, New Mexico, 1940-45.

40. Morton, Louis. THE ATOMIC BOMB AND THE JAPANESE SURRENDER. *Marine Corps Gazette 1959 43(2): 20-28.* An account of the events and decisions which culminated in the use of the atomic bomb on Hiroshima and Nagasaki, Japan, by the United States in August 1945.

41. Mulkin, Barb. LOS ALAMOS—P.O. BOX 1663. *Westways 1977 69(1): 31-34, 72.* History of the town of Los Alamos, New Mexico, centering on its importance as the secret scientific center where the atomic bomb was developed during World War II, 1943-46.

42. Munro, John A. and Inglis, Alex I. THE ATOMIC CONFERENCE 1945 AND THE PEARSON MEMOIRS. *Internat. J. [Canada] 1973/74 29(1): 90-109.* The authors, who have been appointed to edit the remaining volumes of *Mike: The Memoirs of the Right Honourable Lester B. Pearson* (Ontario: U. of Toronto Press, 2 vols., 1972-73), quote here long passages from two documents of 1945 about Canadian-United States discussions on nuclear arms in order to make a point about the late prime minister's character. R. V. Kubicek

43. O'Connor, Raymond G. TRUMAN: NEW POWERS IN FOREIGN AFFAIRS. *Australian J. of Pol. and Hist. [Australia] 1979 25(3): 319-326.* Harry S. Truman added to the authority of the president in international affairs. He increased presidential control over: nuclear weapons; dependence on military force in foreign policy; membership of the United Nations; unilateral protection of the free world; the first of the peacetime alliances (North Atlantic Treaty); promotion of world prosperity (Marshall Plan); new machinery for foreign policymaking (the National Security Council and the Central Intelligence Agency). Most moves were initially controversial, but Truman was an experienced politician and a tradition of bipartisanship had developed during World War II. Truman's reputation as a statesman was greater in the Washington community than with the general public. Based on memoirs and monographs; 21 notes.
W. D. McIntyre

44. Oppenheimer, J. Robert. ON EINSTEIN. *Bull. of the Atomic Scientists 1979 35(3): 36-39.* Reminiscences about Albert Einstein, one written in 1939, the other in 1965, relate his contributions to modern physics and to pacifism, 1919-55.

45. Otis, Cary. ATOMIC BOMB TARGETING: MYTHS AND REALITIES. *Japan Q. [Japan] 1979 26(4): 506-516.* A Japanese myth holds that Kyoto was spared from destruction in the final months before Japan's surrender in 1945 because of the influence of Langdon Warner, the Harvard art historian. In truth, the interim Target Committee, meeting with Secretary of War Stimson and Chief of Staff General George Marshall in Washington, 31 May 1945, did not approve Kyoto as a target, contrary to its earlier recommendations. F. W. Ikle

46. Preda, Eugen. SECRETELE "TUBULUI ALLOYS" [The "tube alloys" secret]. *Magazin Istoric [Rumania] 1975 9(8): 35-37.* Presents the diplomatic history of atomic bomb development, 1942-45. 3 notes.
J. M. McCarthy

47. Reingold, Nathan. WITH HENRY JAMES IN LOS ALAMOS. *Rev. in Am. Hist. 1980 8(4): 541-546.* Review essay of *Robert Oppenheimer: Letters and Recollections,* edited by Alice Kimball Smith and Charles Weiner (Cambridge, Mass.: Harvard U. Pr., 1980); early 1900's-45.

48. Rhodes, Richard. "I AM BECOME DEATH . . . ": THE AGONY OF J. ROBERT OPPENHEIMER. *Am. Heritage 1977 28(6): 70-83.* J. Robert Oppenheimer (1904-67) is portrayed as a brilliant but lonely man. After a sheltered childhood (during which he reveled in books and science), he went to Harvard and to Göttingen. In 1929 he went to Berkeley where he helped establish the reputation of the University of California's school of theoretical physics. In 1942, he was selected to direct the Manhattan Project. Success led him to support the use of the bomb in order that mankind would know of its horrible consequences. By the 1950's, he was convinced that a nuclear stalemate had been reached, and he began to question continued arms development. As a result, he lost his security clearance in 1954. He lived out the remainder of his life in isolation. 8 illus. J. F. Paul

49. Salaff, Stephen. THE DIARY AND THE CENOTAPH: RACIAL AND ATOMIC FEVER. *Can. Dimension [Canada] 1978 13(3): 8-11.* Discusses excerpts from the diary of former Canadian Prime Minister William Lyon MacKenzie King dealing with Canada's participation in the atomic bomb enterprise of the 1940's, and the racist (specifically, anti-Japanese) policies carried out during his 1921-48 administration.

50. Santoni, Alberto. COME SI GIUNSE A HIROSHIMA [How one arrived at Hiroshima—The critical analysis of an historical decision]. *R. Marittima [Italy] 1976 109(2): 77-88.* The public interest which the atomic explosion at Hiroshima has always provoked, both as an historical and moral/political event, has however never been able to prevent analysis of it being distorted by prejudice or by insufficient objectivity. The author therefore, intends to re-examine the determining factors which led the United States to use nuclear weapons against Japan, basing his analysis on the most reliable documentation, and stressing the Japanese economic and military situation at the time, which up to now has been so wrongly interpreted. J

51. Semenishchev, Iu. P. "ISCHEZNOVENIE" NIL'SA BORA V 1943 G. [The "disappearance" of Niels Bohr in 1943]. *Novaia i Noveishaia Istoriia [USSR] 1981 (4): 158-171.* Niels Bohr was one of the few physicists who understood the destructive potential of nuclear power. The US government was aware of his work, and, in order to outpace the research underway in Nazi Germany, decided to bring Bohr to America. In 1943 the Germans ended civilian administration in Denmark and installed military rule. Bohr escaped to Sweden in order to avoid deportation to Germany. On 30 September he was transported to Great Britain at considerable risk, and in December he arrived in the United States. Bohr was opposed to a monopoly on nuclear research by one power and wanted to share it with all nations, including the USSR. Based on the British and American press and secondary sources; 54 notes. V. Sobell Russian.

52. Shapiro, Edward S. THE MILITARY OPTIONS TO HIROSHIMA: A CRITICAL EXAMINATION OF GAR ALPEROVITZ'S *ATOMIC DIPLOMACY*. *Amerikastudien/Am. Studies [West Germany] 1978 23(1): 60-72.* Gar Alperovitz's *Atomic Diplomacy* (1965) is a polemical revisionist tract. At no time was President Harry S. Truman advised by his military chiefs that Japan was on the verge of surrender or that Japanese capitulation would occur without use of the atomic bomb. Of little value in understanding the circumstances surrounding the dropping of the atomic bomb in 1945, *Atomic Diplomacy* reveals much about the outlook and methodology of one representative revisionist diplomatic historian. J/S

53. Sherwin, Martin Jay. THE ATOMIC BOMB AND THE ORIGINS OF THE COLD WAR: U.S. ATOMIC ENERGY POLICY AND DIPLOMACY, 1941-'45. *Am. Hist. R. 1973 78(4): 945-968.* An examination of recently opened atomic energy files indicates that President Franklin Delano Roosevelt left no definitive statement on the postwar role of the atomic bomb. Nevertheless, the policies he chose from among the alternatives of international control, advocated by his science advisors, and an Anglo-American postwar monopoly, urged by Churchill, indicates that the potential postwar diplomatic value of the bomb began to shape his atomic energy policies as early as 1943. The assumption that the bomb could be used to secure postwar diplomatic aims was carried over to the Truman administration. Suggests that historians have exaggerated Roosevelt's confidence in (and perhaps his commitment to) amicable postwar relations with the Soviet Union. In light of this study, the widely held assumption that Truman's attitude toward the atomic bomb was substantially different from Roosevelt's must also be revised. A

54. Sigal, Leon V. BUREAUCRATIC POLITICS AND TACTICAL USE OF COMMITTEES: THE INTERIM COMMITTEE AND THE DECISION TO DROP THE ATOMIC BOMB. *Polity 1978 10(3): 326-364.* With the publication of Graham Allison's *Essence of Decision* the "bureaucratic politics" model added a new dimension to the decision-making theory of foreign policy. Working within the "bureaucratic politics" framework Leon Sigal calls attention to the largely neglected role of committees in the decision-making process. In his analysis of the events leading to the decision to drop the atomic bomb on Japan, he shows how various types of committees can be used tactically to forge interagency agreement on and secure compliance by governmental agencies with policy decisions. The result is a valuable contribution not only to the "bureaucratic

politics" model but also to our understanding of the decision to drop the atomic bomb.
J

55. Smith, Alice Kimball and Weiner, Charles. [ROBERT OPPENHEIMER: LETTERS].
ROBERT OPPENHEIMER: LETTERS AND RECOLLECTIONS. *Bull. of the Atomic Scientists 1980 36(5): 19-27.* Robert Oppenheimer's letters to colleagues and his brother reveal many aspects of his personality.
ROBERT OPPENHEIMER: THE LOS ALAMOS YEARS. *Bull. of the Atomic Scientists 1980 36(6): 11-17.* Robert Oppenheimer's Los Alamos correspondence yields insights into the scientist as well as the private man.

56. Smyth, H. D. THE "SMYTH REPORT." *Princeton U. Lib. Chronicle 1976 37(3): 173-189.* Discusses the author's role in preparing a War Department report on the military feasibility of using atomic energy for strategic purposes against Japan during World War II.

57. Snelders, H. A. M. A. M. MAYER'S EXPERIMENTS WITH FLOATING MAGNETS AND THEIR USE IN THE ATOMIC THEORIES OF MATTER. *Ann. of Sci. [Great Britain] 1976 33(1): 67-80.* In the years 1878 and 1879 the American physicist Alfred Marshall Mayer (1836-1897) published his experiments with floating magnets as a didactic illustration of molecular actions and forms. A number of physicists made use of this analogy of molecular structure. For William Thomson they were a mechanical illustration of the kinetic equilibrium of groups of columnar vortices revolving in circles round their common centre of gravity (1878). A number of modifications of Mayer's experiments were described, which gave configurations which were more or less analogous to Mayer's arrangements. It was Joseph John Thomson who, in publications betwen 1897 and 1907, used Mayer's results to obtain a good deal of insight into the general laws which govern the configuration of the electrons in his atomic model. This article is mainly concerned with Mayer's experiments with floating magnets and their use by a number of physicists. Through his experiments Mayer made a significant, although small, contribution to the theory of atomic structure.
J

58. Steiner, Arthur. BAPTISM OF THE ATOMIC SCIENTISTS. *Bull. of the Atomic Scientists 1975 31(2): 21-28.* Traces the fate of the "Franck Report" to illuminate one aspect of scientists' participation in national policymaking. A seven-member Social and Political Implications Committee, composed of scientists working with the atomic bomb and concerned about its uses, chaired by James Franck, a Nobel laureate in physics, issued a report on 11 June 1945 calling for a "non-lethal demonstration of the soon-to-be-ready atomic bomb." The Franck Report never received "the careful consideration it required" perhaps partly because of a "natural desire to give wider circulation to those views which agree with one's own," and Secretary of War Henry L. Stimson firmly disagreed with the Franck Report's major point. Based on primary and secondary sources; 44 notes.
D. J. Trickey

59. Szilard, Leo. LEO SZILARD: HIS VERSION OF THE FACTS. *Bull. of the Atomic Scientists 1979 35(2): 37-40, (3): 55-59, (4): 28-32, (5): 34-35.* Part I. THE YOUNG SZILARD. Discusses Szilard's childhood in Budapest,

Hungary, and his study of physics in Berlin, 1898-1922. Part II. Chronicles Leo Szilard's years of work in physics in Germany, 1928-33, his escape from Germany in 1933, and his further work in nuclear fission with Albert Einstein, 1933-39. Part III. Discusses the aftermath of a meeting with Washington officials in 1940 explaining the importance of uranium chain reactions and the beginning of nuclear research in the United States, 1940-42. Part IV. Describes Szilard's meeting with James F. Byrnes in 1945 in which Szilard opposed further nuclear testing.

60. Unsigned. TRAINING CAMP FOR THE ATOMIC AGE: WENDOVER FIELD. *Aerospace Historian 1973 20(3): 137-139.* The crew of the *Enola Gay*, the B-29 that dropped the atomic bomb on Hiroshima, Japan, trained at Wendover Field on the Utah-Nevada border, 1944-45. S

61. Villa, Brian Loring. THE ATOMIC BOMB AND THE NORMANDY INVASION. *Perspectives in Am. Hist. 1977-78 11: 463-502.* Franklin D. Roosevelt skillfully refrained from bringing Great Britain into confidence on the atomic bomb project until Winston Churchill openly committed himself to a cross-channel invasion. Although FDR had acted the country squire, his business methods were ingenious and effective. This conclusion contradicts the school of thought which argues that Roosevelt was inept at diplomacy and weak on goals and objectives. Other Roosevelt diplomatic decisions should be reexamined in light of these findings. W. A. Wiegand

62. Villa, Brian Loring. A CONFUSION OF SIGNALS: JAMES FRANCK, THE CHICAGO SCIENTISTS AND EARLY EFFORTS TO STOP THE BOMB. *Bull. of the Atomic Scientists 1975 31(10): 36-42.* Compares James Franck's April 1945 memorandum for Henry A. Wallace with the Franck Committee Report of 11 June 1945, noting that the latter contained contradictory advice. As a result Franck's earlier policy advice on use of the atomic bomb against Japan and the nuclear arms race with the USSR was not superseded at the War Department.

63. Warren, Shields. HIROSHIMA AND NAGASAKI THIRTY YEARS AFTER. *Pro. of the Am. Phil. Soc. 1977 121(2): 97-99.* Medical teams from the military and the Manhattan Project began studying the effects of radiation on Japanese civilian survivors five weeks after the atomic bombs fell. In 1946 President Truman ordered the National Academy of Sciences to form the Atomic Bomb Casualty Commission to care for the 100,000 severely affected survivors. The commission also studied death rates, chromosome anomalies, and cancer incidences. In 1975 the joint US-Japanese Radiation Effects Research Foundation took over from the commission. W. L. Olbrich

64. Wattenberg, Albert. THE BUILDING OF THE FIRST CHAIN REACTION PILE. *Bull. of the Atomic Scientists 1974 30(6): 51-57.* Recalls the construction of the first controlled nuclear chain reaction pile at the University of Chicago's Stagg Field in 1942.

65. Weiner, Charles. CYCLOTRONS AND INTERNATIONALISM: JAPAN, DENMARK AND THE UNITED STATES, 1935-1945. *XIVth International Congress of the History of Science, Proceedings No. 2 (Tokyo and Kyoto: Science Council of Japan, 1974): 353-368.* Examines the process, 1935-45, by which nuclear physics became a major field of research in Japan, Denmark

and the United States following the invention of the cyclotron by Ernest Lawrence in the early 1930's.

66. Wilson, Robert R. A RECRUIT FOR LOS ALAMOS. *Bull. of the Atomic Scientists 1975 31(3): 41-47.* Recounts participation in the construction of a cyclotron at Los Alamos, New Mexico, 1942-43, which preceded the Manhattan Project construction and detonation of the first nuclear bomb.

67. Wolk, Herman S. THE B-29, THE A-BOMB, AND THE JAPANESE SURRENDER. *Air Force Mag. 1975 58(2): 55-61.* Discusses the role of air power in the war against Japan, 1941-45, relying on conversations between the author and General Curtis Lemay.

68. Yavenditti, Michael J. THE AMERICAN PEOPLE AND THE USE OF ATOMIC BOMBS ON JAPAN: THE 1940S. *Historian 1974 36(2): 224-247.* A study of American public opinion in the 1940's on the use of atomic bombs on Japan. The immediate response was largely favorable since it undoubtedly shortened World War II, with little sense of personal responsibility or guilt, especially in view of the secrecy surrounding its development and use. There was adept use of a double standard in evaluation of the conduct of the war, including matters related to the bombs. The controversy which developed was significant but limited in participation. The vocal minority of critics "could not even persuade many members of the intelligentsia to condemn the atomic bombings." 82 notes.
R. V. Ritter

69. Yavenditti, Michael J. ATOMIC SCIENTISTS AND HOLLYWOOD: *THE BEGINNING OR THE END* (1947). *Film and Hist. 1978 8(4): 73-88.* Metro-Goldwyn-Mayer's film portrayed the Manhattan Project's scientists favorably, capitalized on post-Hiroshima curiosity about atomic weaponry, and reinforced public opinion justifying US rights to build and employ the atom bomb against Japan; discusses the scientists' involvement with the film and their concern for the outcome.

70. Yavenditti, Michael J. JOHN HERSEY AND THE AMERICAN CONSCIENCE: THE RECEPTION OF "HIROSHIMA." *Pacific Hist. R. 1974 43(1): 24-49.* "Evaluates aspects of the post-World War II American milieu, the circumstances which led to the writing and publication of *Hiroshima* (by John Hersey, New York, 1946), and the techniques which Hersey employed in order to determine the meaning of his study for the Americans who first encountered it." The objective is "to shed some light on the larger implications of the atomic bombings for the American conscience." Hersey's book contributed to activation of the American conscience in regard to the bomb and laid the groundwork for later assessments, especially from the standpoint of its human victims. 86 notes.
R. V. Ritter

71. York, Herbert F. SOUNDERS OF THE ALARM. *Bull. of the Atomic Scientists 1975 31(10): 43-45.* Discusses the early unsuccessful efforts of scientists to alert American policymakers and the general public to the dangers of atomic warfare and the urgent need for disarmament, 1939-54.

2

THE DEVELOPMENT OF NUCLEAR WEAPONRY

72. Alfven, Hannes. SCIENCE, PROGRESS AND DESTRUCTION. *Bull. of the Atomic Scientists 1979 35(3): 68-71.* Addresses ethical questions concerning production of untested substances, those of possible detriment to the human race, and those involving the arms race.

73. Alkhimenko, A. OPERATIVNOE OBORUDOVANIE OKEANSKIKH I MORSKIKH TEATROV VOENNYKH DEISTVII [Strategic equipment in naval war theaters]. *Morskoi Sbornik [USSR] 1980 (1): 15-20.* Review of the global network of up-to-date bases of all types, communications systems (from numerous underwater cables to satellites), intelligence services (such as acoustic antisubmarine systems), and sophisticated navigational systems available to NATO and allied powers shows that the capitalist powers have the capability to wage war in all naval theaters of operations. Even so, in their preparations for nuclear war, the NATO leadership is devoting massive resources to the further development and perfection of this infrastructure. C. J. Read

74. Ambri, Mariano. CONSIDERAZIONI SULLA "NUOVA STRATEGIA" DEGLI USA [Considerations on the "new strategy" of the United States]. *Civitas [Italy] 1980 31(11-12): 37-45.* Questions the worth and speculates on the ramifications of the new American strategy concerning nuclear warfare that assumes the primary targets in such a war would be industrial, political, and military bases.

75. Anders, Roger M. THE ROSENBERG CASE REVISITED: THE GREENGLASS TESTIMONY AND THE PROTECTION OF ATOMIC SECRETS. *Am. Hist. Rev. 1978 83(2): 388-400.* This article examines how the U.S. Atomic Energy Commission solved the security problem raised by David Greenglass' testimony against the Rosenbergs. It discusses how the Commission verified the technological data in the Greenglass testimony and how it took measures to protect other closely related technological secrets. Using documents recently declassified it shows that Greenglass tried to tell the truth about the technology of the atomic bomb. Refers to earlier article by Michael Parrish (see abstract 152). A

76. Arbab, John. OPPENHEIMER AND ETHICAL RESPONSIBILITY. *Synthesis 1982 5(2): 22-43.* Scientists prior to World War II felt they were responsible only for discovering truth. They were not to be held accountable for how society used their discoveries. With the development of the atomic bomb, however, this assumption began to change. Robert Oppenheimer was not the first to urge scientists to give more attention to their social responsibility, but when he helped formulate the Acheson-Lilienthal Report (or the "Baruch Plan") in 1946, he was playing a leading role in trying to achieve international control of atomic explosives. Mainly secondary sources; 51 notes. M. M. Vance

77. Arkhipov, M. RAZVITIE STRATEGICHESKOGO RAKETNO-IADERNOGO ORUZHIIA SShA POSLE VTOROI MIROVOI VOINY (PO MATERIALAM ZARUBEZHNOI PECHATI) [The development of strategic nuclear missiles in the United States after World War II, according to material from the foreign press]. *Voenno-Istoricheskii Zhurnal [USSR] 1981 23(2): 66-71.* Outlines the development of US nuclear ballistic missiles from 1955 until the present and provides technical data on the following missile systems: Thor, Jupiter, Atlas, Titan, Minuteman, MX, Polaris, Poseidon and Trident. The logistical shortcomings of conventional aircraft with regard to their ability to deliver a nuclear warhead gave the Pentagon the necessary impetus in the postwar period to develop both medium-range and intercontinental ballistic missile systems. 15 notes. A. Brown

78. Aspin, Les. PARLIAMENTARY CONTROL OF DEFENCE: THE AMERICAN EXAMPLE. *Survival [Great Britain] 1973 15(4): 166-170.* Discusses proposed congressional amendments for defense budget cuts in the 1970's, emphasizing the B-l bomber, the *Trident* submarine, and political pressure groups.

79. Ball, Desmond J. THE COUNTERFORCE POTENTIAL OF AMERICAN SLBM SYSTEMS. *J. of Peace Res. [Norway] 1977 14(1): 23-40.* SLBM systems have traditionally been seen as counter-value weapons systems, ideally suited to the support of mutual assured destruction and hence of greater international stability. It is the primary contention of this paper that several quite discrete developments are providing SLBM systems with a potential, at least under some circumstances, for significant counterforce strategic operations. These developments are partly technical (a combination of improved ballistic missile inertial guidance systems, MIRVing, and increases in the navigational accuracy of the FBM submarines), partly strategic, and partly bureaucratic-political. These developments necessitate a rethinking of much of the conventional wisdom on the role of SLBM systems in the American strategic nuclear posture, and of the implications of these systems for arms control. J

80. Ball, Desmond J. and Coleman, Edwin. THE LAND-MOBILE ICBM SYSTEM: A PROPOSAL. *Survival [Great Britain] 1977 19(4): 155-163.* Analyzes US national defense strategies, 1950's-74, by examining the technical and operational feasibility of shelter-based land-mobile ICBM's; describes relevant subsystem technologies and indicates variables and alternatives.

81. Barry, Hamlet J. CARING FOR THE NATIVES. *Washington Monthly 1975 6(11): 59-61.* The Atomic Energy Commission's nuclear testing in the Pacific is over, but the inhabitants of the Marshall Islands are suffering from the effects of radiation exposure. S

82. Barry, Tom. NAVAJOS AND NATIONAL NUCLEAR POLICY. *Southwest Econ. and Soc. 1979 4(3): 21-32.* Discusses the health, economic, cultural, and environmental impact of uranium mining on the Navajo Indians and their reservation in northwestern New Mexico since 1946.

83. Behuncik, John G. NEUTRON WEAPONS AND THE CREDIBILITY OF NATO DEFENSE. *J. of Social and Pol. Studies 1978 3(1): 3-16.* Discusses the properties of the enhanced radiation (fusion-type) weapon and its military advantages over the existing fission-type weapon as a deterrent to attack on NATO countries by the Warsaw Pact.

84. Bellamy, Ian. THE ORIGINS OF THE SOVIET HYDROGEN BOMB: THE YORK HYPOTHESIS. *J. of the Royal United Services Inst. for Defence Studies [Great Britain] 1977 122(1): 56-58.* Most sources hitherto have agreed that the USSR tested its first hydrogen bomb in August 1953. Now evidence suggests that such a test did not take place until November 1955. A former US Director of Defense Research and Engineering, Herbert York (b. 1921), recently stated that the 1953 Soviet test was of an inferior hybrid fission/fusion device, though he did not document the statement. Further indirect evidence is inconclusive, but comparison with the French testing program provides a parallel hybrid stage. The lack of reaction in the United States to the 1953 Soviet claim supports the revisionist view, which has implications for a reevaluation of Soviet foreign policy in the early 1950's. 4 notes. D. H. Murdoch

85. Black, Edwin F. and Cohen, S. T. THE NEUTRON BOMB AND THE DEFENSE OF NATO. *Military Rev. 1978 58(5): 53-61.* Argues that the best defensive weapon against a Soviet attack in Western Europe would be the development of fusion (ER) weapons. The neutron bomb, contrary to press reports, would be designed to neutralize the preponderance of Warsaw Pact armor rather than to kill people without damaging buildings. Although the Soviets have condemned the neutron bomb, it is quite possible they are developing one for their own tactical use. 2 fig., 7 notes. C. Hopkins

86. Blosser, Henry G. CYCLOTRONS AT MICHIGAN STATE UNIVERSITY: A REVIEW OF THEIR SUCCESSFUL DEVELOPMENT. *Centennial Rev. 1981 25(3): 203-224.* Traces the development of the nuclear physics program at Michigan State University (MSU), where recent and current construction will bring two cyclotrons into operation. Since the program's creation in the 1950's MSU has emerged as one of the world's foremost centers for heavy ion nuclear research. Based on the author's first-hand experiences as director of the MSU program; table, 8 fig. A. Hoffman

87. Boitsov, M. PLAN "DROPSHOT" [The "Dropshot" plan]. *Morskoi Sbornik [USSR] 1980 (3): 84-87.* Summarizes, with special reference to naval activity, the Dropshot plan authorized by President Truman in 1949 and ready for use in 1957 for a preventive war against the USSR. The plan envisaged a four-stage war of undefined length beginning with joint US and British strategic

bombing of the USSR, the preparation of all available forces for an all-out attack, the carrying out of that attack, and the establishment of an occupation regime. The subsequent development of nuclear capability lends greater force to the USSR's continuing attempt to limit strategic weapons, 2 tables, note.
C. J. Read

88. Borowski, Harry R. AIR FORCE ATOMIC CAPABILITY FROM V-J DAY TO THE BERLIN BLOCKADE: POTENTIAL OR REAL? *Military Affairs 1980 44(3): 105-110.* Examines how well the US Air Force's atomic monopoly between V-J Day and the Berlin blockade of 1948 translated into real military power. The stockpile of atomic weapons was limited and the Strategic Air Command (SAC), responsible for delivering the weapons, suffered from demobilization, shortages, inadequate leadership, and misguided training. This translated into a widening gap between the potential and real capability of the US strategic air arm. The gap only began to close after General Curtis LeMay took command of SAC and began rebuilding the Command in 1948. Based on Air Force records and other primary sources; 33 notes. A. M. Osur

89. Borst, Gert and Walter, Franz. LANGFRISTIGE TENDENZEN IM RÜSTUNGSWETTLAUF USA-UdSSR [Long-range trends in the U.S.-Soviet armaments race]. *Osteuropa [West Germany] 1973 23(2): 81-104.* A detailed survey of relative strengths in nuclear arms during 1946-72. The USSR has achieved numerical superiority only in rockets, but this is more than offset by the Multiple Independently Targetable Reentry Vehicles (MIRV) of the United States. Since both sides possess a clear second-strike capability, earlier notions of superiority have become obsolete and irrelevant. 6 tables, 7 diagrams, 19 notes.
R. E. Weltsch

90. Brauzzi, Alfredo. AEGIS: IL NUOVO SCUDO DELLA U. S. NAVY [AEGIS: the new shield of the US Navy]. *Riv. Marittima [Italy] 1981 114(10): 11-22.* Examines the new US naval combat system, AEGIS, being developed for service with the Ticonderoga class of missile cruisers, and assesses its performance in trials. Italian.

91. Brauzzi, Alfredo. ENTRA IN SERVIZIO L'*OHIO* [The commissioning of the *Ohio*]. *Riv. Marittima [Italy] 1982 115(1): 113-117.* Describes the launching of the US nuclear submarine in November 1981 and some of its characteristics and capabilities. Italian.

92. Brim, Raymond E. and Condon, Patricia. ANOTHER A-BOMB COVER-UP. *Washington Monthly 1971 12(11): 45-49.* Nuclear testing during the late 1940's and 1950's was open-air, but went underground during the 1960's and 1970's to insure that the testing would be safe; in fact, the risk of fallout continues and the government continues to attempt to cover up the dangers of underground nuclear testing, as a test in Nevada on 25 September 1980 proved.

93. Brown, William D. WHATEVER HAPPENED TO . . . TACTICAL NUCLEAR WARFARE? *Military Rev. 1980 60(1): 46-53.* Today's army is unsure how to train for tactical nuclear warfare and is unable to fight in such an environment. There is also little discussion of tactical nuclear doctrine within the army. Given serious discussion of nuclear warfare in Russia, the United States must begin to actively plan and prepare its forces for this contingency. Based on Army Field Manuals and Soviet writings; chart, 13 notes. D. H. Cline

94. Bundy, McGeorge. THE BEST OF ALL POSSIBLE NUCLEAR WORLDS: A REVIEW ESSAY. *Int. Security 1980 5(1): 172-177.* Reviews Michael Mandelbaum's *The Nuclear Question: The United States and Nuclear Weapons, 1946-1976* (Cambridge: Cambridge U. Pr., 1979).

95. Cavers, David F. THE CARTER EVACUATION PLAN. *Bull. of the Atomic Scientists 1979 35(4): 15-19.* Reviews criticism of President Jimmy Carter's proposed evacuation plan in the event of atomic war—from public opinion and the Arms Control and Disarmament Agency, 1978.

96. Chistiakov, I. KRYLATYE RAKETY SSHA [American winged rockets]. *Voenno-Istoricheskii Zhurnal [USSR] 1979 (3): 73-78.* One aspect of the broadening and speeding up of the arms race by reactionary circles in the United States has been the development of winged rockets derived from German V-1's and V-2's of World War II. Reviews size performance and capabilities of basic types of American winged rockets from the Regulus 1 (SSSM-N-8) of the mid-1950's to the nuclear-armed sea, air, and ground-launched cruise missiles currently being introduced into the American armory. These developments add weight to the Warsaw Pact's view that nuclear energy should be used only for peaceful purposes. 6 photos, table. C. J. Read

97. Cockburn, Andrew and Cockburn, Alexander. THE MYTH OF MISSILE ACCURACY. *Parameters 1981 11(2): 83-89.* The assumption that intercontinental ballistic missiles can hit their targets with accuracy has greatly influenced US and Soviet strategic defense planning over the past 30 years. Concerns have been raised over the use of accurate missiles in a first-strike against each side's own missiles. In addition, the availability of highly accurate missiles seemingly makes nuclear war more acceptable because it may be fought against purely military targets. But, the evidence indicates that missile guidance systems often malfunction, and the existing missiles have never been tested under wartime conditions. There are no certainties about how a wartime missile exchange would end. Based on interviews with defense specialists, US government documents, and secondary sources; 17 notes. L. R. Maxted

98. Cohen, S. T. and Lyons, W. C. A COMPARISON OF U.S.-ALLIED AND SOVIET TACTICAL NUCLEAR FORCE CAPABILITIES AND POLICIES. *Orbis 1975 19(1): 72-92.* Compares the strategic capabilities of US and USSR technology and tactical nuclear arms in the 1970's, emphasizing aspects of US-NATO policy.

99. Coulam, Robert F. NEW DEFENSE SYSTEMS. *Pro. of the Acad. of Pol. Sci. 1982 34(4): 189-200.* Describes the revolutionary changes in the new communications systems including changes in organization. Discusses how communications systems in use were planned for strategic nuclear weapons. It compares the work of the USSR in this area to that of the United States. Primary sources; 7 notes. T. P. Richardson

100. Davis, Lynn Etheridge and Schilling, Warner R. ALL YOU EVER WANTED TO KNOW ABOUT MIRV AND ICBM CALCULATIONS BUT WERE NOT CLEARED TO ASK. *J. of Conflict Resolution 1973 17(2): 207-242.* "This paper 1) explains the variables and methods used to calculate the MIRV threat to the fixed-site land-based ICBM forces of the United States and

the Soviet Union; 2) attempts to reconstruct the calculations presented in the 1969 ABM debate by Albert Wohlstetter, George Rathjens, and John S. Foster about the number of Minutemen which could be expected to survive an attack by forces of various sizes and characteristics; and 3) applies these methods, together with some plausible assumptions about classified variables, to two other security issues: the import of the SALT limitation on the size of the Soviet SS-9 force and the character of the threat to the Soviet fixed-site ICBM force posed by the American MIRV program." J

101. Day, Samuel H., Jr. THE NUCLEAR WEAPONS LABS. *Bull. of the Atomic Scientists 1977 33(4): 21-26, 28-32.* The rival laboratories of Livermore and Los Altos design nuclear warheads. Both have research and development programs in nuclear areas. Both hope to preserve the nation's scientific and technological supremacy in the development of nuclear weapons. President Carter's announced goal of working toward a ban on the testing of all nuclear weapons has raised the question of the future role of the laboratories. Secondary sources; 8 illus., table, graph. D. J. Trickey

102. Dine, Thomas A. MILITARY R&D: CONGRESS' NEXT AREA OF POLICY PENETRATION. *Bull. of the Atomic Scientists 1978 34(2): 32-37.* Creation of a House subcommittee on the Pentagon's research and development budget, reorganization of the congressional budget process, and restructuring of the congressional committee system, 1976-77, all indicate a more aggressive stance by Congress.

103. Dobney, Frederick J. STOCKPILING AND SHORTAGES. *Social Sci. Q. 1976 57(2): 455-465.* Notes that in 1953 the control of strategic-materials stockpiling passed from military to civilian hands, resulting in the politicization of stockpiling decisions. Mining and manufacturing companies have exercised considerable influence since then. Argues that the oil and food shortages of the recent past have given the stockpile new significance and that future decisions should be made only after a coherent, comprehensive consideration of costs and benefits. J

104. Easterbrook, Gregg. FROM SPUTNIK TO THE FLYING SUBMARINE: HOW PENTAGON RIVALRIES GAVE US THE MX. *Washington Monthly 1981 13(8): 10-21.* Rivalries among the US military branches since World War II have revolved around nuclear weapons; the MX missile was designed not so much as a response to world military conditions but as a result of political conditions at the Pentagon; 1960's-81.

105. Eayrs, James. APOCALYPSE THEN: ASPECTS OF NUCLEAR WEAPONS-ACQUISITION POLICY THIRTY YEARS AGO. *Dalhousie Rev. [Canada] 1979-80 59(4): 635-650.* An anecdotal narrative of the development of the hydrogen bomb from 1950 to 1955, emphasizing the opposition within the scientific community to the bomb's development. Based on an address given as part of a study grant at Dalhousie University in 1979; 43 notes.
C. H. Held

106. Fahrney, D. S. GUIDED MISSILES: U.S. NAVY THE PIONEER. *Am. Aviation Hist. Soc. J. 1982 27(1): 15-28.* Reviews the pioneering effort of the navy's Bureau of Aeronautics, which developed the first guided missiles

during the 1930's in the categories of air to air, air to surface, surface to air, and surface to surface; the lack of recognition of the navy's great contributions in remote control airplanes and guided missiles stems from the highly confidential atmosphere of the research.

107. Fahrney, Delmar S. THE BIRTH OF GUIDED MISSILES. *US Naval Inst. Pro. 1980 106(12): 54-60.* The Navy Bureau of Aeronautics developed the world's first air-to-surface guided missile (1938, an N2C-2 radio-controlled airplane drone), the first surface-to-surface guided missile (1942, a TG-2 torpedo plane drone), the first surface-to-air guided missile (1950, a Convair Lark missile), and the first air-to-air guided missile (1952, a Douglas Sparrow missile). The author played a prominent role in the development of the first two, starting in 1936. Because of opposition to the program at naval headquarters in the Pacific, the drone program was never given a fair combat test during World War II, although a number of drone missions were permitted, during which the drones performed well. Many of today's guided missiles, such as the cruise missile and the Regulus missile, can be traced back to the Navy's pioneering efforts. 8 photos.
A. N. Garland

108. Furems, M. DOZY OBLUCHENIIA LICHNOGO SOSTAVA ATOMNYKH KORABLEI I PERSONALA SUBOREMONTNYKH ZAVODOV S.SH.A. [Radiation doses on crews of atomic ships and personnel of ship repair yards in the United States]. *Morskoi Sbornik [USSR] 1980 (3): 69-73.* Summarizes American reports on dosages of radiation received by naval and civilian personnel in shipyards and atomic power stations in the United States published in *Nuclear Safety* (1979 20(3)) and by the Nuclear Regulatory Commission. Doubts the conclusion that doses did not exceed permitted norms and suggests that the findings presented conceal certain situations, such as breakdowns, as a result of which individuals would have been subjected to higher doses than the regulations permit. Covers 1954-79. 5 tables, 4 notes. C. J. Read

109. Gelbond, Florence. THE IMPACT OF THE ATOMIC BOMB ON EDUCATION. *Social Studies 1974 65(3): 109-114.* Analyzes the impact of the nuclear age on American education and illustrates conflicting assessments by various authorities and average Americans in terms of the direction our schools must take. Restructured value systems are strongly suggested with the view of making man more human. 19 notes. L. R. Raife

110. Gerber, Larry G. THE BARUCH PLAN AND THE ORIGINS OF THE COLD WAR. *Diplomatic Hist. 1982 6(1): 69-95.* American security interests and Wilsonian idealism underlay the Baruch plan. Like other policymakers, Bernard Baruch believed that Russia was bent on world domination. Its totalitarian system allowed it to easily violate agreements and react to events more rapidly than the morally superior United States. The American atomic monopoly had to be preserved until enforceable sanctions were in place. As part of world law, the plan would pave the way toward a Wilsonian world of liberal capitalism. Its international controls, which assumed an exchange of information, would liberalize the USSR and make economic blocs unfeasible. Based on the papers of Bernard Baruch and other primary sources; 81 notes. T. J. Heston

111. Giorgerini, Giorgio. I SOTTOMARINI A PROPULSIONE NUCLEARE, LANCIAMISSILI E D'ATTACCO [Nuclear-propelled, missile-firing, and attack submarines]. *R. Marittima [Italy] 1975 108(11): 10-30.* The author reviews the various phases in the development of the attack and missile-firing submarines of the United States Navy, touching as well upon the results obtained in this field by the Soviet, British and French Navies. J

112. Gray, Colin S. MINI-NUKES AND STRATEGY. *Internat. J. [Canada] 1974 29(2): 216-241.* Describes the latest conventional weapons possessed by or available to NATO and enumerates their limitations. Argues for adoption of "mini-nukes," miniaturized nuclear weapons for use on battlefields. 68 notes.
R. V. Kubicek

113. Gray, Colin S. THE MX DEBATE. *Survival [Great Britain] 1978 20(3): 105-112.* Examines the debate over adoption of the MX intercontinental ballistic missile for national security; the greatest drawback is the time necessary to develop such a system.

114. Gray, Colin S. A NEW DEBATE ON BALLISTIC MISSILE DEFENCE. *Survival [Great Britain] 1981 23(2): 60-71.* Reassesses the antiballistic missile debate of 1969-70 in the light of a changed strategic environment, asks whether times have changed to such a degree that 1970 ABM policy positions no longer would be reasonable in 1981, urges reopening a policy debate about ballistic missile defense's possible merit for stabilizing the Soviet-American strategic balance and about the fundamental wisdom of US offence-dominance strategic doctrine.

115. Gray, Robert C. LEARNING FROM HISTORY: CASE STUDIES OF THE WEAPONS ACQUISITION PROCESS. *World Pol. 1979 31(3): 457-470.* This review article presents three case studies of the U.S. strategic weapons acquisition process: the Atlas intercontinental ballistic missile (ICBM), the Polaris fleet ballistic missile (FBM), and the multiple independently targetable reentry vehicle (MIRV). By considering programs that stretch from the 1950's to the 1970's, the essay provides a record of the changing environment of American weapons choices over the past 25 years. After a description of research methods and major arguments, each study is assessed in terms of bureaucratic politics and of its relevance to contemporary policy. Conclusions are drawn about the most fruitful approach to the study of weapons acquisition, about the lessons of these cases for the development of weapons in an age of arms control, and about the challenges of future studies. J

116. Graybar, Lloyd J. BIKINI REVISITED. *Military Affairs 1980 44(3): 118-123.* Analysis of Operation Crossroads, the detonation of two atomic weapons at Bikini Atoll in 1946, to show the origin, conduct, and effect of the tests. Militarily, the operation was a success for the Navy; it had not become obsolete, and it frustrated the Air Force's intention to stage its own nuclear testing against old Japanese ships. The effects on UN negotiations and the peace movement are much more difficult to assess. Primary sources; 28 notes. A. M. Osur

117. Green, Harold P. THE OPPENHEIMER CASE: A STUDY IN THE ABUSE OF LAW. *Bull. of the Atomic Scientists 1977 33(7): 12-16, 56-61.* Executive Order 10450 of April 1953, promulgated by President Eisenhower, was

widely interpreted as an order to rid government service of any person "concerning whom there was any derogatory information that was determined to be true." This was in contrast to the Atomic Energy Commission's standards, under which derogatory information would have to be balanced against favorable information. The appointment of Lewis Strauss, Wall Street banker, as Chairman of the Atomic Energy Commission on 1 July 1953 underscored the changed climate. Strauss had promised J. Edgar Hoover to purge the Commission of several individuals, including Oppenheimer. When it became evident that J. Robert Oppenheimer would demand a hearing and that a board with a predisposition against Oppenheimer was being hand picked, Green resigned his position with the Office of the General Counsel, US Atomic Energy Commission. Based on observations; 5 illus. D. J. Trickey

118. Greenwood, John T. THE AIR FORCE BALLISTIC MISSILE AND SPACE PROGRAM, 1954-1974. *Aerospace Hist. 1974 21(4): 190-205.*

119. Hall, R. Cargill. TO ACQUIRE STRATEGIC BOMBERS: THE CASE OF THE B-58 HUSTLER. *Air. U. Rev. 1980 31(6): 2-20.* Discusses the history of the B-58, including initial studies in 1946, its introduction in 1957, its retirement in 1970, and its impact on the Air Force and the continued development of manned strategic bombers.

120. Hansen, Chuck. NUCLEAR NEPTUNES: EARLY DAYS OF COMPOSITE SQUADRONS 5 & 6. *Am. Aviation Hist. Soc. J. 1979 24(4): 262-268.* During 1949-50, Lockheed's heavy, land-based, twin-engine bomber, the P2V Neptune, served as the US Navy's first carrier-based, nuclear-armed strategic bomber.

121. Harrington, Anne. AFTER THE IMPOSSIBLE HAPPENED. *Synthesis 1982 5(2): 2-21.* Most science fiction before World War II was utopian in the sense that it trusted science and technology to lead mankind into a society of peace, justice, and plenty. Since Hiroshima much of science fiction, like much of fiction in general, has regarded science and technology as ultimately harmful, bearing the seeds of the destruction of civilization. The distinction between science fiction and mainstream fiction has become less clear, much as the lines between jazz and classical music have become blurred. 23 notes.

M. M. Vance

122. Herken, Gregg. A MOST DEADLY ILLUSION: THE ATOMIC SECRET AND AMERICAN NUCLEAR WEAPONS POLICY, 1945-1950. *Pacific Hist. Rev. 1980 49(1): 51-76.* Truman administration atomic energy policy was based on the illusion that America had a monopoly of atomic raw materials and technological expertise and would monopolize nuclear armaments for a generation after Hiroshima. General Leslie R. Groves influenced the making of a policy of excluding other nations from America's atomic secret, and promoted a bogus atomic espionage scare in 1946 to defeat a bill which provided for an all-civilian Atomic Energy Commission. The disclosure in 1949 that the USSR had developed atomic bombs was followed by the popular belief that American traitors had given the atomic secret to Russia. Based on recently declassified official sources, government documents, personal papers, contemporary newspapers and magazines, and secondary sources; 60 notes. R. N. Lokken

123. Herzfeld, Charles M. THE MILITARY R & D PROCESS: A VIEW FROM INDUSTRY. *Bull. of the Atomic Scientists 1978 34(10): 33-40.* Because national security needs to be based on deterrence and on technological and logistic alternatives, suggests a reorganization of military research and development regarding weapons systems acquisition.

124. Holder, William G. STILL GOING STRONG—THE B-52 IN ITS THIRD DECADE. *Air U. R. 1975 26(6): 48-62.* Discusses the career of the B-52 bomber from the flight of the XB-52 in October 1952 through the development of the A through H series and the proposed I series. During this period 742 B-52's were produced, as the bomber underwent many modifications, keeping up-to-date with advancing technology. Although designed to carry a nuclear bomb load, the B-52 proved effective in conventional warfare in Vietnam. 13 illus.
J. W. Thacker, Jr.

125. Huisken, Ron. THE ORIGINS OF THE STRATEGIC CRUISE MISSILE: PERCEPTIONS AND THE STRATEGIC BALANCE. *Australian Outlook [Australia] 1980 34(1): 30-40.* A brief survey of early US missile development from the 1950's, followed by a consideration of why a strategic cruise missile was proposed for 1972. It was not a case of technological determinism, nor pressure from the services, nor to be explained by action-reaction theorem, nor that it filled a void. Suggests that the idea originated in the office of the Secretary of Defense and that it was to obtain bargaining leverage for SALT II and to meet the psychological need for evidence that US technological superiority was being maintained over the more obvious Soviet superiority in numbers. Based on Senate Committee papers; 12 notes. W. D. McIntyre

126. Iklé, Fred C. THE NETHER WORLD OF NUCLEAR MEGATONNAGE. *Bull. of the Atomic Scientists 1975 31(1): 20-25.* Discusses psychological attitudes toward nuclear arms and the threat of fallout 1945-75, including the potential danger of the destruction of the ozone layer in the stratosphere.

127. Johnson, Giff. PARADISE LOST. *Bull. of the Atomic Scientists 1980 36(100): 24-29.* Discusses the controversy over whether it is safe for people to return to Eniwetok atoll in the Marshall Islands after a three-year, $100 million cleanup of the area, which was the site of 43 nuclear tests by the US government from 1946 to 1958.

128. Jones, David C. STRATEGIC ISSUES. *Air Force Mag. 1977 60(4): 30-33.* Discusses strategic issues of US national defense as seen by the Air Force Chief of Staff; article is excerpted from a presentation to the Committee on Armed Services in which David C. Jones introduced three strategic aerospace systems, the B-1, the MX, and the Air-Launched Cruise Missile, 1977.

129. Karp, Walter. WHEN BUNKERS LAST IN THE BACKYARD BLOOM'D: THE FALLOUT SHELTER CRAZE OF 1961. *Am. Heritage 1980 31(2): 84-93.* Relates the short-lived preoccupation with various methods of civil defense, particularly fallout shelters, triggered by the Berlin crisis in 1961, and discusses some of the moral questions which inevitably ensued.

130. Keegan, John. THE HUMAN FACE OF DETERRENCE. *Int. Security 1981 6(1): 136-151.* Suggests some possible scenarios for the ability and willingness of soldiers to perform their duties after nuclear attack, based on the behavior of soldiers after the battles of Cassino and Normandy in 1944. Offers some proposals for governments on how to deal with crises arising from the current hostility between the Eastern Bloc countries and the United States.

131. Keller, Bill. ATTACK OF THE ATOMIC TIDAL WAVE: SIGHTED S.U.M., SANK SAME. *Washington Monthly 1980 12(3): 53-58.* On 25 March 1980, the Defense Department disclosed that Soviet warheads could be exploded in the shallow waters of the continental shelf and create tidal waves to destroy US vessels and cities; the Pentagon made this disclosure to discredit SUMs (Shallow Underwater Mobile submarines), which had been presented as economical alternatives to the Trident submarines and the MX missile system.

132. Kevles, Daniel J. SCIENTISTS, THE MILITARY, AND THE CONTROL OF POSTWAR DEFENSE RESEARCH: THE CASE OF THE RESEARCH BOARD FOR NATIONAL SECURITY, 1944-46. *Technology and Culture 1975 16(1): 20-47.* In February 1945 a stopgap Research Board for National Security was set up as an adjunct of the National Academy of Sciences to "maintain civilian scientific participation in defense research." Opposed by the Bureau of the Budget, it died a year later along with "any independent civilian-controlled agency for defense research," and thus the military became the "dominant patron of academic scientific research outside of atomic energy." Based on archival sources; 79 notes. C. O. Smith

133. Kincade, William H. REPEATING HISTORY: THE CIVIL DEFENSE DEBATE RENEWED. *Int. Security 1978 2(3): 99-120.* Evaluates US and Soviet civil defense preparations and philosophies since 1950; through mutually annihilative power each country holds the other hostage.

134. Kopp, Carolyn. THE ORIGINS OF THE AMERICAN SCIENTIFIC DEBATE OVER FALLOUT HAZARDS. *Social Studies of Sci. [Great Britain] 1979 9(4): 403-422.* The scientific debate over fallout hazards in the 1950's was rooted not in scientific differences but in disciplinary, institutional, and political differences.

135. Krell, Gert. MILITARY DOCTRINES, NEW WEAPONS SYSTEMS, AND ARMS CONTROL: TECHNOLOGICAL DEVELOPMENTS AND POLITICAL ISSUES. *Bull. of Peace Proposals [Norway] 1979 10(1): 38-46.* New technologies such as multiple warheads (MIRV's), cruise missiles, the Backfire bomber, and the SS-20 endanger the stability of the arms race by presenting highly visible signals of attempts to achieve superiority. Furthermore, the new technology threatens arms control. Secondary sources; 17 notes. R. B. Orr

136. Krepon, Michael. A NAVY TO MATCH NATIONAL PURPOSES. *Foreign Affairs 1977 55(2): 355-367.* Reliance on vulnerable supercarriers in high-threat areas underestimates new weapons such as cruise missiles and nuclear submarines. Against low-threat Third World countries, the carrier's military utility is greatest, but its diplomatic utility is most questionable. The new weapons, minicarriers, and land-based aircraft are more flexible alternatives to the supercarrier and better suited to national purposes. Based on congressional records and secondary sources; 20 notes. W. R. Hively

137. Kuter, Laurence S. JFK AND LBJ CONSIDER AEROSPACE DEFENSE. *Aerospace Hist. 1978 25(1): 1-4.* The Commander of the North American Aerospace Defense Command during the Kennedy administration describes a briefing he gave to President John F. Kennedy and Vice President Lyndon B. Johnson in the Oval Office of the White House, 8 February 1962. Having failed to obtain what he considered adequate funds for ballistic missiles in the proposed defense budget, he received permission to present his case to the Chief Executive. In his presentation he spoke of the gaps in our defenses and expressed the need for closing this gap. The President thanked him and said that his recommendations would be considered. He did not receive the funds requested and concluded that he "had done everything compatible with our constitutional government." Were he to have proceeded further he would have had to embark on a personal and probably political crusade. 7 photos. C. W. Ohrvall

138. Kuz'min, I. RAZVEDKA ATOMNYKH RAKETNYKH PODVODNYKH LODOK [Reconnaissance of atomic missile submarines]. *Morskoi Sbornik [USSR] 1979 (5): 66-71.* The Americans regard reconnaissance as vital in combatting nuclear submarines. They observe submarines in and around their home base in order to gather technical information, and they attempt to follow them. One American estimate says that antisubmarine defense is only 20% effective. Existing listening stations are being modernized. International cooperation remains an important factor. 9 notes. B. Holland

139. Leitenberg, Milton. PRESIDENTIAL DIRECTIVE (P.D.) 59: UNITED STATES NUCLEAR WEAPON TARGETING POLICY. *J. of Peace Res. [Norway] 1981 18(4): 309-317.* At least since 1960, US nuclear weapons targeting has been primarily countermilitary and not directed in the first instance against the cities of the USSR. Focuses on the research leading up to Presidential Directive No. 59 of the Carter administration. J/S

140. Levine, Henry D. SOME THINGS TO ALL MEN: THE POLITICS OF CRUISE MISSILE DEVELOPMENT. *Public Policy 1977 25(1): 117-168.* Analyzing Defense Department attitudes toward weapons systems and external political events, the decision to pursue development of the cruise missile is studied. Interservice rivalry, White House pressure, and the cruise missile's role in negotiations with USSR, are explored. Employs a cybernetic model of decisionmaking to explain cruise missile development from the late 1960's to 1976. Implications for defense policymaking are drawn. Based on original research and secondary sources; 100 notes, glossary of terms. J. M. Herrick

141. Macdonald, Hugh. CANADA, NATO AND THE NEUTRON BOMB. *Int. Perspectives [Canada] 1979 (Mar-Apr): 9-11, 16.* The United States developed enhanced-radiation warheads (ERW), generally known as neutron bombs, for use against concentrations of men and tanks. The neutron bomb kills by concentrated massive doses of fast neutron radiation but produces relatively little damage to structures. Intense public controversy has postponed deployment of the weapon, which could correct the imbalance of military strength between NATO and the USSR. The issue has been treated too politically by NATO, too emotionally by the public, and too inconsistently by governments, particularly Canada, which has reacted to the initiatives of others without clear aim or understanding. E. S. Palais

142. Markowitz, Gerald E. and Meeropol, Michael. THE "CRIME OF THE CENTURY" REVISITED: DAVID GREENGLASS' SCIENTIFIC EVIDENCE IN THE ROSENBERG CASE. *Sci. & Soc. 1980 44(1): 1-26.* The US government perpetrated a fraud in the trial that led to the execution of Julius and Ethel Rosenberg in 1953 for allegedly passing on to the USSR the secret to the atomic bomb. The Atomic Energy Commission and the Federal Bureau of Investigation knew that the material passed from David Greenglass to the Rosenbergs was (even if true) insignificant, worthless, and full of gross errors. The testimony of such important scientists as Philip Morrison, Henry Linshitz, Harold Urey, J. Robert Oppenheimer, George Kistiakowsky, and Victor Weiskoff verifies the worthlessness of Greenglass's information. Based on documents won from the US government in a Freedom of Information Act lawsuit; 62 notes. L. V. Eid

143. MccGwire, Michael. WESTERN AND SOVIET NAVAL BUILDING PROGRAMMES 1965-1976. *Survival [Great Britain] 1976 18(5): 204-209.* Compares the resources committed to naval construction by the USSR and by Western nations 1965-76, dividing building programs into 11 categories and giving charts to illustrate each.

144. Meyer, Larry L. THE TIME OF THE GREAT FEVER. *Am. Heritage 1981 32(4): 74-79.* Charles Steen, Joe Cooper, and Vernon Pick were among the most successful early uranium prospectors who made fortunes during the nation's largest bonanza, the uranium rush of the 1950's. Centered in the Colorado Plateau in parts of Utah, Colorado, Arizona, and New Mexico, the boom spilled over into Nevada, California, and Wyoming. 3 illus. J. F. Paul

145. Meyer, Stephen M. ANTI-SATELLITE WEAPONS AND ARMS CONTROL: INCENTIVES AND DISINCENTIVES FROM THE SOVIET AND AMERICAN PERSPECTIVES. *Int. J. [Canada] 1981 36(3): 460-484.* Discusses Soviet and American incentives in developing antisatellite weapons, describes the vulnerabilities of their respective military postures, and assesses the feasibility of arms control in space. Secondary sources; 12 notes. J. Powell

146. Mueller, John E. PUBLIC EXPECTATIONS OF WAR DURING THE COLD WAR. *Am. J. of Pol. Sci. 1979 23(2): 301-329.* Since 1944, various polling organizations have asked the American people some 200 times of their expectation of world war. This paper analyzes these responses—i.e., war expectations—and, by using regression analysis shows how fluctuations in war expectations are associated with content analytic measures of Cold War activities. Investigation of three educational groups reveals education-based differences of war optimism and pessimism. Finally, a "psychology of expectations" is explored in which war expectations are associated with other perceptions. J

147. Nash, Henry T. THE BUREAUCRATIZATION OF HOMICIDE. *Bull. of the Atomic Scientists 1980 36(4): 22-27.* Based on his military intelligence experience in the atomic warfare Air Targets Division of the Air Force at the Defense Department in the 1950's and 1960's, the author discusses factors that tend to isolate planners from the possible human consequences of their planning.

148. Nunn, Jack H. MIT: A UNIVERSITY'S CONTRIBUTION TO NATIONAL DEFENSE. *Military Affairs 1979 43(3): 120-125.* Analyzes the contribution of the Massachusetts Institute of Technology (MIT) to national defense in the training of military personnel, weapons research and development, and policymaking. MIT played an important role, one that has evolved over 100 years, and one that saw the university torn between national duty and academic freedom. Based on MIT records and secondary sources; 33 notes.

A. M. Osur

149. Ognibene, Peter J. THE AIR FORCE'S SECRET WAR ON UNEMPLOYMENT. *Washington Monthly 1975 7(5/6): 58-61.* The lobbyists for the Rockwell International corporation, makers of the B-1 nuclear bomber, portray the B-1 as "the $50-billion solution to the unemployment problem" rather than as a weapon of mass destruction.

150. Paarlberg, Rob. FORGETTING ABOUT THE UNTHINKABLE. *Foreign Policy 1973 (10): 132-140.* Discusses the loss of public interest in the prospect of nuclear war. S

151. Paine, Christopher. PERSHING II: THE ARMY'S STRATEGIC WEAPON. *Bull. of the Atomic Scientists 1980 36(8): 25-31.* Discusses the origins and implications of the 1979 NATO decision to introduce US long-range missiles into Western Europe.

152. Parrish, Michael E. COLD WAR JUSTICE: THE SUPREME COURT AND THE ROSENBERGS. *Am. Hist. Rev. 1977 82(4): 805-842.* In 1953, at the peak of the Cold War, Julius and Ethel Rosenberg were executed by the United States for conspiring to give atomic secrets to the Soviet Union. This reinterpretation of the celebrated "atom spy" case focuses upon the Rosenbergs' many efforts to secure a new trial through appeals to the Supreme Court of the United States. In addition to analyzing the Constitutional and statutory issues raised by the Rosenberg case, this article also explores the conflicts generated by the litigation among members of the Court, including Chief Justice Vinson, Justices Frankfurter, Jackson, and Douglas. A

153. Peck, Earl G. B-47 STRATOJET. *Aerospace Historian 1975 22(2): 61-64.* Describes the B-47 bomber and flying experiences of the author, who was assigned to a Strategic Air Command Wing. Pictures the B-47 as an often admired, respected, and feared aircraft, but never loved. 3 photos.

C. W. Ohrvall

154. Polmar, Norman. SOVIET NUCLEAR SUBMARINES. *US Naval Inst. Pro. 1981 107(7): 31-39.* The first Soviet nuclear-powered submarine was launched in November 1958, some three years after the first US nuclear submarine had been sent to sea. It resulted from a decision by Soviet leaders to scrap earlier plans for a large surface fleet and to rely instead on submarines, missiles, and nuclear weapons. Apparently the Soviets have had less trouble than the United States in manning their boats. Secondary sources; 5 photos, 2 tables.

A. N. Garland

155. Polmar, Norman. USS *NORTON SOUND:* THE NEWEST OLD SHIP: A PICTORIAL. *US Naval Inst. Pro. 1979 105(4): 70-83.* The USS *Norton Sound* is one of the few ships built for the Navy during World War II that is still in the active US fleet. In some respects it is the Navy's most modern ship, because it is now serving as the test ship for the Aegis combat system and the Mark 26/Standard missile system. Launched in November 1943 and commissioned on 8 January 1945, she was built as a large seaplane tender. Over the years she has been modified several times for service as a guided missile test ship and now contains some of the Navy's latest technology. It is planned to use the ship in the years to come to test still more advanced weapons systems and sensors. Secondary sources; 3 notes, 42 photos. A. N. Garland

156. Poss, John R. THE GRAND CANYON'S HIDDEN HOARD. *Westways 1982 74(6): 37-39, 73-74.* Surveys prospecting and mining activities in the Grand Canyon of Arizona since the 16th century, focusing on uranium mining; prospector Daniel Hogan discovered a uranium deposit in 1891 but its utility and value were unknown until the 1950's, and $50 million worth of uranium was mined by 1969.

157. Rabinowitz, Howard N. GROWTH TRENDS IN THE ALBUQUERQUE STANDARD METROPOLITAN STATISTICAL AREA, 1940-1978. *J. of the West 1979 18(3): 62-74.* World War II and the Cold War brought the development of special weapons and atomic research that became the major economic resource of Albuquerque, New Mexico. The city practiced an aggressive policy of annexation to absorb its suburbs and the fleeing middle class. The result has been a conflict between county residents, who want autonomy as well as urban benefits, and the city government, which wants consolidation and control of urban growth. Based on US Census reports, newspapers, and Albuquerque City Commission Minutes; 16 photos, 9 tables, 32 notes. B. S. Porter

158. Rakitin, E. "TRAIDENT" SOVERSHENSTVUETSIA [Trident is being perfected]. *Morskoi Sbornik [USSR] 1980 (6): 79-83.* Traces the development of the Trident missile for nuclear submarines. Foresees the structuring of the new missile Ohio, whose construction began in 1976. The present leaders of the United States, working with the military-industrial complex, have sharply inflated the strategic arms race. Photo, 3 tables, 3 diagrams. J. Mamtora

159. Ramsey, Bill. NOT THE BOMBS, JUST THE PARTS. *Southern Exposure 1979 7(4): 41-43.* Surveys facilities currently being used for the manufacture and assembly of the components used in the construction of US nuclear arms. Four out of the seven operating facilities are now located in the South. More and more production, previously outside, is being transferred to the southern plants. A modernization program is in progress. In addition, nuclear wastes from naval nuclear vessels are handled in the South and components for such vessels are likewise manufactured there. Lists and describes each of the four southern plants and its activities. Map. R. V. Ritter

160. Rodionov, B. and Novikov, N. VNIMANIE: TOMAGAVK! [Attention: Tomahawk!]. *Morskoi Sbornik [USSR] 1980 (5): 77-83.* The Carter administration increased the arms race when, at the December 1979 NATO meeting it proposed deploying cruise missiles in Europe. Submarines are to be armed with

the strategic rocket, Tomahawk, of which there are four types. The cruise missile is not a new weapons system; a similar, less advanced system was available during World War II. However, Tomahawk does possess a number of advantages, notably target accuracy. The Soviet Union is developing cruise missiles. Photo, map, table, 6 diagrams. J. Mamtora

161. Rosenberg, David Alan. AMERICAN ATOMIC STRATEGY AND THE HYDROGEN BOMB DECISION. *J. of Am. Hist. 1979 66(1): 62-87.* Utilizes recently declassified material to analyze the development of American plans for a possible atomic war against the USSR between 1945 and 1950 and the military's role in President Harry S. Truman's decision in 1950 to develop the hydrogen bomb. Concerned over the strength of Soviet conventional forces, the United States by 1949 was firmly committed to a defensive strategy based on atomic weapons. The military's concern over how to counter Soviet conventional capability and possible technological advances prompted the Joint Chiefs of Staff to recommend development of the hydrogen bomb. President Truman accepted the position of the Joint Chiefs on thermonuclear development. 95 notes.
T. P. Linkfield

162. Rosenberg, David Alan. U.S. NUCLEAR STOCKPILE, 1945 TO 1950. *Bull. of the Atomic Sci. 1982 38(5): 25-30.* Recent disclosures about the small size of the US nuclear stockpile during 1946-50 indicate a potential weakness at a time of great international instability; secrecy surrounding the size of the arsenal helped maintain a false view of American nuclear superiority in succeeding years, while handicapping the flexibility of American foreign policy.

163. Rushford, Greg. HOW THE CONDOR WAS KILLED. *Washington Monthly 1976 8(10): 54-60.* Relates the series of events which led up to Congressional rejection of the Condor missile system, a system which the Defense Department badly wanted, 1960's-76.

164. Salaff, Stephen. THE LUCKY DRAGON. *Bull. of the Atomic Scientists 1978 34(5): 21-23.* The crew and catch of a Japanese fishing vessel, the *Lucky Dragon,* were exposed to radioactive ash from the 1 March 1954 hydrogen bomb detonation on Bikini Atoll. US officials refused to admit that deaths and contamination resulted from their exposure.

165. Schneider, Barry. BIG BANGS FROM LITTLE BOMBS. *Bull. of the Atomic Scientists 1975 31(5): 24-28.* Discusses the US nuclear arms system, 1950's-75.

166. Schütze, Walter. LES NOUVELLES ARMES GUIDÉES AVEC PRÉCISION ET LEURS CONSÉQUENCES MILITAIRES ET POLITIQUES [The new precision guided missiles and their military and political consequences]. *Défense Natl. [France] 1976 32(3): 69-86.* Discusses US military technology since 1945, the characteristics of the new precision guided miss1les, and their implications on the conventional military, nuclear, and strategic fronts.

167. Schwartz, Charles. THE BERKELEY CONTROVERSY OVER NUCLEAR WEAPONS. *Bull. of the Atomic Scientists 1978 34(7): 20-24.* Survey of the involvement of the University of California in the production of nuclear arms 1945-78, indicates that control over the labs is military, not civilian, that

there is a substantial interchange of high level personnel between the labs and the military, lab managers tend to take an aggressive political role in promoting particular weapons systems as well as military strategy and national policy, supervision by national leaders is minimal, and information provided to the public is likewise minimal.

168. Schwellen, Joachim. ULMS-PROGRAMM TROTZ SALT-GESPRÄCHEN [The Undersea Long-range Missile System program in spite of SALT]. *Aussenpolitik [West Germany] 1972 23(5): 257-268.* The program for the establishment of an Undersea Long-range Missile System (ULMS) that has been planned as an answer to the numerical superiority of Soviet nuclear power in the late 1960's is carried on in spite of the beginning of the Strategic Arms Limitation Talks.

169. Slay, Alton D. MX, A NEW DIMENSION IN STRATEGIC DETERRENCE. *Air Force Mag. 1976 59(9): 44-49.* Discusses the modernization of the Intercontinental Ballistic Missile Force, and in particular, the fourth generation ICBM, known as Missile X or MX, 1957-76.

170. Smirnov, A. and Novikov, N. ZA DVADTSAT' LET OT "IUPITERA" K "TRAIDENTU" [The 20 years from Jupiter to Trident]. *Morskoi Sbornik [USSR] 1980 (4): 82-84.* Reviews the development of ballistic missiles from the Jupiter in 1955 to the Trident in 1980, including the problems of size, liquid fuel, and the fire hazard. Since the United States developed the Polaris in April 1960, it has been modified every two years. Every six years a new rocket has appeared. This intense development has increased the firing distance fourfold. The power of the head has doubled. The Soviet Union regards these developments as a threat, and demands more of the Soviet Armed Forces for military readiness. 2 tables.
J. Mamtora

171. Smit, Wim A. and Boskma, Peter. LASER FUSION. *Bull. of the Atomic Scientists 1980 36(10): 34-38.* Discusses the lack of regulations regarding research on laser fusion and the problems that arise as nations are able to develop thermonuclear weapons without restriction; 1960-79.

172. Smith, Datus C., Jr. THE PUBLISHING HISTORY OF THE "SMYTH REPORT." *Princeton U. Lib. Chronicle 1976 37(3): 191-200.* Discusses the publishing history of H. D. Smyth's 1945 War Department report on the military and strategic feasibility of using atomic energy, 1946-73, including translation rights.

173. Steinbruner, John and Carter, Barry. ORGANIZATIONAL AND POLITICAL DIMENSIONS OF THE STRATEGIC POSTURE: THE PROBLEMS OF REFORM. *Daedalus 1975 104(3): 131-154.* Due to the complicated organizational procedures of the defense establishment, the political management of weapons acquisition is poorly developed. Certain organizational changes would lessen the problems involved: specify the desired military requirements and integrate these into general defense needs; end concurrent scheduling of research, development, and production; improve managerial control of budget procedures; and analyze the domestic implications of weapons programs. Based on federal documents and secondary sources; 34 notes.
E. McCarthy

174. Sussman, Leonard R. INFORMATION CONTROL AS AN INTERNATIONAL ISSUE. *Pro. of the Acad. of Pol. Sci. 1982 34(4): 176-188.* Describes what is revolutionary about new defense communication systems and how the Defense Department has changed its organization and the communications systems for strategic nuclear weapons. Compares the system of the USSR to that of the United States. Primary sources; 7 notes. T. P. Richardson

175. Tell, Geoffrey. THE SAFEGUARD DEBATE: IMAGE AND REALITY. *J. of Royal United Services Inst. for Defence Studies [Great Britain] 1974 119(4): 40-46.* The ballistic missile defense (Safeguard) debate, 1968-72, may well be regarded as one of the most significant of the era in the field of US defense policy. Two levels of debate were involved: technical argument and differences in the judgmental assumptions of the decision-makers. Concludes that "many of the differences that arose between the two sides in the Safeguard debate seem to have been fueled, not so much by disputes over the Safeguard system itself, but by different positions on the workings of the international political system." 34 notes. D. H. Murdoch

176. Terzibaschitsch, Stefan. AMERIKAS ATOM-UNTERSEEFLOTTE IN DER ZWANGSJACKE DER WERFTKAPAZITÄT [US atomic submarine fleet in the strait jacket of shipyards' capacity]. *Marine Rundschau [West Germany] 1975 72(7): 385-403.* The *USS Nautilus* was 20 years old in 1975. The author reviews the development of atomic submarines during 1955-75 and notes that shipyards do not have the capacity for future development. 7 pictures, 3 tables. G. E. Pergl

177. Terzibaschitsch, Stefan. DIE SITUATION DER AMERIKANISCHEN FLUGZEUGTRÄGERWAFFE [The situation of the American aircraft carrier weapon]. *Marine Rundschau [West Germany] 1974 71(2): 69-92.* Explains the strategic value of aircraft carriers in the US Navy, and the tasks of this weapon since World War II. Focuses on conventional and nuclear propulsion for carriers, registers deactivated ships, and presents the actual strength of US active carriers from *Hornet* (1943) to *Nimitz* (1974). 11 illus., 7 tables.
G. E. Pergl

178. Terzibaschitsch, Stefan. WAFFENSCHUL- UND ERPROBUNGSSCHIFFE DER U.S. NAVY 1930-76 [Weapons training and test ships of the US Navy, 1930-76]. *Marine Rundschau [West Germany] 1976 73(6): 351-367.* Details the course of naval weapons development through the histories of the four gunnery training and weapons test ships of the US Navy since 1930. Three were old battleships—*Utah, Wyoming,* and *Mississippi*—and the last a converted seaplane tender, *Norton Sound.* The first two served to train antiaircraft gunners at sea until *Utah* was lost in 1941 and *Wyoming* scrapped in 1947. *Mississippi* replaced them in 1948 and also served as evaluation ship for the new antiaircraft missile system until stricken in 1956. The Navy rebuilt *Norton Sound* as its first true floating weapons laboratory in 1948. She serves with continuing modifications today as a primary test ship for nearly all naval missile systems. 20 photos.
K. W. Estes

179. Ulsamer, Edgar. AT EGLIN AFB ADTC ADDS NEW DIMENSIONS TO TAC AIR. *Air Force Mag. 1974 57(3): 30-35.* Discusses the revolutionary impact of guided weapons on the scope and effectiveness of tactical air power, with reference to the programs of the Armament Development and Test Center at Eglin Air Force Base, Florida, 1960's-70's.

180. Ulsamer, Edgar. DNA'S BUSINESS: THINKING THE UNTHINKABLE. *Air Force Mag. 1976 59(9): 50-54.* Discusses the activities of the Defense Nuclear Agency, which investigates the effects and side effects of nuclear arms, 1963-76.

181. Ulsamer, Edgar. INTERVIEW WITH GEN. GEORGE S. BROWN: AT STAKE USAF'S ECONOMIC SURVIVAL. *Air Force Mag. 1973 56(9): 50-57.* General George S. Brown, US Air Force's Chief of Staff, discusses how to overcome the problem of rising costs of developing, maintaining, buying, and manning aircraft, missiles, and weapon systems, and inflation, with particular reference to the Acquisition Cost Evaluation project.

182. Ulsamer, Edgar. THE SOVIET DRIVE FOR AEROSPACE SUPERIORITY. *Air Force Mag. 1975 58(3): 44-49.* Discusses the Soviet creation of Intercontinental Ballistic Missile (ICBM) forces, the US military research and development (R & D) efforts, and the difference in the level of military investment in the two states.

183. Ulsamer, Edgar. USAF'S NEW FLEXIBLE CAPABILITIES. *Air Force Mag. 1975 58(4): 26-31.* Discusses the capabilities of the US Air Force's weapons systems in the 1970's, including the mobile-based Intercontinental Ballistic Missile, maneuvering warheads, and all-weather modular standoff weapons of the Air Force Systems Command.

184. Weisgall, Jonathan M. THE NUCLEAR NOMADS OF BIKINI ATOLL. *Foreign Policy 1980 (39): 74-98.* In 1946, the United States evacuated the residents of the Pacific atoll of Bikini (in the Marshall Islands) in order to use their homeland as a nuclear arms testing site. Since then, the record of American dealings with the refugees has been one of "neglect, thwarted hopes, and unkept promises." The United States is planning to terminate its trusteeship over the area, and still has not reached a settlement with the Bikinians. Map.

T. L. Powers

185. Werrell, Kenneth P. THE CRUISE MISSILE: PRECURSORS AND PROBLEMS. *Air U. Rev. 1981 32(2): 36-50.* Sketches the history of the cruise missile, "a dispensable, pilotless, self-guided, continuously powered, air-breathing warhead-delivery vehicle"; 1914-80.

186. Whitmore, William F. THE ORIGIN OF POLARIS. *US Naval Inst. Pro. 1980 106(3): 55-59.* The US Navy did not become actively involved in a ballistic missile program until 1955. But late in that year, as a result of the submission of the Killian Report to President Dwight D. Eisenhower, the Navy was ordered to develop a sea-based system that used the Army's Jupiter missile. The Navy organized a Special Projects Office and put Rear Admiral William F. Rabon in charge of it. The author of this article was the Office's chief scientist and senior operations research advisor. By late 1956, it was apparent that the

Jupiter could not be used, and the Navy was given the go-ahead for its Polaris program. The first fleet ballistic missile submarine, the USS *George Washington* (SSBN-598), was commissioned on 30 December 1959. In July 1960, the *George Washington* fired her first Polaris missile from underwater. Secondary sources; 3 photos, 3 notes. A. N. Garland

187. Wit, Joel S. AMERICAN SLBM: COUNTERFORCE OPTIONS AND STRATEGIC IMPLICATIONS. *Survival [Great Britain] 1982 24(4): 163-174.* Discusses the historical development of sea-launched ballistic missiles (SLBM) from 1969-82, especially the strategic and arms-control implications of their improved accuracy.

188. Wrenn, Catherine; West, Ronald E.; and Kreith, Frank. ATTITUDES OF COLORADO CITIZENS TOWARD UNDERGROUND NUCLEAR EXPLOSIONS. *Colorado Q. 1974 23(2): 159-172.*

189. Wright, Frank. NEVADA HUMANISTS VIEW MX. *Nevada Hist. Soc. Q. 1981 24(2): 176-182.* Since 1979, Nevada humanists have attempted to assess the impact of federal proposals to preempt large areas of Nevada for a roving MX missile weapons complex. In a film, "Battle Born: MX in Nevada," and a booklet, *MX in Nevada: A Humanistic Perspective* (1980), humanists have attacked the picture of Nevada as a wasteland in which a weapons complex would create few adverse consequences. Secondary sources; 17 notes. H. T. Lovin

190. York, Herbert F. and Greb, G. Allen. MILITARY RESEARCH AND DEVELOPMENT: A POSTWAR HISTORY. *Bull. of the Atomic Scientists 1977 33(1): 12-26.* Overview of the technological arms race focuses on the development of managerial and supervisorial positions, specifically those held by scientists and engineers, 1945-75.

191. York, Herbert F. THE ORIGINS OF THE LAWRENCE LIVERMORE LABORATORY. *Bull. of the Atomic Scientists 1975 31(7): 8-14.* Herbert York, the first director of the Lawrence Livermore laboratory, recalls the circumstances surrounding the establishment of that laboratory, the rivalry between it and Los Alamos, and the accelerated rate of progress in nuclear arms design that resulted. Based in part on a chapter of the author's forthcoming book, *The Advisors: Oppenheimer, Teller and the Super Bomb*; 6 illus., 6 notes.
D. J. Trickey

192. Zacharias, Jerrold R. PRAGMATISM, SECRECY AND MORAL VALUES. *Bull. of the Atomic Scientists 1976 32(10): 57-59.* Review article prompted by Herbert F. York's *The Advisors: Oppenheimer, Teller and the Superbomb* (W. H. Freeman, 1975) which analyzes the decisions and decision-making involved in producing the American hydrogen bomb.

3

THE BALANCE OF TERROR

193. Ackley, Richard T. THE STRATEGIC BALANCE IS TIPPING. *J. of Social and Pol. Studies 1978 3(3): 203-216.* Argues that there is a growing imbalance between US and Soviet strategic forces and programs; drawn from 1969-77 statistics.

194. Adams, Gordon. WHAT DO WEAPONS SECURE? *Bull. of the Atomic Sci. 1982 38(4): 8-10.* Examines the buildup of the US military in the 1980's, the role of the Defense Department, congressional supporters, and the defense industry; discusses the growing criticism of the buildup among some elements of US society.

195. Alexander, Charles C. NOT-SO-HAPPY DAYS. *Rev. in Am. Hist. 1979 7(3): 427-432* Reviews Robert A. Divine's *Blowing on the Wind: The Nuclear Test Ban Debate, 1954-1960* (New York; Oxford U. Pr., 1978) and discusses Soviet-American foreign relations, the escalation of nuclear arms, and the formation of public policy during the Cold War.

196. Allen, Scott. SANCHO! MY ARMOR! MY SWORD! *US Naval Inst. Pro. 1977 103(10): 18-23.* Since the early 1960's, the US defense policy has depended on the so-called mutual assured destruction (MAD) concept. Unfortunately, our strategic planners misinterpreted Soviet intentions when they developed the MAD concept. The Soviet doctrine of "war fighting" calls for an increase in both Soviet conventional and nuclear forces to a point where they can win any war. In short, the Soviet doctrine states simply "Dictate—or take!" The Soviets are not concerned with equal forces; they want dominant forces. The only thing that will stop the Soviets is a clear sign from us that we will match their best, and better it. Drawing. A. N. Garland

197. Altherr, Marco. LES ORIGINES DE LA GUERRE FROIDE: UN ESSAI D'HISTORIOGRAPHIE [The origins of the Cold War: a historiographical essay]. *Relations Int. [France] 1977 (9): 69-81.* Notes the extent and the passion of the debate among historians about the origins of the Cold War. Divides opinions into traditional, realist, revisionist, and postrevisionist schools, the revisionists in turn being classified as "hard" and "soft." Reviews leading points of controversy, including eastern Europe, the role of Harry Truman, the atomic bomb, and economic determinism. A conflict of generations with different historical experiences helps explain the historiographical disagreements.
R. Stromberg

198. Andersen, A. E. FLEXIBLE RESPONSE AND THE MARINE CORPS. *Marine Corps Gazette 1972 56(1): 40-44.* Describes the change in US military strategy after 1953, modifying the commitment to all-out nuclear war with the concept of a limited response to enemy aggression.

199. Appleby, Charles A. NUCLEAR STRATEGY AT THE CROSS-ROADS. *SAIS Rev. 1981 (1): 69-82.* Discusses "punishment" and "denial" schools of nuclear deterrence among American strategists.

200. Arbatov, Georgi. THE DANGERS OF A NEW COLD WAR. *Bull. of the Atomic Sci. 1977 33(3): 33-40.* Discusses the arms race, detente, nuclear arms control arrangements, SALT, Soviet compliance with human rights provisions of the Helsinki Agreement, and Soviet treatment of dissidents, 1975-76.

201. Aspaturian, Vernon D. THE USSR, THE USA AND CHINA IN THE SEVENTIES. *Military R. 1974 54(1): 50-63.* Discusses the foundations of a post-Cold War world order being established by the United States, the USSR, and China in the 1970's, emphasizing the function of nuclear stalemate.

202. Bailey, Martin J. DETERRENCE, ASSURED DESTRUCTION, AND DEFENSE. *Orbis 1972 16(3): 682-695.* Considers military strategy since the 1950's and offers defense policy suggestions. S

203. Ballard, William T. STRATEGIC POWER: ON BALANCE. *Air U. R. 1973 24(4): 89-94.* Reviews Joseph I. Coffey's *Strategic Power and National Security* (U. of Pittsburgh Press, 1971), which analyzed U.S., Soviet, and Communist Chinese military strength. Concludes that U.S. military superiority has diminished in recent years. Note. J. W. Thacker, Jr.

204. Barlow, William J. SOVIET DAMAGE-DENIAL: STRATEGY, SYSTEMS, SALT, AND SOLUTIONS. *Air U. Rev. 1981 32(6): 2-20.* Examines Soviet nuclear objectives and the principal historical elements of Soviet military strategy, emphasizing the relationship between US and Soviet nuclear doctrine and their respective positions on the Strategic Arms Limitation Talks.

205. Barnaby, Frank. ARMAMENTS AND DISARMAMENT. *Bull. of the Atomic Scientists 1976 32(6): 25-32.* Details quantitatively the extent, nature, and significance of wars, armaments development, the arms trade, and nuclear proliferation since World War II.

206. Barnaby, Frank. WORLD ARSENALS IN 1978. *Bull. of the Atomic Scientists 1979 35(7): 18-26.* Reviews the extent of military spending, existing arsenals, and approaches toward disarmament.

207. Beres, Louis René. TILTING TOWARD THANATOS: AMERICA'S "COUNTERVAILING" NUCLEAR STRATEGY. *World Pol. 1981 34(1): 25-46.* America's current nuclear strategy seeks to improve deterrence with a counterforce targeting plan that exceeds the requirements of mutual assured destruction. This "countervailing" nuclear strategy codifies an enlarged spectrum of retaliatory options. J/S

208. Bernstein, Barton J. THE CHALLENGES AND DANGERS OF NUCLEAR WEAPONS: FOREIGN POLICY AND STRATEGY, 1941-1978. *Maryland Hist. 1978 9(1): 73-99.* Reviews US nuclear policies since 1941. The arms race, founded upon mutual distrust, will continue, but nuclear war is unlikely unless an amoral technologist errs. Primary and secondary sources; 79 notes. G. O. Gagnon

209. Bernstein, Barton J. THE WEEK WE ALMOST WENT TO WAR. *Bull. of the Atomic Scientists 1976 32(2): 12-21.* Discusses military strategy and the actions of President John F. Kennedy and USSR Premier Nikita Khrushchev during the 1962 Cuban Missile Crisis, questioning whether the threat of nuclear war was as great as it was thought to be.

210. Beukel, Erik. DEN NYE KERNEVÅBENDEBAT OG DEN AMERIKANSKE REGERING [The new debate on nuclear arms and the American government]. *Internasjonal Politikk [Norway] 1980 (3): 419-447.* In the sixties and the first half of the seventies there was a strong support for Mutual Assured Destruction (MAD) and bilateral American-Soviet arms control. Scholars and writers today emphasize the moral failure of MAD and the necessity of strategic defense. A strong suspicion of the USSR is common, and unilateral American defense measures are advocated. The Carter administration supports a moderate "war fighting" doctrine today. Considers the ABM treaty, the proliferation of nuclear weapons, and the psychological barrier against the use of nuclear weapons. J/S

211. Black, Edwin F. NATIONAL SECURITY POLICY: GIVE THE PUBLIC A VOICE. *US Naval Inst. Pro. 1977 103(6): 18-24.* In May 1972, President Richard M. Nixon broke with the past concerning US foreign policy operations and sent to the Congress the three policy documents that made up the so-called Moscow Peace Package. Until then, such documents would have been highly classified, and few people in or out of our highest government policy circles would ever have known their contents. Much has happened since 1972 to that package of documents, as well as to our doctrines of mutual assured destruction (MAD) and "essential equivalence." The United States is falling behind as a world power. President Jimmy Carter and his administration can do much to correct the present situation and should open to the public full and frank discussions of our national security requirements conducted at the National Security Council and Congressional levels. 2 photos. A. N. Garland

212. Blechman, Barry M. and Kaplan, Stephen S. ARMED FORCES AS POLITICAL INSTRUMENTS. *Naval War Coll. Rev. 1978 30(4): 80-87.* During 1946-75 the United States used its armed forces for political purposes 215 times. A show of force is most effective in delaying adverse political decisions in foreign countries and in supporting the policies of allies rather than blunting hostile reactions. When Soviet forces are also present, US gestures are less efficacious, although the presence of land-based tactical aircraft and strategic nuclear forces tends to diminish the Soviets' demonstration. This may indicate that the Air Force, rather than the Navy, works better when a show of armed strength is used. Displays by the military have been effective, but they cannot adequately substitute for long-range policies or consistent goals. J. W. Leedom

213. Blechman, Barry M. and Hart, Douglas M. THE POLITICAL UTILITY OF NUCLEAR WEAPONS: THE 1973 MIDDLE EAST CRISIS. *Int. Security 1982 7(1): 132-156.* Analyzes past nuclear threats, particularly the one that took place between the United States and the USSR during the October War.

214. Bloomfield, Lincoln P. AMERICAN APPROACHES TO MILITARY STRATEGY, ARMS CONTROL, AND DISARMAMENT: A CRITIQUE OF THE POSTWAR EXPERIENCE. *Policy Studies J. 1979 8(1): 114-119.* Development of US strategic thinking is based on perception of a hostile USSR, global turbulence, and a nuclear arsenal, 1945-70's.

215. Borst, Gert. USA UND UdSSR—WETTRÜSTEN UND GLEICHGEWICHT [USA and USSR—arms race and balance]. *Osteuropa [West Germany] 1974 24(9): 627-643.* Surveys US and Soviet nuclear armaments 1965-73. Though internal political pressures prolong the arms race on both sides, the two superpowers face a rapidly diminishing probability of changing the existing "balance of terror." To continue the nuclear race at the present rate will invite domestic instability, weaken alliance systems, and lead to nuclear proliferation elsewhere. 3 tables, 3 diagrams, 18 notes. R. E. Weltsch

216. Bowman, Richard C. NATO IN A TIME OF CRISIS. *Air Force Mag. 1975 58(4): 49-54.* Considers problems of NATO, 1950's-70's, particularly the crises in energy and economics, and the threat of atomic warfare, and discusses ways to preserve NATO's viability.

217. Božić, Nemanja. RAVNOTEŽA SILE U VOJNO-POLITIČKOJ MISLI U SAD [Balance of power in military-political thought in the United States]. *Medjunarodni Problemi [Yugoslavia] 1979 31(2): 93-101.* Examines US doctrine on nuclear arms and disarmament in the context of US views of the balance of military and diplomatic power. Serbocroatian.

218. Brach, Hans Günter. AMERIKANISH-SOWJETISCHE BEZIEHUNGEN: VOM KALTEN KRIEG ZUR ENTSPANNUNG? AMERIKANISCHE ENTSPANNUNGSPOLITIK GEGENÜBER DER SOWJETUNION UND CHINA (I) [US-USSR relations: from cold war to detente? US policies of detente toward the USSR and China: Part I]. *Neue Politische Literatur [West Germany] 1979 24(4): 513-548.* While immediate postwar armaments developments were controlled by the United States and the USSR, the explosion of the Chinese nuclear bombs and the increasing ideological differences between China and the USSR have changed the balance of power. Richard Nixon's foreign policy changed the modalities and style of US foreign policy but not US goals. 127 notes. R. Wagnleitner

219. Brodie, Bernard. THE DEVELOPMENT OF NUCLEAR STRATEGY. *Int. Security 1978 2(4): 65-83.* Reviews advances in weaponry since 1945 and concludes that, while the United States no longer has unchallenged superiority in nuclear armaments, the vast destructive power of such weapons is still the greatest deterrent to their possible use between the United States and the USSR, and that preemptive or first-strike strategies for either side are unreasonable.

220. Brooks, Harvey. THE MILITARY INNOVATION SYSTEM AND THE QUALITATIVE ARMS RACE. *Daedalus 1975 104(3): 75-97.* Provided that the Strategic Arms Limitation Treaty (1972) remains in force, the qualitative arms race is not likely to lead to major strategic instabilities. The most promising control procedures are mutually agreed upon test limitations and a comprehensive nuclear test ban. Technology has helped to fuel the arms race, yet a selective reduction in the rate and areas of innovation would contribute to international security. Secondary sources; 17 notes. E. McCarthy

221. Bruce, Geoffrey F. SALVATION OR ARMAGEDDON? *Int. Perspectives [Canada] 1975 (3): 31-36.* Nuclear proliferation poses the dilemma of nuclear use: as an energy source or for armament.

222. Burke, Gerard K. THE METAPHYSICS OF POWER REALITIES AND NUCLEAR ARMAMENTS. *Military Rev. 1975 55(9): 14-24.* Except in terms of raw numbers of launch vehicles and nuclear warheads, there is no effective way to configure the base equation upon which the all-important metaphysical perception of power is calculated. Therefore, the United States should not relinquish its international stature by accepting anything less than true parity with the USSR. Primary and secondary sources; 3 illus., 18 notes.
J. K. Ohl

223. Burshop, E. H. S. SCIENTISTS AND SOLDIERS. *Bull. of the Atomic Scientists 1974 30(9): 4-8.* Discusses attitudes of American scientists from the Jason group of the US Institute of Defense Analysis toward their work and its repercussions for US involvement in the Vietnam War, 1966-73.

224. Buteux, Paul. THEATRE NUCLEAR WEAPONS AND EUROPEAN SECURITY. *Canadian J. of Pol. Sci. [Canada] 1977 10(4): 781-808.* Evaluates to what extent present arguments about the strategic and political function of nuclear armaments deployed by NATO are comparable to those made in the 1950's and 1960's. 52 notes. R. V. Kubicek

225. Canby, Steven L. DAMPING NUCLEAR COUNTERFORCE INCENTIVES: CORRECTING NATO'S INFERIORITY IN CONVENTIONAL MILITARY STRENGTH. *Orbis 1975 19(1): 47-71.* Questions the ability of tactical nuclear arms to substitute for conventional weapons in US and NATO military strategy toward the USSR in the 1970's, including issues of economic efficiency.

226. Canby, Steven L. NATO MUSCLE: MORE SHADOW THAN SUBSTANCE. *Military R. 1973 53(2): 65-74.* NATO's posture and operating procedures are inappropriate to its strategic requirements. By emphasizing nuclear deterrence and conventional forces that are structured for a long war it is ill-equipped to fight a short war in the crucial central region. The Warsaw Pact, meanwhile, has bought more security for less money by opting for high initial combat forces which can potentially swamp NATO before it completes mobilization. Reprinted from *Foreign Policy* 1972 8 (see abstract 11A:4945). 4 illus.
J. K. Ohl

227. Carlin, Robert J. A 400-MEGATON MISUNDERSTANDING. *Military R. 1974 54(11): 3-12.* Explores the existing relation between war fighting and war deterrence and argues that continued effort be given to finding standards for an effective strategic deterrent force.

228. Carreras Matas, Narciso. ASUNTOS DE PLANEAMIENTO DE DEFENSA [The business of defense planning]. *Rev. General de Marina [Spain] 1978 194(1): 7-32.* Describes the organizations and process of short- and long-range defense planning in the US Department of Defense. An elaborate planning system is designed to ensure that an effective deterrence, a readiness to fight anywhere under any conditions, and a capability for rapid mobilization will always be available despite constantly changing circumstances. 10 tables.
W. C. Frank

229. Carter, Barry. FLEXIBLE STRATEGIC OPTIONS: NO NEED FOR NEW STRATEGY. *Survival [Great Britain] 1975 17(1): 25-31.* Evaluates the flexibility of US military strategy and nuclear arms potential toward the USSR in the 1970's, emphasizing the policies of Secretary of Defense James Schlesinger.

230. Clark, Wesley K. GRADUALISM AND AMERICAN MILITARY STRATEGY. *Military Rev. 1975 55(9): 3-13.* Examines gradualism, also known as graduated response or graduated pressure, and its relationship to the tension derived from employing military power to achieve political ends. This policy failed in Vietnam because it was adopted without a clear plan of how the conflict was to be terminated. Yet the concept is not dead, and the problem posed by its use must be resolved if the United States is to maintain an effective military strategy. Primary and secondary sources; 27 notes. J. K. Ohl

231. Cohen, S. T. U.S. STRATEGIC NUCLEAR WEAPON POLICY—DO WE HAVE ONE? SHOULD THERE BE ONE? *Air U. R. 1975 26(2): 12-25.* Examines present US strategic nuclear arms policy in order to determine who makes the policy, what forces would be needed, and what circumstances would cause such forces to be used. Advocates the exploitation of advanced technology for new strategic systems rather than the suppression of technology and new systems through agreements like the Strategic Arms Limitation Talks (SALT). Based on Defense Department reports; 2 notes. J. W. Thacker, Jr.

232. Coles, Harry L. STRATEGIC STUDIES SINCE 1945: THE ERA OF OVERTHINK. *Military R. 1973 53(4): 3-16.* A vast outpouring of strategic studies during 1945-68 persuaded many American military intellectuals that application of military power to political ends could be reduced to something of an exact science. The Vietnam war, however, has shown that too many of these studies were based on hypothetical models and schemes that did not adequately take into account historical particularity. Based on primary and secondary sources; 4 illus., 43 notes. J. K. Ohl

233. Collins, John M. MANEUVER INSTEAD OF MASS: THE KEY TO ASSURED STABILITY. *Orbis 1974 18(3): 750-762.* Discusses US strategic policy concerning nuclear arms since the 1960's. S

234. Commager, Henry Steele. COMMITMENT TO POSTERITY: WHERE DID IT GO? *Am. Heritage 1976 27(5): 4-7.* The United States has maintained the institutions of the Founding Fathers while largely betraying their principles. The Founding Fathers looked to posterity for vindication, as the future belonged to America. Today Americans do not look to posterity, perhaps because with the threat of nuclear holocaust, they foresee no posterity. Illus.

J. F. Paul

235. Cowhey, Peter F. and Laitin, David D. BEARING THE BURDEN: A MODEL OF PRESIDENTIAL RESPONSIBILITY. *Int. Studies Q. 1978 22(2): 267-296.* Presidential responsibility for decisions in foreign policy varies according to the political structure of the international system, the degree of consensus among the countries involved, and the degree of consensus with the United States. Suggests that a model that incorporates these factors can provide a set of decisionmaking rules for future presidents. Demonstrates how the model works by applying it retrospectively to decisions involving the Marshall Plan, the Multilateral Nuclear Force, the Cuban Missile Crisis and the Vietnam War. Fig., 31 notes. E. S. Palais

236. Crosby, Ralph D., Jr. THE CUBAN MISSILE CRISIS: SOVIET VIEW. *Military Rev. 1976 56(9): 58-70.* Discusses the USSR's desire to gain a strategic balance of power with the United States in placing nuclear missiles in Cuba in 1962, and political factors in the Soviets' decision to withdraw them following President John F. Kennedy's announcement of a naval blockade.

237. DeForth, Peter W. U.S. NAVAL PRESENCE IN THE PERSIAN GULF: THE MIDEAST FORCE SINCE WORLD WAR II. *Naval War Coll. R. 1975 28(1): 28-38.* Nuclear stalemate and its most recent manifestation, detente, have led the United States to adopt a policy of "conflict avoidance" as the primary emphasis of national strategy. In response, the Navy has given the presence role a far higher priority in its consideration of the four principal mission areas. One example of the success of this role is the small and relatively weak force in the Persian Gulf—ill-equipped for a major combat role, but nonetheless a model of presence which can be useful in planning future US strategy. J

238. Denis, Jacques. LES HORIZONS DE LA COMPÉTITION STRATÉGIQUE SOVIÉTO-AMÉRICAINE [Horizons of Soviet-American strategic competition]. *Défense Natl. [France] 1980 36(8): 41-52.* Discusses the idea of equilibrium between the armed forces of the USSR and the United States in light of fundamental political and cultural differences between the two countries.

239. DePuy, William E. TACTICAL NUCLEAR WARFARE. *Marine Corps Gazette 1981 65(4): 64-66.* Reviews John P. Rose's *The Evolution of U.S. Army Nuclear Doctrine* (Boulder, Colo.: Westview Pr., 1980), which views the evolution of US Army nuclear doctrine with dismay because it has been developed in accordance with political preferences rather than the real nature of the threat and the rigors of the nuclear battlefield.

240. Divine, Robert A. EISENHOWER, DULLES, AND THE NUCLEAR TEST BAN ISSUE: MEMORANDUM OF A WHITE HOUSE CONFERENCE, 24 MARCH 1958. *Diplomatic Hist. 1978 2(3): 321-330.* Discusses General Andrew Goodpaster's memorandum, recently declassified, of a White

House conference on nuclear testing held on 24 March 1958. The Soviets at this time were staging an apparently accelerated series of tests, with the purpose, as the CIA correctly surmised, of following up their tests with the highly embarrassing announcement of a unilateral suspension of testing. Secretary of State John Foster Dulles convened the meeting to discuss the possibility of stealing a march on the Soviets by announcing a two-year test suspension before the USSR's expected announcement. Dulles dropped the idea in response to adverse comment from Atomic Energy Commission Chairman Lewis Strauss and various defense spokesmen. Discussion based on archival, contemporary press and periodical, other primary and secondary sources; 14 notes. L. W. Van Wyk

241. Donato, Alberto. IL BILANCIO DELLA DIFESA U. S. A. PER IL 1974-75 [US defense estimate 1974-75]. *R. Marittima [Italy] 1974 107(6): 41-45.* After recalling the reasons which lead the U.S.A. to keep a visible and conventional nuclear balance with the U.S.S.R. in order to support all their very wide moral interests in the world, the author reviews the criteria according to which the Defense Estimate has been planned and the objectives to be achieved. J

242. Dornan, Robert K. EXPORTING AMERICAN TECHNOLOGY: A NATIONAL SECURITY PERSPECTIVE. *J. of Social and Pol. Studies 1977 2(3): 131-142.* The commercial exportation of technology to Communist countries might diminish US hegemony and threaten national security; discusses export of computer technology associated with nuclear weapons, the Export Administration Act of 1969, and the operations of the International Export Control Coordinating Committee, 1949-78.

243. Dumas, Lloyd J. THIRTY YEARS OF THE ARMS RACE: THE DETERIORATION OF ECONOMIC STRENGTH AND MILITARY SECURITY. *Peace and Change 1976 4(2): 3-9.* Maintenance of the military establishment 1945-75 has been a drain on the US economy, and continued strength in military systems creates a deterioration in the military aspect of national defense.

244. Dunn, Lewis A. PROLIFERATION WATCH: HALF PAST INDIA'S BANG. *Foreign Policy 1979 (36): 71-88.* India's detonation of a nuclear device in May 1974 seemed to portend widespread nuclear arms proliferation among developing nations. Thanks largely to American restraining efforts, that threat has been contained, but it is by no means dead. T. L. Powers

245. Ellsberg, Daniel. CALL TO MUTINY. *Monthly Rev. 1981 33(4): 1-26.* Discusses US threats to use nuclear arms to settle international disputes since 1948.

246. Emel'ianov, V. KTO UGROZHAET MIRU I STABIL'NOSTI NA PLANETE? [Who threatens peace and stability on the planet?]. *Aziia i Afrika Segodnia [USSR] 1982 (6): 2-5.* The nuclear arms race threatens annihilation; cites Leonid Brezhnev's proposals for disarmament and notes the warlike course of the United States under Ronald Reagan. Russian.

247. Enthoven, Alain C. U.S. FORCES IN EUROPE: HOW MANY? DOING WHAT? *Foreign Affairs 1975 53(3): 513-532.* Critics of the US role in NATO say that the allies are not doing their fair share financially or militarily. Actually the US burden has become more equalized, and the potential savings

are insignificant by comparison with the importance of what is at stake. NATO land forces are not now outnumbered by those of the Warsaw Pact, but any US withdrawal weakens NATO's political position and could lead to Russian dominance and the "Finlandization" of Europe. NATO should change its emphasis on tactical nuclear weapons toward stronger conventional defenses. Table, 24 notes. C. W. Olson

248. Epstein, William. THE INEXORABLE RISE OF MILITARY EXPENDITURES: DESPITE DETENTE, THE BUDGETS ALWAYS GO UP. *Bull. of the Atomic Scientists 1975 31(1): 17-19.* Discusses the rise of military expenditures in the federal budget in spite of detente and nuclear arms limitation agreements with the USSR, 1959-74.

249. Etzold, Thomas H. GEWALT UND DIPLOMATIE IM NUKLEAREN ZEITALTER [Force and diplomacy in the nuclear age]. *Frankfurter Hefte [West Germany] 1974 29(2): 93-100.* Reviews American defense strategy since 1945.

250. Feiveson, H. A. THINKING ABOUT NUCLEAR WEAPONS. *Dissent 1982 29(2): 183-194.* Discusses the tragic and unreliable thinking of nuclear weapons experts in the United States, who lack relevant decisionmaking experience, are uncertain about Soviet plans for nuclear war, and are awed by nuclear weapons but are willing to fight a nuclear war.

251. Fialka, John J. THE AMERICAN CONNECTION: HOW ISRAEL GOT THE BOMB. *Washington Monthly 1979 10(10): 50-58.* In the late 1960's the Central Intelligence Agency proved that the private Nuclear Materials and Equipment Corporation in Apollo, Pennsylvania, was responsible for Israel's obtaining the enriched uranium necessary to become a nuclear power.

252. Flores Pinel, Fernando. LA GRAN PARADOJA DE LA POLÍTICA INTERNACIONAL CONTEMPORÁNEA [The great paradox of international contemporary politics]. *Estudios Centro Americanos [El Salvador] 1977 32(350): 875-892.* Explains changes in the international political scene since World War II from the viewpoint of power politics rather than ideological considerations. The existence of enormous power vacuums in the postwar world due to the debilitation of certain Western democracies encouraged belief in a Soviet menace. In addition to its superior atomic power, the United States created a system of alliances to militarily surround the USSR. The Soviet response was expansion in Eastern Europe and acceleration of the arms race. Both powers tinkered in the internal politics of the weaker nations in the opposite camp. Territorial dimensions lost significance as power and security became the decisive factors. By 1974, relative nuclear equality of the superpowers led to the initiation of detente due to world fears of autodestruction. Care must be taken with sources as most have been written by participants. 49 notes. Article to be continued.
J. M. Walsh

253. Freedman, Lawrence. THE ATLANTIC CRISIS. *Int. Affairs [Great Britain] 1982 58(3): 395-412.* The crisis in the Atlantic Alliance during 1980 is due in part to the economic recession, the changed nuclear deterrent power of the United States, the US global perspectives and faulty consultation within the alliance.

254. Freedman, Lawrence. NATO MYTHS. *Foreign Policy 1981-82 (45): 48-68.* The NATO doctrine of flexible response—theater nuclear weapons (TNF) in Europe to counterbalance superior Soviet conventional forces—no longer expresses the Atlantic alliance's consensus on defense strategy. NATO must convince the Soviets that an invasion of Western Europe risks certain nuclear retaliation, without alarming Western Europeans uneasy about nuclear arms and indecisive in their support of TNF modernization. In this context, the myth of flexible response offers controversy, not reassurance, for the NATO allies. A credible deterrent strategy of conventional forces is the answer to this problem. M. K. Jones

255. Friedberg, Aaron L. A HISTORY OF THE U.S. STRATEGIC "DOCTRINE": 1945 TO 1980. *J. of Strategic Studies [Great Britain] 1980 3(3): 37-71.* Discusses US strategic nuclear doctrine, focusing on the character, structure, and purpose of US plans for nuclear war and how those plans have evolved. The United States has never adhered to a doctrine of mutually assured destruction, and indeed has never had a strategic nuclear doctrine. Instead of a single integrated set of ideas, values, and beliefs, the United States has had a complex and sometimes contradictory mixture of notions, principles, and policies. And, without a more unified and coherent doctrine, it will be extremely difficult for the United States to make progress toward redressing an increasingly unfavorable strategic balance. Based on Defense Department and other primary sources; 109 notes. A. M. Osur

256. Gaboury, L. R. and Etzold, Thomas H. PROBLEM AND PARADOX: TACTICAL NUCLEAR WEAPONS IN NATO. *Marine Corps Gazette 1976 60(4): 46-50.* Discusses US military strategy regarding the stockpiling of tactical nuclear arms in NATO for the potential defense of Western Europe in the 1970's and the role of US-USSR foreign relations.

257. Gall, Norman. ATOMS FOR BRAZIL, DANGER FOR ALL. *Foreign Policy 1976 (23): 155-201.* West Germany's agreement to provide Brazil with the largest transfer of nuclear technology ever given a developing country is an initial step in worldwide nuclear proliferation. US readiness in the past to provide nuclear technology to both its allies and to private American corporations dictates that the technology is available to almost any country willing to pay the high price. One solution to eliminate competition and provide controls would be an international confederation of suppliers which through its directorate could set the standards and controls for peaceful nuclear development. Primary and secondary sources; 54 notes. C. Hopkins

258. Garigue, Philippe. LES DÉVELOPPEMENTS ET PROBLÈMES DE LA PENSÉE STRATÉGIQUE CONTEMPORAINE [Developments and problems in contemporary strategic thought]. *Can. J. of Pol. Sci. 1975 8(2): 235-253.* Notes particularly the contribution that conflict theory has made as applied to the examination of the strategy of deterrence between the United States and the USSR. 70 notes. R. V. Kubicek

259. Garthoff, Raymond L. THE DEATH OF STALIN AND THE BIRTH OF MUTUAL DETERRENCE. *Survey [Great Britain] 1980 25(2): 10-16.* Within the context of nuclear and thermonuclear weapons possessed by the

United States and the Soviet Union, perhaps the most important change in Soviet foreign policy that took place with Stalin's death originated in Stalin's own Politburo where, despite his opposition, there was an increasing awareness of the political and military importance of nuclear weapons and their significance for Marxist-Leninist theory and for Soviet policy. 17 notes. V. Samaraweera

260. Geneste, Marc. LA "BOMBE À NEUTRONS," LA DÉFENSE DE L'EUROPE ET LA FLEXIBLE [The "neutron bomb," the defense of Europe, and the "flexible response"]. *Défense Natl. [France] 1977 33(12): 43-57.* Assesses the importance of the US neutron bomb as an international threat and as an element to dissuade the pursuit of nuclear armament, with reference to the experiences of the 1960's.

261. Gnevushev, N. "NOVAIA IADERNAIA STRATEGIIA" SSHA I RAZVIVAIUSHCHIESIA STRANY [The US "new nuclear strategy" and the developing nations]. *Aziia i Afrika Segodnia [USSR] 1981 (9): 11-13.* Discusses the impact of Directive 59, issued by President Jimmy Carter in 1979 and implemented by President Ronald Reagan, on political and military stability in the world, focusing on the directive's strategy involving US military bases and nuclear arms. Russian.

262. Goldhamer, Herbert. THE US-SOVIET STRATEGIC BALANCE AS SEEN FROM LONDON AND PARIS. *Survival [Great Britain] 1977 19(5): 202-207.* Analyzes attitudes expressed in Great Britain's *The Economist* and France's *Le Monde* pertaining to strategic arms balance between the United States and the USSR, 1948-73, finding that when the Soviet Union seemed to have hegemony, *The Economist* viewed this as a military threat while *Le Monde* saw it as an increased political advantage for the USSR.

263. Gontaev, A. AMERIKANSKAIA VOENNAIA DOKTRINA [American military doctrine]. *Morskoi Sbornik [USSR] 1978 (2): 110-117.* Traces the evolution of American military strategy from World War II to the 1970's, arguing that the concept of "mass nuclear retaliation" in the 1950's, that of "flexible response" in the 1960's, and that of "realistic deterrence" in the 1970's are part of the same overriding strategy of "containing communism."

264. Gorshkov, S. G. NAVIES IN WAR AND IN PEACE. *US Naval Inst. Pro. 1974 100(10): 56-65.* After World War II "American policy and strategy" were aimed at only one goal, "the achievement of rule over the entire world by American monopolistic capital." The only thing that "forced the latter-day pretenders to world supremacy to restrain their aggressive zeal" was the creation and testing of nuclear weapons in the Soviet Union. Even though US longing for world domination has not abated, "the U.S.S.R. and the other Socialist countries have stood as an immovable force in the path of these aspirations of American imperialism." In this latter endeavor, "navies . . . have played and are playing the role of an instrument of state policy in peacetime." 2 photos, 10 notes.
A. N. Garland

265. Gray, Colin S. ACROSS THE NUCLEAR DIVIDE-STRATEGIC STUDIES, PAST AND PRESENT. *Int. Security 1977 2(1): 24-46.* Presents the case for the relevance of pre-nuclear strategic thought and practice in assessing the policy problems arising since the 1945 arrival of the nuclear age.

266. Gray, Colin S. FOREIGN POLICY AND THE STRATEGIC BALANCE. *Orbis 1974 18(3): 706-728.*

267. Gray, Colin S. HOW DOES THE NUCLEAR ARMS RACE WORK? *Cooperation and Conflict [Denmark] 1974 9(4): 285-295.* Many analysts are convinced that they understand how arms races *work*. However, very few arms race studies rest upon a disciplined use of historical data, or upon a willingness to consider seriously more than one or two of the more prominent candidates for the role of *the* driving factor in an arms race. This article suggests that any analysis of the dynamics of the nuclear arms race cannot afford to neglect the operation of the following factors: interstate action-reaction processes; interarmed service action-reaction processes; intraarmed service action-reaction processes; bureaucratic political games; the very individual structures and processes of each political system; military-industrial complexities; foreign policy goals and strategies; and finally, technological innovation. Synergistic combinations of the operation of these factors produce and sustain what we term arms races. Policy prescription for arms control, if it is to be relevant, must rest upon a deep understanding of the processes to be controlled. An appreciation of the true complexity of arms race phenomena is one important intellectual step towards the ability to control. J

268. Gray, Colin S. NATIONAL STYLE IN STRATEGY: THE AMERICAN EXAMPLE. *Int. Security 1981 6(2): 21-47.* Attempts to characterize the unique American orientation to nuclear strategy which is based on its history, geopolitical position, and dominant religion, and assesses the impact of that orientation on our present relationship with the USSR.

269. Gray, Colin S. THE NUCLEAR CONNECTION. *Military Rev. 1974 54(9): 3-13.* Discusses steps Europe and NATO should take in the 1970's in preparation for potential nuclear war, emphasizing the ambiguity of US detente with the USSR.

270. Gray, Colin S. NUCLEAR STRATEGY: THE DEBATE MOVES ON. *J. of the Royal United Services Inst. for Defence Studies [Great Britain] 1976 121(1): 44-50.* Discusses developments in nuclear deterrence theory in the 1970's, specifically the "Schlesinger doctrine" of greater flexibility in US strategic nuclear options and creation of a hard target counterforce capability for the United States to match that of the USSR. Opponents of the notion of nuclear parity have long tended to ignore the unchanged Soviet belief in the possibilities for political exploitation of nuclear imbalance. The core of the Schlesinger doctrine is the creation of limited strategic options below the level of the Single Integrated Operations Plan; however, "not all credible threats deter" and the United States may find that the Schlesinger strategy will prove inadequate in terms of operational realities. 17 notes. D. H. Murdoch

271. Gray, Colin S. STRATEGIC STABILITY RECONSIDERED. *Daedalus 1980 109(4): 135-154.* Criticizes concepts of strategic stability which have led Western policymakers into neglecting the operational dimensions of military strategy and insists that the West needs new concepts of stability that are adequately responsive to Soviet developments and that will give theoretical underpinning to the determination of the military requirements that will enable the West to defend its vital interests.

272. Gray, Colin S. and Payne, Keith. UNDER THE NUCLEAR GUN: VICTORY IS POSSIBLE. *Foreign Policy 1980 (39): 14-27.* Atomic warfare need not end in complete catastrophe; it can be won or lost at any level. US strategists must take account of this and plan accordingly if US nuclear power is to support US foreign policy objectives. T. L. Powers

273. Gray, Colin S. THE URGE TO COMPETE: RATIONALES FOR ARMS RACING. *World Politics 1974 26(2): 207-213.* Arms races among nations normally result from seemingly good and sufficient reasons. Elements of the rationale behind competitive armament include deterrence, defense, diplomacy, the functional threat, vested interests, reputation, and technology. Proponents of arms control consider not only the negative but also the positive consequences for international security of competitive armament.
D. C. Richardson

274. Greico, Joseph M. and Clarke, Duncan L. NATIONAL SECURITY AFFAIRS: A SELECTIVE BIBLIOGRAPHY. *Policy Studies J. 1978 7(1): 157-164.* Presents a bibliography of books, articles, and dissertations published during the 1960's-70's pertaining to national security.

275. Griffith, Robert. TRUMAN AND THE HISTORIANS: THE RECONSTRUCTION OF POSTWAR AMERICAN HISTORY. *Wisconsin Mag. of Hist. 1975 59(1): 20-50.* Reviews the historical literature on President Harry S. Truman's foreign policy, including changing interpretations and evaluations of the Marshall Plan, Truman Doctrine, Point Four, atomic diplomacy, the Korean War, containment and the origins of the Cold War. Also deals with Truman's domestic policy, highlighting his handling of inflation, domestic security, McCarthyism, housing, civil rights, anti-trust enforcement, and public power programs. Revisionist historians view Truman's administration as a failure because it represented mid-century liberalism which "served only to rationalize an aggressive and militaristic foreign policy, while betraying the cause of social justice at home." Presented at the annual meeting of the Organization of American Historians in 1972. 9 illus., 84 notes, biblio. N. C. Burckel

276. Griffith, William E. BONN AND WASHINGTON: FROM DETERIORATION TO CRISIS? *Orbis 1982 26(1): 117-133.* Examines the deteriorating relationship between the United States and West Germany during 1976-81. The West German peace and ecological movement, US attitude toward the use of intermediate nuclear weapons, and the growing strains on the US economy are discussed in detail. The United States should concentrate on rebuilding the solidarity of NATO, which would improve relations with West Germany. Based on published sources; 19 notes. J. W. Thacker, Jr.

277. Gurney, Ramsdell, Jr. ARMS AND THE MEN. *Bull. of the Atomic Scientists 1975 31(10): 23-33.* National leaders in the 20th century have ignored the historical lesson that armament buildups lead to war despite their disarmament efforts; accounts for the continued arms race and unsuccessful disarmament efforts of the US and the USSR since 1945.

278. Haakonsen, Per. NATOS TNF-VEDTAK OG UTSPILLET OM RUSTNINGSKONTROLL [NATO's TNF decision and arms control proposals]. *Internasjonal Politikk [Norway] 1980 (3): 449-458.* In December 1979 the

Allies decided to deploy 572 medium-range missiles in Western Europe. The weapons—part of NATO's Theater Nuclear Forces (TNF)—were intended to counter a massive Soviet deployment of similar rockets (SS 20) aimed at Europe. Part of the so-called integrated TNF decision was a call for negotiations on different arms control issues. The neutron weapon debate in 1977 heavily influenced the decision to attach to the deployment of a new generation of nuclear missiles in Europe an elaborate package of arms control measures. The Allied disarmament proposals had little or no end in themselves but were initiated primarily to serve the greater purpose of securing the TNF in the decisionmaking processes in the respective NATO countries. More modern nuclear arms are needed in Western Europe, but this is being achieved in a way that in fact compromises the cause of disarmament. J

279. Hagan, Kenneth J. and Kipp, Jacob W. U.S. AND U.S.S.R. NAVAL STRATEGY. *US Naval Inst. Pro. 1973 99(11): 38-44.* Contains a review of the Soviet perception of alternative American naval strategies, demonstrating that a strategy based on submarines, especially those with nuclear power and ballistic missiles, is frighteningly credible to the Soviets, while a strategy based on attack aircraft carriers and amphibious assault forces lacks credibility. 2 illus.
J. K. Ohl

280. Halperin, Morton. CLEVER BRIEFERS, CRAZY LEADERS AND MYOPIC ANALYSTS. *Washington Monthly 1974 6(7): 42-49.* Examines the bureaucratic politics of nuclear war, and detects three potentially vulnerable points: a "clever" but deranged pro-nuclear war advisor, a mentally unstable leader, and a myopic strategy analyst. S

281. Hamburg, Roger. MASSIVE RETALIATION REVISITED. *Military Affairs 1974 38(1): 17-23.* Suggests that the Nixon administration's post-Vietnam policies involve a revival of massive retaliation. Based on public statements and special studies; 38 notes. K. J. Bauer

282. Harrington, Michael. NUCLEAR THREAT. *Society 1980 18(1): 16-21.* Examines US foreign policy with regard to developing nations and focuses on views of detente and attitudes toward the USSR, represented by the author on one side and Carl Gershman on the other.

283. Head, Richard G. TECHNOLOGY AND THE MILITARY BALANCE. *Foreign Affairs 1978 56(3): 544-563.* Presents doctrinal and technological factors that shape the US-Soviet military balance. The United States, for example, appreciates the element of uncertainty in battle and relies on broad-based research and development that produces sophisticated, multipurpose weapons systems. The Soviets, however, consider war a precise science; their highly structured research and development efforts are weighted more toward numerical superiority. American technological superiority will continue only if appropriate investment strategy is combined with the realization that US-Soviet competition will continue for many years. M. R. Yerburgh

284. Helms, Robert F., II. A NEW STRATEGY FOR THE US ARMY. *Military Rev. 1975 55(8): 49-55.* Describes some of the more significant changes affecting military strategy today, discusses their possible effects, on strategy, and offers an alternative for the US Army during 1975-90. After World War II,

American strategy grew out of the Cold War policy of containment, a "police the world" strategy. Now strategy must be based on the maintenance of a credible conventional force in Europe, the protection of US interests abroad, the protection of the homeland, and the preservation of domestic tranquility. 11 notes.
J. K. Ohl

285. Heurlin, Bertel. RYKKER ATOMKRIGEN NAERMERE? EN ANALYSE AF NYE TENDENSER I AMERIKANSK STRATEGI [Is an atomic war closer? An analysis of new tendencies in American strategy]. *Økonomi og politik [Denmark] 1974 48(4): 304-342.* The change in American policy from defense to strike capability with MIRV brings atomic war closer. Another change has been from civilian targets to military installations. The strategy combines the military (weapons) and political (doctrine and international support), with alteration from massive retaliation to flexible strategy. New relations with the Soviet Union are to prevent an arms race. Abandonment of balance with the USSR in favor of flexibility raises the argument of increased risks of war. The conflict of Henry A. Kissinger and James Schlesinger prompted debate about whether political factors are more important than strategic. Schlesinger holds that the USSR must realize that lessened arms supply does not give an opportunity for international political action. R. E. Lindgren

286. Hoeber, Amoretta M. SOME MYTHS ABOUT THE STRATEGIC BALANCE. *Air U. R. 1975 26(5): 85-91.* Examines several widely held opinions about the strategic balance, including the ideas that numbers of strategic weapons does not count, limiting damage from a nuclear attack is not a matter of policy choice, US qualitative advantage in weaponry can affect numerical disadvantages, and the US must not build new weapons because it would just intensify the arms race. Concludes that these are incorrect assumptions, and that assured destruction of the enemy is insufficient. There are many reasons why increasing accuracy may be beneficial; and arms control agreements that give quantitative advantages to the USSR could be highly disadvantageous to the US. Based on published sources; 10 notes. J. W. Thacker, Jr.

287. Hoeber, Francis P. MYTHS ABOUT THE DEFENSE BUDGET. *Air U. Rev. 1978 29(6): 2-17* Examines the arguments for cutting defense spending and analyzes myths used for justification. The myths are: 1) defense is nonproductive, 2) we need to shift our priorities, 3) the arms race, usually meaning the strategic nuclear arms race, comprises the major part of the defense budget, 4) the defense budget mostly goes for weapons, 5) that the Russian's military threat is a perennial Defense Department budget-time trick, and 6) that we cannot afford more for defense. Concludes that the amount of "national defense the U.S. can afford is how much it needs and has the political will to provide." Published sources; 3 tables, 29 notes. J. W. Thacker, Jr.

288. Hoeber, Francis P. and Hoeber, Amoretta M. THE SOVIET VIEW OF DETERRENCE: WHO WHOM? *Survey [Great Britain] 1980 25(2): 17-24.* The mutual deterrence approach only makes sense when it is mutual. The views attributed to the Soviets are not the real core of Soviet thought and if it is recognized that the Soviet thrust, both politically and militarily, is toward building credible threats while denying credibility to US threats, it would be clear that mutual deterrence is not a Soviet objective. 12 notes. V. Samaraweera

289. Hollist, W. Ladd. AN ANALYSIS OF ARMS PROCESSES IN THE UNITED STATES AND THE SOVIET UNION. *Int. Studies Q. 1977 21(3): 503-528.* Examines alternative explanations, or models, of arms build-ups in the USSR and the United States from 1948-70. The United States reacts more to changes in Soviet arms expenditures than to levels of Soviet arms, while the Soviet Union tends to react more to the level of US arms than to changes. In addition to the insights gained into the arms race process, this research, based on computer simulation, represents a contribution to a comprehensive international relations theory. Based on published statistics and secondary works; 4 tables, 39 notes.
E. S. Palais

290. Huff, Rodney L. NUCLEAR REACTORS AND FOREIGN POLICY: CHALLENGES OF A GLOBAL TECHNOLOGY. *Policy Studies J. 1974 3(2): 181-185.* Discusses the policy questions posed by US commitments to provide nuclear power to other nations. S

291. Huntington, Samuel P. AFTER CONTAINMENT: THE FUNCTIONS OF THE MILITARY ESTABLISHMENT. *Ann. of the Am. Acad. of Pol. and Social Sci. 1973 (406): 1-16.* "After World War II, the United States reconstituted its military policy for the third time in its history. A strategy of deterrence was adopted as the military counterpart to a foreign policy of containment. This strategy involved military alliances, overseas deployments, larger and diversified military forces, higher levels of readiness, and development of programs for strategic retaliation, continental defense, European defense, and limited war. By 1972, the original basis for this strategy was disappearing; public support for military burdens had decreased; the Soviet Union had achieved military parity with the United States; Europe, Japan, China were independent centers of economic and political power; local hegemonic powers had emerged in the Third World. For the foreseeable future, only the Soviet Union is in a position to aspire to global preeminence and thus pose a significant threat to U.S. security. Hence the U.S. must aim to avoid: military inferiority vis-à-vis the Soviet Union, diplomatic isolation among the major powers, and exclusion by the Soviet Union from political or economic access to any major portion of the Third World. These goals require military forces to support diplomacy as well as to maintain deterrence. More specifically, they require: a redefinition of the role of the strategic retaliatory force, recognizing its diplomatic as well as deterrent functions; the adaptation of U.S. forces deployed in and designed for the defense of Europe to the more general purpose of great power reinforcement; and the conversion of limited war forces into counterintervention forces to deter Soviet military intervention in the Third World. While civilians played a major role in developing the strategy of deterrence, the principal responsibility for elaborating these changes in strategy will rest with military officers." J

292. Iklé, Fred Charles. BOMBS AND REACTORS: THE NUCLEAR DIVIDE. *Bull. of the Atomic Scientists 1980 36(1): 38-42.* Discusses a direct relationship between the spread of atomic energy and the arms control question, focusing on changing attitudes of US policymakers since 1945.

293. Jackson, William D. SOVIET IMAGES OF THE U.S. AS NUCLEAR ADVERSARY, 1969-1979. *World Pol. 1981 33(4): 614-638.* Images of the United States as nuclear adversary presented in official Soviet commentary pro-

vide useful clues in the analyses of Soviet strategic policy. Hard, high-threat images stressing the continuing danger of nuclear war are functionally associated with conservative policies emphasizing the need for efforts to improve war-fighting capabilities. Less militant adversary images appear associated with more moderate defense policies. In the 1970's, sharp divergences in adversary images appeared in official Soviet commentaries, indicative of disagreement within the USSR on the defense policy implications of the Strategic Arms Limitation Talks. The policy implications of shifts in adversary images and the location of the political leadership in terms of conflicting moderate and conservative images are examined for the period 1969-79. J

294. Jackson, William D. THE SOVIETS AND STRATEGIC ARMS: TOWARD AN EVALUATION OF THE RECORD. *Pol. Sci. Q. 1979 94(2): 243-262.* Discusses the controversy in the explanation of Moscow's behavior in strategic arms competition. Contends that the USSR has not consistently sought to achieve nuclear superiority and that Soviet interest in curbing the strategic competition has been substantial. J/S

295. Jeffries, Vincent. POLITICAL GENERATIONS AND THE ACCEPTANCE OR REJECTION OF NUCLEAR WARFARE. *J. of Social Isssues 1974 39(3): 119-136.* "The paper examines differential attitudes toward the use in war of nuclear weapons among age strata, analyzed from the perspective of political generations. Data for the study came from a probability sample of 477 adults living in a metropolitan area. On the basis of historical events and the differing climates of public opinion pertaining to war, three political generations are discriminated: Dissent (those born 1943-1949), Cold War (those born 1927-1942), and World War II (those born before 1927). Results suggest the viability of political generations thus defined. The generation of Dissent is most likely to reject nuclear war, while those of the generation of World War II are most likely to accept nuclear warfare. The basic relationship between age and attitudes toward nuclear war is examined within the context of occupation, sex, fear of communism, and patriotism." J

296. Jenson, John W. NUCLEAR STRATEGY: DIFFERENCES IN SOVIET AND AMERICAN THINKING. *Air U. Rev. 1979 30(3): 2-17.* Discusses the differences in present thinking about nuclear strategy in the United States and the USSR, with a history of the development of their nuclear strategies and a discussion of the factors that influence each. The Soviets regard nuclear war in more traditional terms than the US concept of mutual destruction. US options include a return to reliance on the assured destruction concept, the pursuance of balance-of-power politics, and development of a decisive counterforce capability. Based on published sources; 17 notes. J. W. Thacker, Jr.

297. Jervis, Robert. WHY NUCLEAR SUPERIORITY DOESN'T MATTER. *Pol. Sci. Q. 1979-80 94(4): 617-633.* Examines the current debate over the capability of [Soviet and] American nuclear weapons. He argues that the implicit assumption behind the call for adding warheads to develop a strategy of "flexible response" including a counterforce capability is wrong. Jervis concludes that the real issue is not the number of warheads to be developed but the resolve and willingness to use them. J

298. Johnson, M. Glen. INTEREST STRUCTURES, DECISION-MAKING PROCESSES, AND UNITED STATES FOREIGN POLICY. *Int. Studies [India] 1979 18(4): 595-614.* Examines various considerations and interests that underlie US foreign policy and analyzes their relationship to the decision-making process. Although the United States defines order and stability as its main international objectives, it has developed the concept of regionally dominant powers (like India in the Asian subcontinent), maintaining order under the protection of US nuclear deterrence. Domestic issues such as energy and food have acquired interconnections with foreign policy considerations. 16 notes.
T. P. Linkfield

299. Jones, Kenneth Macdonald. THE ENDLESS FRONTIER. *Prologue 1976 8(1): 35-46.* The wartime performance of science convinced most Americans by the closing months of World War II that some sort of program to encourage postwar scientific inquiry sponsored by the federal government was desirable. The debate over the kind of governmental assistance to be made available was furthered through the publication of *Science, The Endless Frontier* by Dr. Vannevar Bush, wartime director of the Office of Scientific Research and Development. He and others felt that government support of science was not incompatible with scientific freedom. After July 1945, the public's rather vague, open-ended image of science gradually narrowed and became more utilitarian as affected by the atomic bomb and the onset of the Cold War. Based on primary and secondary sources. N. Lederer

300. Kaplan, Fred. A STRATEGY FOR DISARMAMENT. *Working Papers for a New Society 1977 5(1): 18-28.* Examines the growing nuclear arsenal in the United States despite attempts to limit arms stockpiling; compares American holdings with those of the USSR and discusses foreign policy.

301. Kapur, Ashok. NUCLEAR PROLIFERATION IN THE 1980S. *Int. J. [Canada] 1981 36(3): 535-555.* First wave, or "vertical" nuclear proliferation, concerns the quantity of weapons presently stockpiled by members of the nuclear club. Second wave, or "horizontal" proliferation, concerns the acquisition of nuclear capabilities by developing nations, such as India, Pakistan, Israel, Argentina, Brazil, and South Africa. These nations are not merely interested in nuclear arms but in the downward mobility of the superpowers and the upward mobility of themselves in the global hierarchy. Discusses the argument against proliferation as formulated in the 1960's as well as subsequent arguments. Secondary sources; table, 11 notes. J. Powell

302. Kelleher, Catherine McArdle. THE PRESENT AS PROLOGUE: EUROPE AND THEATER NUCLEAR MODERNIZATION. *Int. Security 1981 5(4): 150-168.* Discusses the Western Alliance's problems in balancing the conflicting requirements of theater and broader East-West equilibrium at a time of strategic parity, American-European differences in risk assessment vis-à-vis the Soviet Union, the impact of these differences on coordination in arms procurement and arms limitation, and how Alliance decisionmaking will accommodate increasing European assertiveness on issues of conventional-nuclear balance within the West.

303. Kinnard, Douglas. PRESIDENT EISENHOWER AND THE DEFENSE BUDGET. *J. of Pol. 1977 39(3): 596-623.* President Dwight D. Eisenhower accepted Harry S. Truman's position that the United States should lead the non-Communist world. However, Eisenhower urged that American military strategy be based on strategic deterrence, not on balanced forces. His 1952 campaign promises to end the Korean War, balance the federal budget, and reduce taxes led to pressure to reduce defense expenditures throughout his two terms, even in the post-Sputnik era. The National Security Council was a forum for achieving consensus, not a decisionmaking body. Continental air defense and conventional forces yielded to the development of resources capable of massive retaliation on the Soviet Union. Based on primary and secondary sources; 58 notes.
A. W. Novitsky

304. Kirgis, Frederic L., Jr. NATO CONSULTATIONS AS A COMPONENT OF NATIONAL DECISIONMAKING. *Am. J. of Int. Law 1979 73(3): 372-406.* Identifies the extent to which the NATO allies have used consultation as a policymaking device, based on documented instances. Develops the framework for NATO consultations within North Atlantic Treaty provisions. Action within the alliance affecting NATO defense of Europe has included decisions to limit military commitments, use or production of tactical nuclear weapons, and the appointment of the US Supreme Allied Commander, Europe. 143 notes.
R. J. Jirran/S

305. Klare, Michael T. DE LA DISUASIÓN A LA CONTRAFUERZA: ESTRATEGIA NUCLEAR DE LOS ESTADOS UNIDOS EN LOS AÑOS 70'S [Counterforce deterrence: nuclear strategy of the US in the 1970's]. *Rev. Mexicana de Ciencias Pol. y Soc. [Mexico] 1975 21(81): 33-60.* Secretary of Defense James R. Schlesinger announced in January 1974 that US nuclear policy encompassed "assured destruction"; that is, should the USSR launch a nuclear attack the United States could only retaliate in kind.

306. Klein, Jean. VENTES D'ARMES ET D'ÉQUIPEMENTS NUCLÉAIRES: LES POLITIQUES DES ETATS-UNIS ET DES PAYS D'EUROPE OCCIDENTALE DEPUIS LA GUERRE D'OCTOBRE 1973 [Sales of arms and nuclear equipment: the policies of the United States and the countries of Western Europe since the October War of 1973]. *Politique Étrangère [France] 1975 40(6): 603-620.* Discusses the effects of the petroleum crisis coupled with the October War of 1973 on policies regarding the sales of arms and nuclear equipment, particularly to countries in the Middle East, by Western Europe and the United States.

307. Knight, Jonathan. RISKS OF WAR AND DETERRENCE LOGIC. *Can. J. of Pol. Sci. 1973 6(1): 22-36.* The logic that says World War II was caused by appeasement and disarmament and that therefore we must negotiate from strength in order to prevent another large-scale war has a serious flaw when applied, because it does not assure genuine security and lasting peace. 38 notes.
R. V. Kubicek

308. Kolowicz, Roman. US AND SOVIET APPROACHES TO MILITARY STRATEGY: THEORY VS. EXPERIENCE. *Orbis 1981 25(2): 307-329.* "Soviet and American approaches to strategic problems in the nuclear era

appear to be similar: the military technologies are similar, both sides generally understand the qualitative and quantitative aspects of the other's weapons systems, and both sides have been engaged in protracted diplomatic and technical negotiations on strategic arms limitations. . . . " They diverge on many "issues regarding the uses, limitations, and purposes of military power as well as the rules of the game. . . . " Soviet and US approaches to the use of military force are at odds. The "instrumental, strategic, and highly politicized Soviet approach contrasts sharply with the more emotional, political, and moralizing American approach." 70 notes. J. W. Thacker, Jr.

309. Korb, Lawrence J. THE ISSUES AND COSTS OF THE NEW UNITED STATES NUCLEAR POLICY. *Naval War Coll. R. 1974 27(3): 28-41.* "The changes in strategic defense policy recently articulated by Secretary of Defense Schlesinger have given rise to considerable speculation on the relative strengths found in United States and Soviet nuclear arsenals. The intrinsic complexities of measuring throw weight, accuracy, and both warhead and launch vehicle number, further complicated by the difficulty of President Nixon and Premier Brezhnev in achieving any significant progress toward limiting the new technology, have made such speculation relatively subjective and open to a wide variety of opinion. Unfortunately, any disequilibrium in the nuclear equation is bound to be one that, although created in the name of national security, makes the actors on both sides feel progressively less secure." J

310. Krell, Gert. DIE KRITIK DER AMERIKANISCHEN RÜSTUNG UND DIE DEBATTE UM DIE "NATIONAL PRIORITIES" [The criticism of the American armaments industry and the debate about the "national priorities"]. *Politische Vierteljahresschrift [West Germany] 1973 14(4): 527-566.* The armaments industry has been criticized severely since the Vietnam War. Previously it was assumed that weapon-producing was necessary for employment and for fulfilling America's commitments. But the counterargument runs that arms production is self-generating and wasteful ($2 billion a year): Ralph Nader has drawn up a list of six subeconomies, from many of which the arms industry suffers. The most recent hopes for arms limitation are SALT; but there is a fear that they may lead to arms expansion. Concludes that America should adopt a new direction. Based on secondary works; 6 tables, 107 notes. A. Alcock

311. Kublig, Bernd W. NUKLEARENERGIE UND NICHTVERBREITERUNG VON KERNWAFFEN [Nuclear energy and the nonproliferation of nuclear weapons]. *Neue Politische Literatur [West Germany] 1979 24(4): 487-512.* The Indian nuclear experiments of 1974 showed the clear context between peaceful use of nuclear atomic energy and nuclear arms, already mentioned in the concept "Atoms for Peace" of President Eisenhower in the 1950's. The Nuclear Nonproliferation Treaty (1968) did not prevent the development of nuclear weapons by nonsignatory states in Asia, Africa, and Latin America in the 1960's and 1970's. R. Wagnleitner

312. Kugler, Jacek; Organski, A. F. K.; and Fox, Daniel J. DETERRENCE AND THE ARMS RACE: THE IMPOTENCE OF POWER. *Int. Security 1980 4(4): 105-138.* The United States and the USSR are building nuclear arms but, despite what they say, they are not competing with one another and mutual deterrence is not taking place. Nuclear arms races, in contrast with nonnuclear arms races, are chimeras in international relations.

313. Kupperman, Robert H.; Behr, Robert M.; and Jones, Thomas P., Jr. THE DETERRENCE CONTINUUM. *Orbis 1974 18(3): 728-750.* Discusses the current relation between strategic response and economic policy. S

314. La Tournelle, Guy de. ARMES NUCLÉAIRES ET POLITIQUE ÉTRANGÈRE (D'APRÈS HENRY A. KISSINGER, EN 1957) [*Nuclear Weapons and Foreign Policy* by Henry A. Kissinger: published in 1957]. *Études Polémologiques [France] 1975 5(15): 19-48.* Published in 1957, the book must be placed in the context of that time. But it still offers treble interest: it sheds some light on American psychology; it demonstrates the intellectual investigation and intuitive insight of a man thereafter called to carry the highest responsibility in the world. It shows the difficulty in our nuclear age of establishing a non-bellogenic relationship between force and diplomacy. J

315. Lambelet, John C. TOWARDS A DYNAMIC TWO-THEATER MODEL OF THE EAST-WEST ARMS RACE. *J. of Peace Sci. 1973 1(1): 1-38.* Presents a primitive East-West arms race model which compares both strategic and conventional forces in the postwar arms race between the United States and the USSR. Suggests that the model is not yet adequate for simulation and forecasting, and improvement will require interdisciplinary research. Based on theoretical models; 14 tables, 24 notes, biblio. M. I. Elzy

316. Lauren, Paul Gordon. THEORIES OF BARGAINING WITH THREATS OF FORCE: DETERRENCE AND COERCIVE DIPLOMACY. Lauren, Paul Gordon, ed. *Diplomacy: New Approaches in History, Theory, and Policy* (New York: Free Pr., 1979): 183-211. Examines the use of the threat of force as a part of diplomacy and the behavior of those reacting to it. Before the advent of strategic bombing and the possibility of nuclear holocaust, war itself was regarded as a legitimate instrument of foreign policy. Enumerates the large amount of US literature in the Cold War era on the theory of bargaining with threats of force. Coercive diplomacy tries to initiate behavior by fear, deterrence to inhibit behavior by fear. Historical examples are President Paul Kruger of the Transvaal against Great Britain, Nikita Khrushchev and his threats of the use of Medium Range Ballistic Missiles, and Britain and France against Nazi Germany after the invasion of Poland. 98 notes. S

317. Lebedev, Nikolai Ivanovich and Kortunov, Sergei Vadimovich. PROBLEMA RAZORUZHENIIA I IDEOLOGICHESKAIA BOR'BA: KRITIKA APOLOGETOV GONKI VOORUZHENII [Problems of disarmament and ideological struggle: a critique of arms race apologists]. *Novaia i Noveishaia Istoriia [USSR] 1980 (4): 3-22.* Stresses the good intentions of the USSR and all socialist countries in promoting international peace and mutual trust, and the USSR's contributions, including SALT II, to these aims. Argues against the reasons put forward by some for continuing the arms race: the "Soviet threat" and the claim that the USSR is spending increasingly on arms, the notion of a "balance of terror" to maintain peace, the claim that the armaments industry is vital to Western economies, the "technical problems" in achieving arms limitation or disarmament, and the Chinese claim that a third world war is inevitable and that the arms race must continue. 83 notes. J. S. S. Charles

318. Lodal, Jan M. DETERRENCE AND NUCLEAR STRATEGY. *Daedalus 1980 109(4): 155-176.* Deterrence has been the primary mission of US nuclear weapons, but the continued Soviet buildup of offensive nuclear arms, particularly when combined with their buildup of conventional forces, has made the maintenance of stable deterrence ever more difficult, a difficulty surmountable by a revised military strategy which returned to a partial dependence on active defenses accompanied by a serious ABM development program.

319. Lugato, Giuseppe. LA NUOVA CORSA AGLI ARMAMENTI EL-'AGGIORNAMENTO DELLE STRATEGIE [The new race toward armaments and the updating of strategy]. *Civitas [Italy] 1974 25(3/4): 3-20.* "Analyzes two subjects: the new race toward armaments, in which the big countries are engaged in spite of relaxation and dialogue; and the strategic revolution being carried on by the USA. These subjects are considered as far as the connections among the big powers and their effects on the remaining part of the world are concerned. The balance of forces is no longer the same as it was two years ago. The Soviet Union has in fact completed MIRV tests and new armaments are being developed. The United States' answer is the appropriation for the defence of the record-sum of 80.6 milliards of dollars, and the complete modification of their nuclear strategy: after the 'massive retaliation,' they have shifted to 'flexible response,' that is a gradual use of atomic arms. Can we therefore hope for relaxation to go on? Is bipolarism bound to become more and more imperfect and precarious? This is the essential question which one cannot easily answer." J

320. Luttwak, Edward N. NUCLEAR STRATEGY: THE NEW DEBATE. *Commentary 1974 57(4): 53-59.*

321. Lynn, Laurence E., Jr. ARM WAVING AT THE ARMS RACE. *Int. Security 1979 4(1): 117-126.* Reviews the Boston Study Group's *The Price of Defense: A New Strategy for Military Spending* (New York, 1979), which recommends drastic decreases in defense spending, and finds it a misleading, invidious, and specious work which does not permit objective comparison with such careful and complete studies of the problem as the Brookings Institution's *Selling National Priorities, the 1979 Budget.*

322. Lyth, Einar. SUPERMAGTERNES DOKTRINER [The doctrines of the superpowers]. *Svensk Tidskrift [Sweden] 1975 62(9): 386-393.* Examines the defense policy of the United States and the USSR, 1945-74. Both countries' doctrines have moved from an aggressive stance emphasizing atomic power to a defensive position stressing spheres of interest. This development is due to their growing military similarity. Sweden's position in the power system lies geographically between the two spheres and is therefore exposed, because the USSR is expanding militarily at the Kola Peninsula. U. H. Bartels

323. Maddox, Robert James. ATOMIC DIPLOMACY: A STUDY IN CREATIVE WRITING. *J. of Am. Hist. 1973 59(4): 925-934.* Gar Alperovitz's *Atomic Diplomacy: Hiroshima and Potsdam: The Use of the Atomic Bomb and the American Confrontation with Soviet Power* (New York: Simon and Schuster, 1965) has become a staple of New Left historiography, but Alperovitz' use of his sources distorts and misrepresents the evidence. Arguments cited in support of his thesis in fact refer to other subjects, time sequences are altered, key words are

deleted which would invalidate the arguments, certain words are given a sinister weight they do not deserve, and some statements are contradicted by their very sources. The uncritical reception of this lamentable scholarship points to shortcomings in the critical mechanisms of the profession. 43 notes. K. B. West

324. Mandelbaum, Michael. THE BOMB, DREAD, AND ETERNITY. *Int. Security 1980 5(2): 3-23.* Explores the psychological impact of nuclear armaments, and the resulting threat of nuclear annihilation, on American society since 1945.

325. Marawitz, Wayne L. NUCLEAR PROLIFERATION AND U.S. SECURITY. *Air U. Rev. 1977 28(2): 19-28.* Analyzes the probability of future proliferation of nuclear arms and the possible effects on US security. The relative availability of the technology necessary to build nuclear weapons and the increasing availability of nuclear materials from nuclear reactors is discussed. Possible nuclear powers of the future are examined in terms of capabilities and motivation. Suggests that the United States develop the intelligence capacity to keep track of nuclear proliferation, and that the United States publicly state that any nuclear threat, blackmail, or attack will be answered with extreme measures. Based on published sources; 3 tables, 18 notes. J. W. Thacker, Jr.

326. Martin, J. J. NUCLEAR WEAPONS IN NATO'S DETERRENT STRATEGY. *Orbis 1979 22(4): 875-895.* In the mid-1970's the focus of NATO strategy began to shift. The United States stressed the need for strong military forces emphasizing a conventional defense. The European members, although improving their conventional defenses, have continued to insist on the potential use of nuclear arms as the primary means to deter a Warsaw Pact attack. "The threat of escalation to total war must be maintained in both cases, but other options should be available as well." The strategy of near-total reliance on nuclear weapons "suffers from three fatal flaws: it is militarily unsound; it has major political defects; and it would weaken deterrence." Believes we have no choice but to deal with the possibility that our adversaries may use nuclear weapons or that an overwhelming conventional attack may threaten the loss of Western Europe. 18 notes. E. P. Stickney

327. Martin, Laurence. CHANGES IN AMERICAN STRATEGIC DOCTRINE—AN INITIAL INTERPRETATION. *Survival [Great Britain] 1974 16(4): 158-164.* Discusses US military strategy and nuclear arms potential in 1974, emphasizing Secretary of Defense James Schlesinger's policies toward ICBM's and the Strategic Arms Limitation Talks with the USSR.

328. Mayer, Laurel A. and Stupak, Ronald J. THE EVOLUTION OF FLEXIBLE RESPONSE IN THE POST-VIETNAM ERA: ADJUSTMENT OR TRANSFORMATION? *Air U. R. 1975 27(1): 11-21.* Examines the development of US national strategic doctrine of deterrence since 1950 from massive retaliation and flexible response to the present situation. "Despite all the dialogue devoted to a re-evaluation of strategic theory, especially massive retaliation and the literature dealing with limited war and coercive diplomacy, . . . Much of strategic thinking has failed to keep pace with the rapidly changing international situation." In effect, US strategic thinking has focused too exclusively on situations of superpower confrontation and attempted to apply such things to problems that are not applicable. Based on printed sources; illus., 41 notes.
J. W. Thacker, Jr.

329. Maynes, Charles William. OLD ERRORS IN THE NEW COLD WAR. *Foreign Policy 1982 (46): 86-104.* In the context of increasingly effective first strike capabilities, long-standing US-USSR tensions threaten catastrophic nuclear war. Since WWII, Soviet concern over American military power spurred Moscow to strengthen itself. This, in turn, alarmed Washington into increasing its military might. Compounding this was a tradition of preserving absolute American security and a frozen view of the Soviets as bent on world domination. Detente and arms limitations have floundered in this atmosphere.

M. K. Jones

330. McCormick, Gordon H. and Miller, Mark E. AMERICAN SEAPOWER AT RISK: NUCLEAR WEAPONS IN SOVIET NAVAL PLANNING. *Orbis 1981 25(2): 351-367.* Analyzes Soviet naval strategy since the 1950's and the role that nuclear arms would play in a war. 17 notes.

J. W. Thacker, Jr.

331. McDonald, W. Wesley. WHAT FUELS THE SOVIET-AMERICAN STRATEGIC ARMS RACE?: AN EXAMINATION OF THE INTERSTATE ACTION-REACTION MODEL AND ITS ADVOCATES. *Towson State J. of Int. Affairs 1978 13(1): 29-42.* Discusses the action-reaction model, one of several current theories of arms-race behavior which deal with the dynamics of the Soviet-American strategic arms race, 1970's.

332. McGlinchey, Joseph J. and Seelig, Jakob W. WHY ICBMS CAN SURVIVE. *Air Force Mag. 1974 57(9): 82-85.* Discusses the US program for nuclear defense, with particular reference to the long-standing question of the survivability of US land-based missiles in the event of nuclear attack by the USSR.

333. McNeill, William H. THE PURSUIT OF POWER: A HISTORIAN REFLECTS. *Bull. of the Atomic Sci. 1982 38(4): 19-23.* Discusses the nature of political and military power in the world from 1945 to 1982, focusing on efforts to bring about disarmament, on conflict between the United States and the USSR, and on the necessity of a new world government.

334. Melby, Svein. DEN AMERIKANSKE DOKTRINEENDRINGEN: HVA OG HVOR MYE ER NYTT? [The change in American nuclear doctrine: what and how much is new?]. *Internasjonal Politikk [Norway] 1980 (4): 719-735.* Denies that the change in American nuclear strategy represents something fundamentally new, and that the risks of nuclear war have increased. Both the theories behind the "new" strategy and the efforts to develop it can be traced back as far as the 1950's.

J/S

335. Mikulín, Antonín. OZBROJEŃE SÍLY USA—HLAVNÍ NÁSTROJ GLOBÁLNÍ STRATEGIE IMPERIALISMU [US Armed Forces: the main instrument of imperialist global strategy]. *Hist. a vojenství [Czechoslovakia] 1978 27(5): 137-162, (6): 119-149.* Part I. A study of the system of American military strategy since 1945 divides the past decades in some historical epoques: 1945-53; the time of "balanced armed forces": the years 1954-60 were influenced by the idea of "massive revanche"; 1960-69 was the time of "elastic reaction"; and in 1970 the military circles began the strategy of "realistic deterrent." Dissects US military planning and the strategic thinking of top ranking leaders at that time. 27 notes. Part II. Outlines further development of American strategic thinking

and examines basic military conceptions as they appear backed by the armed potential of the United States. Compares actions taken by the Nixon, Ford, and Carter administrations and the volume of armament of all capitalist countries. Views US conceptions of warfare in all dimensions as too unrealistic and unbalanced. Based on published printed sources; 42 notes. G. E. Pergl

336. Miller, Kenneth G. BLUEPRINT FOR ABDICATION. *Air U. Rev. 1978 29(4): 31-38.* Examines US actions and Soviet responses since 1965 and presents a possible blueprint for Soviet covert and overt actions in the future. Possible covert efforts are: 1) supporting any individuals or groups whose activities contribute to a reduction of US defense spending, 2) supporting all extremist groups, 3) exacerbating economic problems at every possible opportunity, and 4) giving the widest possible coverage of all US problems and failures in foreign affairs. Some of the possible overt actions listed are: 1) encouraging trade at all levels, especially for high technology items, 2) a stronger effort to surpass US military capability, and 3) focusing attention on US intelligence activity. The author is not fearful that the Soviets will overpower the US, but that the US will abdicate its position in world affairs. Published sources; 30 notes.
J. W. Thacker, Jr.

337. Millett, Stephen M. THE CAPABILITIES OF THE AMERICAN NUCLEAR DETERRENT 1945-1950. *Aerospace Hist. 1980 27(1): 27-32.* Reviews the historical record of US nuclear deterrence capabilities immediately after World War II. Covers the nuclear armaments available, the delivery capabilities, and a scenario of possible atomic war. Concludes that America did not have a nuclear capability to deter Soviet aggression in July, 1948, but that capability was acquired at least by 1950. Based on secondary sources; 5 photos, 33 notes.
C. W. Ohrvall

338. Molineu, Harold. THE IMPACT OF STRATEGIC AND TECHNOLOGICAL INNOVATIONS ON NUCLEAR DETERRENCE. *Military Rev. 1978 58(1): 7-16.* Recent developments in strategy and technology could have serious consequences on the fragile peace between the USSR and the United States since 1969.

339. Monin, M. LOZH' STOLETIIA I PRAVDA ISTORII [The lie of the century and the truth of history]. *Voenno-Istoricheskii Zhurnal [USSR] 1980 (1): 63-71.* Despite the signing of SALT II, the US government continues to stockpile arms and follow its aggressive policies toward the USSR and the other Communist countries. Since 1945, the US military has drafted a series of plans for an attack on the USSR which have been seen as inevitably leading to World War III: Charioteer and Fleetwood in 1948, Trojan and Dropshot in 1949. The general shift in the balance of world forces in favor of socialism and in particular the economic and military developments of the USSR have prevented the United States from carrying through these plans. Secondary sources; 6 notes, 2 plans.
L. Waters

340. Moore, William C. COUNTERFORCE: FACTS AND FANTASIES. *Air Force Mag. 1974 57(4): 49-52.* Discusses the change in US nuclear strategy in the 1970's with reference to the adoption of the concept of Counterforce by the United States and its implications.

341. Morgenthau, Hans J. HENRY KISSINGER, SECRETARY OF STATE: AN EVALUATION. *Encounter [Great Britain] 1974 43(5): 57-61.* Discusses Henry A. Kissinger's ability to adjust intellectual conviction to political exigencies, both as Secretary of State since 1969 and even earlier when he published *Nuclear Weapons and Foreign Policy* (1957).

342. Murry, William V. CLAUSEWITZ AND LIMITED NUCLEAR WAR. *Military Rev. 1975 55(4): 15-28.* Examines the applicability of nine principles of war, defined in United States Army Field Manual 100-5, to limited atomic warfare. Only the three principles of mass, economy of force, and maneuver enhance success in a limited nuclear war. The two principles, objective and offensive, do not enhance success. The principles of security and simplicity require renaming; and unity of command and surprise are not true principles of war. Based on secondary sources; 10 illus., 2 figs., 11 notes. J. K. Ohl

343. Myrdal, Alva Reimer. THE ARMS RACE IS AN INTERNATIONAL EPIDEMIC. *Center Report 1973 6(1): 18-21.* Monopoly over military technology is giving rise to "a similar monopoly over new technologies of immense importance in the economic life of nations," extending to foreign relations, in general, and concentrated in the two superpowers. S

344. Nacht, Michael. THE FUTURE UNLIKE THE PAST: NUCLEAR PROLIFERATION AND AMERICAN SECURITY POLICY. *Int. Organization 1981 35(1): 193-212.* An examination of the past relationships between nuclear proliferation and American security policy substantiates several propositions. The political relationship between the United States and each new nuclear weapon state was not fundamentally transformed as a result of nuclear proliferation. With the exception of the USSR, no new nuclear state significantly affected US defense programs or policies. American interest in bilateral nuclear arms control negotiations has been confined to the USSR. A conventional conflict involving a nonnuclear ally prompted the United States to intervene in ways it otherwise might not have in order to forestall the use of nuclear weapons. The intensification of the superpower rivalry and specific developments in their nuclear weapons and doctrines, the decline of American power generally, and the characteristics of nuclear threshold states all serve to stimulate nuclear proliferation. J/S

345. Nacht, Michael. THE UNITED STATES IN A WORLD OF NUCLEAR POWERS. *Ann. of the Am. Acad. of Pol. and Social Sci. 1977 (430): 162-174.* There is little consistency in American policy toward those states that have obtained independent nuclear weapons capabilities. Bilateral relations between the United States and the new nuclear state prior to weapons acquisition have proven to be far more accurate indicators of future trends in U.S. policy than the acquisition by the state of nuclear weapons per se. In the future, five basic options confront the United States: malign neglect, nuclear realignment, confrontation politics, equality promotion, and adaptive continuity. The last option, which involves the implementation of a variety of political-military and energy-related strategies, is the most likely one to be adopted. Major shocks to the international system, however, will drive the United States toward greater use of sanctions against the new nuclear states. J

346. Neal, Fred Warner. THE NEW AMERICAN-SOVIET COLD WAR. *Korea & World Affairs [South Korea] 1981 5(4): 529-536.* Discusses Soviet-US relations since the election of Ronald Reagan, commenting on the developing Cold War as evident in the failure of détente, the issue of Soviet SS-20 missiles in Europe, the European pacifist movement, and the volatile political situation in Eastern Europe.

347. Neumann, Gerd. OST-WEST-HANDEL ODER RÜSTUNGSKONJUNKTUR? FRIEDLICHE KOEXISTENZ ODER ESKALATION? ZUR WIRTSCHAFTLICHEN LAGE DER USA IM JAHRE 1949 [East-West trade or arms race? Peaceful coexistence or escalation? Concerning the economic situation of the USA in the year 1949]. *Jahrbuch für Wirtschaftsgeschichte [East Germany] 1976 (4): 31-50.* Industrial production in the United States reached its height in October 1948 and thereafter declined markedly. The ensuing economic crisis lasted only 11 months, but evoked comparisons with the economic situation just before the Great Depression. The Truman Administration reacted in September and October 1949 by increasing US military expenditures and offering assistance to American allies overseas, thus initiating the Cold War. Although the economic crisis ended immediately, Truman's policies ruined East-West trade relationships and precipitated the arms race. US actions thus caused more problems than they solved. Secondary sources; 6 tables, 4 graphs, 86 notes.
R. J. Bazillion

348. Nitze, Paul H. ASSURING STRATEGIC STABILITY IN AN ERA OF DÉTENTE. *Foreign Affairs 1975 54(2): 207-232.* "There is every prospect that under the terms of the SALT agreements the Soviet Union will continue to pursue . . ." nuclear superiority. "It is urgent that the United States take positive steps to maintain strategic stability and high-quality deterrence." 20 notes.
R. Riles

349. Norton, Augustus R. MYOPIC VISIONS OF THE ARMS RACE: THE IMMORTALITY OF METAPHORS. *Naval War College Rev. 1978 31(2): 48-53.* Misunderstanding and misapplication of some enduring arms race metaphors obscure distinctions that should be made and can lead to conclusions not supported by logic.

350. Nyström, Sune. AMERIKANSKA PROBLEM [The American problem]. *Kungliga Krigsvetenskaps Akademiens Handlingar och Tidskrift [Sweden] 1969 173(8): 433-449.* Discusses US security policy, especially nuclear strategy and the development of new biochemical weapons. The latter has been discussed in the Packard study (the report by the National Security Council with David Packard as spokesman in 1969). The United States will urge Western Europe to increase contributions to the Western defense budget. The relationship between the Defense Department and Congress has deteriorated since Eisenhower, culminating in a study from July 1969 by 45 Democrats in a congressional conference on the military budget and national priorities. Notes the problems of the overkill capacity of the biochemical weapons. U. G. Jeyes

351. Nyström, Sune. OM KRIGETS NYA ANSIKTE [The new face of war]. *Kungliga Krigsvetenskaps Akademiens Handlingar och Tidskrift [Sweden] 1968 172(5): 167-174.* Examines how attitudes to warfare changed concurrently

with technical and economic developments between World War I and the Vietnam War, and assesses the impact of nuclear arms on military strategies since 1945. U. G. Jeyes

352. Obrador Serra, Francisco. EL COMPONENTE TERRESTRE DE LAS FUERZAS ARMADAS NORTEAMERICANAS [The land component of American armed forces]. *Rev. General de Marina [Spain] 1981 201(Sept): 145-160.* The present primary theater of the US military is Europe. In the event of war with the USSR, the United States plans to deploy ground forces by sea and air to Europe, where their equipment and supplies will be waiting. The risks of strategic nuclear exchanges and of slow, vulnerable sea transport introduce elements of uncertainty into American strategy. Yet the United States is continually improving its weapons systems. This key aspect of the American war plans is a fundamental element in the strategy of NATO. 2 tables. W. C. Frank, Jr. Spanish.

353. Osgood, Robert E. US SECURITY INTERESTS IN OCEAN LAW. *Survival [Great Britain] 1975 17(3): 122-128.* Discusses the role of maritime law in US defense, military strategy, and nuclear arms policy in the 1970's, including the Caribbean region, the Indonesian straits, and the Arctic.

354. Outrey, Georges. LA DIRECTIVE PRÉSIDENTIELLE NO. 59 [Presidential executive order no. 59]. *Défense Natl. [France] 1980 36(10): 125-130.* Discusses US presidential directives on defense from 1973 and particularly President Jimmy Carter's executive order no. 59. 8 notes.

355. Phillips, Dennis. COLD WAR TWO: IMPLICATIONS FOR THE STRATEGIC ARMS RACE. *Australian Q. [Australia] 1980 52(2): 144-151.* Studies the deterioration in Soviet-American relations consequent upon the USSR's invasion of Afghanistan and says that, even in the face of this act, it is the continual escalation of the arms race and the militarization of the economy and public opinion that is the major present problem.

356. Pipes, Richard. WHY THE SOVIET UNION THINKS IT COULD FIGHT AND WIN A NUCLEAR WAR. *Commentary 1977 64(1): 21-34.* Prevalent strategic theory in the United States holds that there would be no winner in an all-out atomic war, but Soviet doctrine holds that the better-prepared contestant would win such a contest. American refusal to acknowledge this Soviet view is arrogant, ignorant, and dangerous. Insularity, technological superiority, and industrial capacity have contributed to the lack of an ingrained strategic tradition in America. American and Soviet strategies differ because of dissimilar views of the role of conflict and violence, and of the function of the military establishment. D. W. Johnson

357. Polyanov, N. NATO: THE ANATOMY OF AN ANACHRONISM. *Int. Affairs [USSR] 1977 (2): 88-97.* The growth of detente, and especially the conference at Helsinki, have created a new set of conditions for NATO. The cold warriors who run the alliance are unwilling to adapt to the new situation, and yet are being forced to do so in a number of ways, for example through the disarmament talks at Vienna with Warsaw Pact leaders. Thus NATO practice and rhetoric betray many contradictions. Western militarists have stepped up their propaganda about the Soviet threat, and have spoken of the need to bring

Spain into the alliance along with former members France and Greece, and to extend NATO activities to the Indian Ocean. Nonetheless, a more realistic view is emerging among farsighted Western leaders. L. W. Van Wyk

358. Porro, Jeffrey D. THE POLICY WAR: BRODIE VS. KAHN. *Bull. of the Atomic Sci. 1982 38(6): 16-19.* Discusses the contrasting nuclear strategies of Bernard Brodie and Herman Kahn, assessing their impact on presidential administrations from Kennedy to Reagan.

359. Posen, Barry R. and VanEvera, Stephen W. OVERARMING AND UNDERWHELMING. *Foreign Policy 1980 (40): 99-118.* The mistaken impression of a shift in world power from the United States to the USSR has led to growing cries for a dramatic increase in US defense spending. In fact, NATO already outspends the Warsaw Pact. Such Western weaknesses as now exist derive from errors in doctrine, force structure, and weapons choice. Increased spending without strategic reforms, far from boosting US power, will only perpetuate existing problems, while the consequent damage to the US economy could result in a genuine decline in national security. Note. T. L. Powers

360. Postbrief, Sam. DEPARTURE FROM INCREMENTALISM IN U.S. STRATEGIC PLANNING: THE ORIGINS OF NSC-68. *Naval War Coll. Rev. 1980 33(2): 34-57.* Recently declassified documents and other materials make reexamination of the Truman administration national security planning apparatus a valuable example of policy formation in crises. The arguments formed and advanced in the present strategic debate are strikingly similar to those offered in the debate on NSC-68, the first major official American policy planning document of the Cold War. 76 notes. J

361. Power, Paul F. THE INDO-AMERICAN NUCLEAR CONTROVERSY. *Asian Survey 1979 19(6): 574-596.* Indo-US relations have improved since 1977, but not without controversy over the US nuclear nonproliferation policy and law and Indian reaction to them. President Carter and Prime Minister Desai have met to try to work out differences, but India will not agree to US demands for full-scope safeguards before delivering fuel supplies and continuing nuclear cooperation with India. Soviet offers of assistance to India with fewer strings attached complicate the issue, as does Pakistan's nuclear development project. 33 notes. M. A. Eide

362. Preston, Adrian. THE "NEW" CIVIL-MILITARY RELATIONS: RETROSPECT AND PROSPECT. *Air U. R. 1973 24(3): 51-53.* Examines the effect of the nuclear guerrilla age on civil-military relations and concludes that the demands are greater than ever for precise and intelligent regulation of military power; recommends a comprehensive system of interpenetration. Based on secondary sources; 3 photos, 2 notes. J. W. Thacker, Jr.

363. Quaroni, Pietro. MALE I SREDNJE SILE U SVETSKOJ POLITICI [The small and medium powers in world politics]. *Medjunarodni Problemi [Yugoslavia] 1970 22(4): 133-136.* There is a "balance of nuclear fear" between the Great Powers, but at the "sub-nuclear level" the desire to win over the small- and medium-sized countries foments rivalry between the United States and the USSR, both in Europe and in the developing nations.

364. Quester, George H. CAN DETERRENCE BE LEFT TO THE DETERRENT? *Polity 1975 7(4): 523-529.* A review of seven publications concerned with problems of nuclear deterrence: Arthur D. Larson, *National Security Affairs: A Guide to Information Sources* (Detroit: Gale Research Company, 1973); James E. Dougherty, *How To Think About Arms Control and Disarmament* (New York: Crane Russak and Company, 1973); Robert M. Lawrence, *Arms Control and Disarmament: Practice and Promise* (Minneapolis: Burgess Publishing Company, 1973); Ralph Sanders, *The Politics of Defense Analysis* (New York: Dunellen, 1973); Alton H. Quanbeck and Barry M. Blechman, *Strategic Forces: Issues for the Mid-Seventies* (Washington: Brookings Institution, 1973); Wynfred Joshua, *Nuclear Weapons and the Atlantic Alliance* (New York: National Strategy Information Center, 1973); and Morton A. Kaplan, ed., *Strategic Thinking and its Moral Implications* (Chicago: University of Chicago Center for Policy Study, 1973). These books reflect the decline in interest in strategic studies, which phenomenon may disastrously return warfare preparation solely into the hands of the generals. 11 notes. V. L. Human

365. Quester, George H. THE SUPERPOWERS AND THE ATLANTIC ALLIANCE. *Daedalus 1981 110(1): 23-40.* Examines whether the erosion of US strategic nuclear forces and the weaker and less competent post-Vietnam image of the United States have raised doubts in Western Europe about the US commitment to its defense. Europeans by and large still seem to believe that the US deterrent remains credible.

366. Rathjens, G. W. FLEXIBLE RESPONSE OPTIONS. *Orbis 1974 18(3): 677-687.* Discusses James R. Schlesinger's strategic arms proposals in 1975. S

367. Ravenal, Earl C. UNDER THE NUCLEAR GUN: DOING NOTHING. *Foreign Policy 1980 (39): 28-39.* An interventionist mentality, such as that which led the United States into Korea and Vietnam, is reemerging in the wake of the 1979 Soviet invasion of Afghanistan. Interventionism, in a world faced with atomic warfare, is bankrupting, counterproductive, and potentially fatal. The middle options having been stripped away, the only rational alternative is to acknowledge the limits of America's power, to attempt to compartmentalize rather than to link the world's troubles, and to adopt a generally isolationist foreign policy. T. L. Powers

369. Record, Jeffrey. TO NUKE OR NOT TO NUKE: A CRITIQUE OF RATIONALES FOR A TACTICAL NUCLEAR DEFENSE OF EUROPE. *Military Rev. 1974 54(10): 3-13.* NATO must maintain and improve its conventional arms rather than rely on tactical nuclear weapons.

370. Reed, Robert H. ON DETERRENCE: A BROADENED PERSPECTIVE. *Air U. R. 1975 26(4): 2-17.* Examines the problem of developing a military strategy of deterrence and supporting regional strategies in today's complex world. Believes that the maintenance of international stability will be a key concept in guiding US strategy at the regional and local level. Military aid and sales with little likelihood of direct US involvement will be the chief means of supporting this strategy. J. W. Thacker, Jr.

371. Roschlau, Wolfgang. DER HERAUSBILDUNG DER FLEXIBLEN MILITÄRSTRATEGIE UND DIE HAUPTENTWICKLUNGSRICHTUNGEN DER NATO STREITKRÄFTE IN DEN SECHZIGER JAHREN [The development of flexible military strategy and the directions of development of NATO forces in the 60's]. *Militärgeschichte [East Germany] 1979 18(2): 182-192.* As a result of a military and strategic situation in Europe that had changed in favor of socialist states, the NATO leadership developed a strategy stressing flexibility in types of possible warfare. The basic goals were preparation for nuclear war, more effective organization and closer relations among the separate forces, creation of more effective forms of troop preparedness, and the improvement of recruitment, the numerical expansion of forces, and the creation of a ready reserve. 20 notes. H. D. Andrews

372. Rosecrance, Richard. DÉTENTE OR ENTENTE? *Foreign Affairs 1975 53(3): 464-481.* Believes "the superpower relationship must be preserved, and that if it collapses a much greater tension and polarization in world politics will follow." Choices made now could cause the unraveling of world politics. Draws historic parallels with German policy under Otto von Bismarck, who built up a carefully maintained network of alliances prior to his dismissal in 1890. After the dissolution of the Russian connection the Triple Alliance and Triple Entente were formed, leading to World War I. Similarly, the United States and Russia were trapped into a cold war after World War II. In today's era of nuclear weapons and economic disruption in the industrialized countries, a partial entente is necessary to stabilize the world. The United States needs a positive policy or detente will jeopardize relations with Japan and Western Europe without positive gains from Russia. 13 notes. C. W. Olson

373. Rosenberg, David Alan. "A SMOKING RADIATING RUIN AT THE END OF TWO HOURS": DOCUMENTS ON AMERICAN PLANS FOR NUCLEAR WAR WITH THE SOVIET UNION, 1954-1955. *Int. Security 1981 6(3): 3-38.* Introduces a navy memorandum on an air force briefing of Strategic Air Command plans and capabilities for war and a summary of the findings contained in a Defense Department Weapons Systems Evaluation Group report; they shed considerable light on American plans and preparations for nuclear war in the first years of the Eisenhower administration, illuminating discrepancies between strategy and operational planning in the bomber era of the 1950's.

374. Rumsfeld, Donald H. THE STATE OF AMERICAN DEFENSE. *Orbis 1980 23(4): 897-910.* Examines US defense between 1977 and 1980, focusing on strategic strength. Details the decline of defense spending, changes in strategic doctrine, the SALT process, and the military strength of the USSR. US defense is too weak and must be strengthened immediately. 3 notes.
J. W. Thacker, Jr.

375. Russett, Bruce M. ASSURED DESTRUCTION OF WHAT? A COUNTERCOMBATANT ALTERNATIVE TO NUCLEAR *MAD*NESS. *Public Policy 1974 22(2): 121-138.* The targets of atomic war should be shifted from civilian populations to military targets to strengthen our foreign policy of nuclear deterrence. S

376. Salomon, Michael D. NEW CONCEPTS FOR STRATEGY PARITY. *Survival [Great Britain] 1977 19(6): 255-262.* Strategic thought of the 1950's-60's, epitomized and formed by three theorists, Bernard Brodie, Herman Kahn, and Thomas Schelling, incorporated the realization that military effectiveness and maintenance of national security were competing objectives, a key factor missing in current strategic thought, represented by Mutual Assured Destruction (MAD) and limited counterforce.

377. Schilling, Warner R. U.S. STRATEGIC NUCLEAR CONCEPTS IN THE 1970S: THE SEARCH FOR SUFFICIENTLY EQUIVALENT COUNTERVAILING PARITY. *Int. Security 1981 6(2): 49-79.* Analyzes the reasons why the United States did not contest the buildup of Soviet strategic capabilities, which began in 1964; describes the strategic response of the United States in light of this buildup; comments on the current Soviet-American state of parity and surveys future options.

378. Schlesinger, James R. THE EVOLUTION OF AMERICAN POLICY TOWARD THE SOVIET UNION. *Int. Security 1976 1(1): 37-48.* US foreign policy since World War II developed as a counter to the power of the USSR and as a result of international responsibilities; mentions later issues of deterrence, detente, and continued Soviet emphasis on global struggle.

379. Schlesinger, James. THE POWER TO DETER. *Center Mag. 1976 9(2): 45-47.* Advocates a strong military deterrence force and defense policy for the success of detente with the USSR.

380. Schneider, Mark B. NUCLEAR FLEXIBILITY AND PARITY. *Air Force Mag. 1974 57(9): 76-78.* In the light of SALT I and Robert S. McNamara's defense policies in the 1960's, particularly the concept of Mutual Assured Destruction (MAD), discusses Defense Secretary James Schlesinger's objectives for restoring US nuclear parity with the USSR.

381. Schroeder, Pat. SHOOTING AT EMPTY SILOS. *Washington Monthly 1974 6(3): 43-47.* One of three articles on "Three Ways to Increase Security— And Save Money Too." S

382. Schulze, Ludwig. DIE SPIELTHEORIE UND DIE AMERIKANISCHE MILITÄRSTRATEGIE [Game theory and US military strategy]. *Wehrkunde [West Germany] 1968 17(2): 87-93.* US military strategies after World War II, especially those based on the theory of escalation of Hermann Kahn, do not include the strategic situation and military security of Central Europe.

383. Schwartz, Charles. HELPING THE PENTAGON AIM RIGHT: A REPORT ON THE SCIENTIFIC TARGET PLANNING STAFF. *Bull. of the Atomic Scientists 1974 30(9): 9-13.* Examines technical, military, and political considerations in determining targets of US nuclear missiles; assesses the Single Integrated Operations Plan of the Joint Chiefs of Staff and the impact of MIRV and ABM systems on the Strategic Arms Limitation Talks, 1972-74.

384. Scott, William F. ARE WE UNDERRATING SOVIET MILITARY MANPOWER? *Air Force Mag. 1974 57(4): 26-31.* Presents an analysis of Soviet military manpower which is above the generally accepted estimates, and considers the implications of this for the United States in the light of strategic arms limitations.

385. Scoville, Herbert, Jr. FLEXIBLE MADNESS? *Foreign Policy 1974 (14): 164-177.* Criticizes the "flexible response" strategy of nuclear policies of the Nixon administration. S

386. Sharp, Jane M. O. NUCLEAR WEAPONS AND ALLIANCE COHESION. *Bull. of the Atomic Sci. 1982 38(6): 33-36.* Discusses the status of NATO with respect to nuclear capabilities and responsibilities, with special attention to European anxieties over deployment of US nuclear systems in Europe.

387. Singh, L. P. REGIONAL POWER VS. GLOBAL POWER IN ARMS CONTROL: INDIA, AMERICA AND NUCLEAR AFFAIRS. *India Q. [India] 1979 35(3): 351-361.* Examines India's role as a regional power determined to maintain its flexibility on nuclear policy by resisting nuclear safeguards that are proposed by any great power—but especially by the United States. Accordingly, India has retained its nuclear autonomy by following an independent course of action mainly to ensure its security in south Asia. By working closely with Canada to exchange peaceful nuclear technology, India alerts both China and America to the reality that India must remain a nuclear regional power. Covers 1951-78. US documents, secondary sources; 22 notes.
W. R. Johnson

388. Skogan, John Kristen. NATOS FØRSTEBRUKSDOKTRINE: BAKGRUNN, INNVENDINGER OG ETTERTANKE [NATO's first use doctrine: background, remonstrances, and reflections]. *Int. Pol. [Norway] 1982 (1B): 141-171.* Discusses in some detail objections, expostulations, and arguments that can and have been raised regarding the current NATO doctrine of first use of nuclear arms. J/S Norwegian.

389. Smart, Ian. THE GREAT ENGINES: THE RISE AND DECLINE OF A NUCLEAR AGE. *Int. Affairs [Great Britain] 1975 51(4): 544-553.* Examines the nature and development of the nuclear age, its implications for international politics, and differing judgements regarding the value and effect of nuclear weapons. This review article centers on nine recent publications (1973-75) on various aspects of nuclear weapons; 26 notes. P. J. Beck

390. Smith, Gaddis. WHAT WE GOT FOR WHAT WE GAVE: THE AMERICAN EXPERIENCE WITH FOREIGN AID. *Am. Heritage 1978 29(3): 64-71, 76-81.* Considers the value of US foreign aid which has totaled over $300 billion since the start of World War I. Limited in scale before World War II, foreign aid increased in scale in 1947 with the aid for Greece and Turkey, quickly followed by the Marshall Plan and Point 4. Setbacks in China and Korea and the Russian explosion of an atomic bomb in 1949-50 led to an almost complete militarization of foreign aid. After John F. Kennedy instituted the Peace Corps and the Alliance for Progress, the Vietnam disaster again led to serious questioning of the aid programs. No clear answer appears to the question: has the United States received its money's worth? 19 illus. J. F. Paul

391. Smoke, Richard. THEORIES OF ESCALATION. Lauren, Paul Gordon, ed. *Diplomacy: New Approaches in History, Theory, and Policy* (New York: Free Pr., 1979): 162-182. Nuclear weaponry sparked interest in escalation as a subject of study only recently separated from diplomacy and war. The Cuban missile crisis made "escalation" part of everyone's vocabulary, but the sources of

escalation theory in the United States occurred after World War II. Sketches the roles of 1) Otto von Bismarck and the Franco-Prussian War of 1870 and 2) John F. Kennedy, Robert McNamara and the Cuban missile crisis of 1962 in order to illustrate the significance and implications of escalation decisionmaking. 31 notes. S

392. Staudenmaier, William O. CIVIL DEFENSE IN SOVIET AND AMERICAN STRATEGY. *Military Rev. 1978 58(10): 2-14.* Differences in attitude toward atomic warfare have led the USSR to construct a massive and extensive civil defense strategy while the United States has a relatively less developed one. Psychological implications of such preparedness could lead the Soviets to a false sense of security about safely surviving a US second strike.

393. Steinbruner, John D. NUCLEAR DECAPITATION. *Foreign Policy 1981-82 (45): 16-28.* The command structure—the organizational and technical network of communication and command centers—of US strategic nuclear forces in the 1980's is extremely vulnerable to Soviet nuclear attack. Though President Ronald Reagan's defense policy has recognized this and called for strengthening the security of the command structure, technical solutions to command structure vulnerability probably are not possible at an affordable cost. M. K. Jones

394. Steinbruner, John D. and Garwin, Thomas M. STRATEGIC VULNERABILITY: THE BALANCE BETWEEN PRUDENCE AND PARANOIA. *Int. Security 1976 1(1): 138-181.* Vulnerability of US land-based missile forces to those of the USSR is "more a state of mind than a physical condition."

395. Struck, Myra. THEATER NUCLEAR FORCES: EUROPE'S NEW MAGINOT LINE. *SAIS Rev. 1981 (2): 113-129.* Examines the historical development of the theater nuclear forces debate within NATO; any reevaluation of NATO's current force posture and operational doctrine must begin with a look at how and why it emerged in the 1950's.

396. Sviatov, G. I. O POLITIKE SSHA V OBLASTI STROITEL'STVA VOORUZHENNYKH SIL I ORGANICHENIIA VOORUZHENII [US policy in the sphere of armaments]. *Voprosy Istorii [USSR] 1978 2: 66-92.* The article examines US policy in the field of armaments. The author analyzes a number of state documents (Presidential messages, reports by Defence Secretaries, etc.) and research works by American scientists. These and other materials enable the author to draw the conclusion that the Soviet Union's consistent peace policy, the tangible relaxation of international tension and the signing of Soviet-American agreements on the limitation of armaments have brought about a certain quantitative lessening of the armaments race in the U.S.A. in the first half of the seventies, although the improvement and perfection of armaments is continuing at a high rate. Owing to the efforts of the enemies of détente the US military budget has begun to soar again in real terms during the past two or three years. This is coupled with the further intensification of the armaments race in the U.S.A. All the peace-loving forces both in the United States and beyond its borders must redouble their efforts and work persistently to curb this dangerous race in armaments. J

397. Svyatov, G. and Kokoshin, A. NAVAL POWER IN THE US STRATEGIC PLANS. *Int. Affairs [USSR] 1973 (4): 56-62.* Considers the role of naval forces, the merchant marine and Poseidon ballistic missiles in US and NATO naval strategy in the 1970's; emphasizes strategic considerations in the Mediterranean and Indian Ocean areas.

398. Thomas, James A. COLLAPSE OF THE DEFENSIVE WAR ARGUMENT. *Military R. 1973 53(5): 35-38.* In World War II the United States successfully utilized the defensive war argument, anchored in the American tradition of property rights and fervor for universal moral values, to justify its participation. The Korean War and Vietnam War, however, demonstrated that the defensive war argument is not suitable for an era of protracted political conflict, and "the time to begin thinking of an alternative is now."

J. K. Ohl

399. Thompson, Kenneth W. THE COLD WAR: THE LEGACY OF MORGENTHAU'S APPROACH. *Social Res. 1981 48(4): 660-676.* Discusses Hans J. Morgenthau's views on the relation of Soviet imperialism and Communism in the context of the Cold War, his analysis of nuclear war and the Middle East, and his belief that diplomacy, not force, must be used to defuse East-West power conflicts in the 20th century.

400. Toner, James H. NATIONAL SECURITY POLICY: THE UNSPOKEN ASSUMPTIONS. *Air. U. Rev. 1977 28(5): 36-43.* Examines the unspoken assumptions involved in making national security policy. The five unspoken assumptions are: 1) American society is worthy of protection; 2) there is a real threat to American security; 3) the United States has the means to establish its national security; 4) the United States has the will to establish its national security; and 5) security is a flexible, adjustable concept. Concludes with a discussion of the objectives of defense policy. They are: to ensure the survival of the nation; to avoid nuclear war; and to promote the peace and prosperity of the world. Diagram; 19 notes. J. W. Thacker, Jr.

401. Tow, William T. CHINA'S NUCLEAR STRATEGY AND US REACTIONS IN THE "POST-DÉTENTE" ERA. *Military Rev. 1976 56(6): 80-90.* Discusses the implications of China's nuclear strategy for arms control agreements and foreign policy toward the United States in the 1970's; examines China's desire to weaken detente between the United States and the USSR.

402. Treverton, Gregory F. NUCLEAR WEAPONS AND THE "GRAY AREA." *Foreign Affairs 1979 57(5): 1075-1089.* A primary issue for the United States and its European allies is ". . . what to do about the nuclear threat to Western Europe, and to NATO's deterrent, posed by Soviet systems targeted on Western Europe—the SS-20 mobile missile and other Soviet weapons in the 'gray area' between the strategic and the tactical." As senior partner, the United States will initiate some sensitive, policy decisions, e.g., whether to beef up NATO's nuclear arms systems and take the brunt of the ensuing political heat.

M. R. Yerburgh

403. Trofimenko, G. A. EVOLIUTSIIA VOENNO-POLITICHESKOI STRATEGII SSHA POSLE VTOROI MIROVOI VOINY [The evolution of US military-political strategy after the Second World War]. *Voprosy Istorii*

[USSR] 1976 (3): 64-90. Drawing on a number of official US documents, the author closely examines the stages of the evolution which occurred in America's military-political strategy after the Second World War. Under the influence of objective factors the US Administration was adapting its strategy to the changing world situation, going over from the tough strategy of "containment" at the height of the cold war to the more flexible military-strategic conceptions. The author analyzes the most up-to-date trends in US strategy: special emphasis on the ocean strategy, the attempts to galvanize the idea of a "limited counter-force war." The article subjects to a critical examination the Pentagon's allegation about the systematic reduction of the US military budget during the last few years. The author also examines the objective causes which brought about a relaxation in Soviet-American relations. J

404. Trofimenko, Henry A. COUNTERFORCE: ILLUSION OF A PANACEA. *Int. Security 1981 5(4): 28-48.* Discusses Soviet strategic capacities as perceived by the Pentagon, as a rationale for the development of rough parity in the nuclear arsenal of the United States.

405. Tucker, Robert W. ISRAEL AND THE UNITED STATES: FROM DEPENDENCE TO NUCLEAR WEAPONS? *Commentary 1975 60(5): 29-43.* Discusses possible effects of recent changes in US foreign policy in the Middle East on Israel's defense policy, emphasizing the danger of nuclear arms.

406. Ulsamer, Edgar. ADJUSTING TRIAD TO MOUNTING SOVIET THREATS. *Air Force Mag. 1974 57(4): 54-60.* Considers the survivability of the US Air Force's ICBM's and ways in which this can be increased significantly, with reference to the American Aeronautics and Astronautics Associations meeting in Washington in 1974.

407. Ulsamer, Edgar. A BLUEPRINT FOR SAFEGUARDING THE STRATEGIC BALANCE. *Air Force Mag. 1976 59(8): 68-74.* Discusses the conclusions of the Air Force Association's symposium "Tomorrow's Strategic Options" held in April 1976.

408. Ulsamer, Edgar. GENERAL BROWN LOOKS AT US DEFENSE NEEDS. *Air Force Mag. 1975 58(8): 33-37.* General George S. Brown, Chairman of the Joint Chiefs of Staff, discusses ways of maintaining US strategic nuclear forces on a par with those of the USSR.

409. Ulsamer, Edgar. INCREASING MOMENTUM IN SOVIET STRATEGIC SYSTEMS. *Air Force Mag. 1973 56(3): 62-64.* Recent developments in Soviet military technology have produced a new, improved generation of strategic air weapons, 1970's.

410. Ulsamer, Edgar. THE PENTAGON LOOKS AT NEW STRATEGIC OPTIONS. *Air Force Mag. 1974 57(2): 52-55.* An interview with Deputy Secretary of Defense William P. Clements, Jr., in which, in the context of Strategic Arms Limitation Talks, he discusses the need for the United States to match the strategic capabilities of the USSR.

411. Ulsamer, Edgar. A REPORT ON AEC RESEARCH: WARHEAD DESIGN AND NUCLEAR STRATEGY. *Air Force Mag. 1974 57(6): 55-59.* An interview with Major General Edward B. Gillet, the Atomic Energy Commission's Assistant General Manager for National Security, on nuclear warhead design as it affects US nuclear deterrence strategy.

412. Ulsamer, Edgar. SAC'S COMMANDER LOOKS AT OUR STRATEGIC NEEDS. *Air Force Mag. 1973 56(12): 46-50.* In an interview, General John C. Meyer, Commander in Chief Strategic Air Command, discusses the balance of strategic power between the United States and the USSR.

413. Ulsamer, Edgar. THE SOVIET ICBM THREAT IS MOUNTING. *Air Force Mag. 1973 56(11): 34-37.* Discusses what detente means to the United States and the USSR, and considers the threat of the USSR's research and development programs, ranging from ballistic missiles to strategic bombers.

414. Ulsamer, Edgar. SOVIET OBJECTIVE: TECHNOLOGICAL SUPREMACY. *Air Force Mag. 1974 57(6): 22-27.* Discusses the USSR's efforts, in particular the policies of Defense Minister Marshal Andrei Grechko, to outdistance the United States in the introduction of new offensive missile systems and beamed energy weapons.

415. Ulsamer, Edgar. STRATEGIC WARNING: CORNERSTONE OF DETERRENCE. *Air Force Mag. 1974 57(5): 40-47.* Discusses the capabilities of the North American Air Defense Command and its systems for nuclear surveillance, warning, and attack.

416. Ulsamer, Edgar. UPGRADING USAF'S ICBMS FOR THE COUNTERFORCE ROLE. *Air Force Mag. 1974 57(2): 56-60.* Discusses ways of maintaining the United States ICBM force as a major deterrent to nuclear war, and developments intended to counter the USSR's high rate of technological advance in strategic weapons systems.

417. Ulsamer, Edgar. URGENT US R&D REQUIREMENTS. *Air Force Mag. 1974 57(4): 36-40.* An interview with Malcolm R. Currie, Director of Defense Research and Engineering, on national policy for deterring nuclear war and maintaining US military superiority.

418. Urban, Josef. VOJENSKOPOLITICKÉ ZÁMĚRY AMERICKÉHO IMPERIALISMU NA POČÁTKU 80. LET [Military and political intentions of American imperialism at the beginning of the 1980's]. *Hist. a Vojenství [Czechoslovakia] 1981 30(5): 109-127.* The Carter administration retreated from the principles of peaceful coexistence in 1977. An intensive ideological and psychological offensive against socialist lands was triggered. Its theme was the defense of human rights. A continuous effort was made to isolate the socialist world both economically and culturally. When Reagan entered the White House, military escalation took place, especially in the area of ballistic missiles. Even when there are some explanations for the defensive doctrine of the "new nuclear strategy," it is obvious that constant American aggression cannot be camouflaged by slogans. Only the strength of the socialist camp impedes American plans for world dominance. 25 notes. Czech. G. E. Pergl

419. Van Cleave, William R. and Barnett, Roger W. STRATEGIC ADAPTABILITY. *Orbis 1974 18(3): 655-676.* Approves Secretary of Defense James R. Schlesinger's strategic flexibility proposals concerning the deployment of nuclear arms during the 1970's. S

420. Vardamis, Alex A. NUCLEAR WEAPONS AND FOREIGN POLICY: FAULTY RATIONALE FOR CURRENT PRACTICE. *Naval War Coll. Rev. 1978 31(2): 89-103.* Questions conclusions of Henry A. Kissinger in his book *Nuclear Weapons and Foreign Policy,* published in 1957.

421. Vernant, Jacques. ARMEMENTS, STRATÉGIES ET POLITIQUES DANS LA COEXISTENCE ET LA RIVALITÉ [Arms, strategies and policies in coexistence and rivalry]. *Défense Natl. [France] 1975 31(10): 103-109.* Since the 1960's, international relations have been dominated by the peaceful coexistence policy of the USSR and the United States, which excludes direct military conflict and leads them to compromise.

422. Vernant, Jacques. PROLIFÉRATION ET DÉVELOPPEMENT [Proliferation and development]. *Défense Natl. [France] 1977 33(7): 115-121.* Briefly surveys current international relations, focusing on East-West relations and Jimmy Carter's foreign policy, on the occasion of the so-called "north-south" conference held in Paris (30 May-2 June 1977). Secondary sources; 5 notes.

423. Vidich, Arthur J. PROSPECTS FOR PEACE IN THE NUCLEAR WORLD. *J. of Pol. and Military Sociol. 1980 8(1): 85-97.* Analyzes US-Soviet relations as a dyadic interaction which is both symbiotic and dialectical in nature. In this relationship the possibility of mutual simultaneous destruction has encouraged restraint in the use of force and stimulated the development of new forms of diplomatic etiquette. Seen in these terms, the chances for peace rest much more in the evolving relationship—in the as-yet-unfulfilled passing present—than in the process of rational planning. The emergence of China as an autonomous actor creates a triadic situation which alters both codes of etiquette and the conditions for nuclear peace. 8 notes, 12 ref. J

424. Wallace, J. F. NUCLEAR DETERRENCE DOES NOT MAKE CIVIL DEFENSE UNNECESSARY. *Military Rev. 1979 59(5): 50-59.* The concept of mutual and universal destruction in the event of nuclear war has led the American government to pooh-pooh the idea of civil defense. The idea is popular with the public, which cares not to even think about a nuclear war. Conversely, the USSR has gone ahead with both intensive military development and civil defense, which has left the United States in a highly vulnerable position. This situation must be remedied if the nation is not to lose even more of its bargaining power. 2 photos, 25 notes. V. L. Human

425. Wallace, Michael D. ARMAMENTS AND ESCALATION: TWO COMPETING HYPOTHESES. *Int. Studies Q. 1982 26(1): 37-56.* A comparative empirical study of disputes between the Great Powers over a period of 150 years demonstrates that international conflicts that were accompanied by arms races were much more likely to result in war than those in which an arms race did not occur. The test confirms the "arms race" hypothesis, which states that rapid bilateral militarization is an important link in the escalation process. Military preparedness appears to have little or no capacity to prevent war. Data

obtained from the Correlates of War Project at the University of Michigan; 7 tables, 18 notes. E. S. Palais

426. Wallace, Michael D. ARMS RACES AND ESCALATION: SOME NEW EVIDENCE. *J. of Conflict Resolution 1979 23(1): 3-16.* Examines the relationship between the arms race and international military escalation, based on an arms race index, and discusses them in light of SALT II.

427. Walters, Ronald W. APARTHEID AND THE ATOM: THE UNITED STATES AND SOUTH AFRICA'S MILITARY POTENTIAL. *Africa Today 1976 23(3): 25-35.* American foreign policy will continue to support racist rule in South Africa and seek to undermine the revolutionary leadership of Mozambique and Angola. Perceived American economic and strategic considerations will dictate continued opposition to majority rule unless American blacks and others oppose the policies which may permit South Africa to become one of the nuclear powers. Based on secondary sources; 23 notes. G. O. Gagnon

428. Walters, Ronald W. URANIUM POLITICS AND U.S. FOREIGN POLICY IN SOUTHERN AFRICA. *J. of Southern African Affairs 1979 4(3): 281-300.* Considers the role of uranium procurement, processing, and pricing in US-South African relations since the early 1960's. The precise degree of uranium interdependence between these two countries is difficult to determine, but it is a growing factor in America's political economy. South African control over substantial uranium reserves and their international marketing through a cartel formed in 1972 is seen as a large consideration in current US policy toward Southern Africa. Based on newspapers and secondary sources; 66 notes.
L. W. Truschel

429. Webb, B. H. THE ORIGINS OF THE COLD WAR. *Kleio [South Africa] 1974 6(1): 13-26.* The Western Allies' weakness in delaying the opening of a Second Front was the situation which developed into the Cold War, aggravated for the USSR by the atom bomb and the Marshall Plan, and cooled when it manufactured its own bomb.

430. Wells, Samuel F., Jr. THE ORIGINS OF MASSIVE RETALIATION. *Pol. Sci. Q. 1981 96(1): 31-52.* Examines the popular understanding of the concept of massive retaliation and its refinement in the early months of the Eisenhower administration. The Truman administration had developed both the doctrine and the defense programs that formed the basis of its successor's new declaratory policy. J

431. Werner, Roy A. DOWN THE ROAD TO ARMAGEDDON? *Military Rev. 1975 55(7): 30-40.* Future American security will be endangered because elite youth will draw "lessons" from the recent past which will be misapplied later. These youth, the policymakers of tomorrow, have concluded that military solutions are no longer viable and, in effect, will psychologically disarm the United States. By doing so, they may make international conflict more likely than if the nation maintained the stance of the period 1945-1973—"the will to exercise power learned from Munich." 19 notes. J. K. Ohl

432. Willhelm, Sidney M. THE RISE OF STATE RULE: AN EXPLORATORY AND INTERPRETIVE ESSAY. *Catalyst [Canada] 1977 (9): 1-58.* Examines the decline of capitalism as a political control in the federal government due to the increasing importance of the Central Intelligence Agency and growing reliance on nuclear arms as a method of maintaining national security, 1940's-70's.

433. Wohlstetter, Albert. IS THERE A STRATEGIC ARMS RACE? *Survival [Great Britain] 1974 16(6): 277-292.* Questions the existence of a nuclear arms race between the United States and the USSR in the 1970's, emphasizing military strategy, missile capability, and the role of defense spending. Reprinted from *Foreign Policy* 1974 (15); see abstract 446.

434. Wohlstetter, Albert. RACING TOWARD OR AMBLING BACK? *Survey [Great Britain] 1976 22(3-4): 163-217.* The Soviet-American "arms race" continues to be fueled by an "action-reaction" cycle. However, the United States has not driven the cycle, for the United States has, since the 1960's, underestimated Soviet force and decreased defense spending. Perceptions of the current "arms race" resemble doctrines which developed in England between the two World Wars. Contrary to popular belief, the United States consistently underestimated rather than overestimated Soviet strategic forces. A steady decline in the US defense budget and strategic forces belies the existence of an arms race. Secondary sources; table, 35 charts, 61 notes. D. R. McDonald

435. Wohlstetter, Albert. SPREADING THE BOMB WITHOUT QUITE BREAKING THE RULES. *Foreign Policy 1976-77 (25): 88-96, 145-179.* Contends that the possibility of nuclear arms proliferation has increased greatly because of the almost uncontrolled export of nuclear technology. This can be avoided through the implementation of nuclear weapons support guarantees to certain key nations, restrictions on the export of readily fissionable materials, the denial of assistance for critical experiments in foreign countries, regulation of the export of enrichment and separation facilities, and deferment of the decision to separate plutonium in the United States. Primary and secondary sources; 2 graphs, 30 notes. E. Hopkins

436. Wohlstetter, Albert. THREATS AND PROMISES OF PEACE: EUROPE AND AMERICA IN THE NEW ERA. *Orbis 1974 127(4): 1107-1144.* Discusses foreign relations, the use of nuclear arms, and military strategy of the United States, Europe, and the USSR, 1950's-70's.

437. Wolfe, Gary K. *DR. STRANGELOVE*, *RED ALERT*, AND PATTERNS OF PARANOIA IN THE 1950'S. *J. of Popular Film 1976 5(1): 57-67.* Discusses patterns of paranoia in American thought and focuses on Stanley Kubrick's *Dr. Strangelove* and Peter George's 1958 novel *Red Alert*, which inspired it, as a representative of the 1950's fear of communism, assassination, and nuclear holocaust.

438. Wood, Robert S. FROM COLD WAR TO DETENTE: CHANGE AND CONTINUITY. *Soviet Union 1975 2(2): 198-204.* Reviews *Caging the Bear: Containment and the Cold War* (New York: Bobbs, 1974) by Charles Gati; *The Role of Nuclear Forces in Current Soviet Strategy* (U. of Miami, 1974) by Leon Goure et al.; and *The US-USSR Nuclear Weapons Balance* by Edward

Luttwak. It has become accepted that the dominant features of post-World War II international politics and the assumptions underlying American foreign policy are undergoing substantial changes. Consequently the recent detente policy of Henry A. Kissinger is aimed at the encouragement of "an environment in which competitors can regulate and restrain their differences and ultimately move from competition to cooperation." G. E. Snow

439. York, Herbert F. NUCLEAR PROLIFERATION: AN UNHAPPY HISTORY. *Center Mag. 1976 9(3): 71-77.* Discusses nuclear proliferation, 1945-75, and the diplomatic attempts by the United States and other nations to restrict the possession of nuclear arms.

440. York, Herbert. REDUCING THE OVERKILL. *Survival [Great Britain] 1974 16(2): 70-74.* Discusses the threat of overkill in US and USSR nuclear stockpiling in the 1960's and 70's, emphasizing strategic principles in nuclear deterrence and balance between the two nations.

441. Zhigalov, I. I. OBSHCHESTVENNOST' VELIKOBRITANII I PLANY RAZMESHCHENIIA NA EE TERRITORII AMERIKANSKOGO NEITRONNOGO ORUZHIIA I KRYLATYKH RAKET [British public opinion on deployment of the US neutron weapon and cruise missiles on British territory]. *Voprosy Istorii [USSR] 1981 (11): 66-80.* Analyzes the development of the British antiwar movement in 1978-81, especially against the deployment of the US neutron weapon and cruise missiles. The author shows the factors behind the antimissile movement, its character, specifics, the emergence of new antiwar organizations, new forms and methods of struggle, and against this background, the positions and activity of major political parties, trade unions, and influential social organizations. The conclusion he draws from his analysis is that the struggle against nuclear weapons is again in the forefront of British political life, and that the antiwar movement is gaining momentum and mass support. Russian. J

442. Zumwalt, Elmo R. TOTAL FORCE. *US Naval Inst. Pro. 1979 105(5): 88-107.* The United States and the Soviet Union have been, and are today, "engaged in a gigantic struggle whose end is not in sight." A country's total force is made up of eight components—nuclear military, conventional military, political, ideological, economic, technological, social, and diplomatic forces. Today, the Soviets have superiority in the first three, while the United States is superior in the economic, technological, and social components. The countries appear even in the other two. The United States must find the leadership and the will to support any effort to stand up to the USSR, and in these efforts, sea power will play an important role. Unless the United States is prepared to make that effort over the long run, it will encounter increasing difficulties in fending off the increasing Soviet challenge to its freedom and independence. Secondary sources; 22 notes, 12 photos. A. N. Garland

443. —. AFTER VIETNAM: RESURGENT U.S. MILITARISM. *Social Policy 1977 7(4): 8-15.* Through examination of defense spending, nuclear arms proliferation, atomic proliferation, and a general attitude among policy makers toward proliferation of militarism, the Bay Area Group of the Inter-University Committee to Stop Funding War and Militarism concludes that foreign policy is gradually moving toward a similar militaristic attitude.

444. —. [DETENTE]. *Pol. Étrangère [France] 1977 42(3-4): 273-290.*

Ikle, Fred Charles. LA DÉTENTE ET LES RELATIONS OUEST-EST [Detente and East-West relations], *pp. 273-283.* Discusses the evolution of European internal political structures, conventional and nuclear military forces, and relations with the United States and within European alliances since the 1950's.

Ruehl, Lothar. INTERVENTION [Comment], *pp. 285-290.* Discusses the causes behind the increase in the USSR's conventional and nuclear forces in terms of Soviet doctrine and history since World War II.

445. —. IS THERE A STRATEGIC ARMS RACE? *Foreign Policy 1974 (16): 48-92.*

Wohlstetter, Albert. RIVALS, BUT NO "RACE," pp. 48-81. Challenges the opinion that the United States is engaged in a spiralling strategic arms race with the USSR, for the following reasons: 1) US strategic spending has decreased, 2) the relative destructiveness of US strategic forces has declined, 3) the decreased vulnerability of US strategic forces has increased their stability, 4) the controllability of US force has been enhanced by improved protective methods, which have made hair-trigger response unnecessary, and 5) improvements in quality have made quantitative increases less necessary. However, there still is military competition with the Soviet Union, particularly in quality. Generally, the United States has overestimated future Soviet military potential. Primary and secondary sources; 6 graphs, 20 notes.

Nitze, Paul H. COMMENT, pp. 82-83. Agrees that the United States has tended to overestimate Soviet military capabilities. Indicates that the usual oversimplified stereotypes fail to explain the complex interrelationships of the United States and the Soviet Union.

Alsop, Joseph. COMMENTS, pp. 83-88. Although the United States has greatly overestimated the Soviet defense effort, an adequate defense insurance policy requires such overestimation; to underestimate might be fatal. Favors an arms race in which the United States leads in all categories for the sake of security. Secondary sources.

Halperin, Morton H. and Stone, Jeremy J. COMMENTS, pp. 88-92. The United States does not overestimate Soviet military capabilities, but has invariably set the rate and scale for most of the individual steps in the strategic arms race. Secondary sources. R. F. Kugler

446. —. [IS THERE A STRATEGIC ARMS RACE?]

Wohlstetter, Albert. IS THERE A STRATEGIC ARMS RACE? *Foreign Policy 1974 (15): 3-20.*

Nitze, Paul H. THE STRATEGIC BALANCE BETWEEN HOPE AND SKEPTICISM. *Foreign Policy 1974/75 (17): 136-156.*

Aaron, David. A NEW CONCEPT. *Foreign Policy 1974/75 (17): 157-165.*

Warnke, Paul C. APES ON A TREADMILL. *Foreign Policy 1975 (18): 12-29.*

Myrdal, Alva. THE HIGH PRICE OF NUCLEAR ARMS MONOPOLY. *Foreign Policy 1975 (18): 30-43.*

Holst, Johan Jorgen. WHAT IS REALLY GOING ON? *Foreign Policy 1975 (19): 155-177.*

Nacht, Michael L. THE DELICATE BALANCE OF ERROR. *Foreign Policy 1975 (19): 163-177.*

Wohlstetter, Albert. HOW TO CONFUSE OURSELVES. *Foreign Policy 1975 (20): 170-198.*
A debate on the myths and realities of the arms race. Wohlstetter argues that the United States is, in effect, disarming unilaterally, while the Soviet Union is not. His opponents argue that, while the United States did not always attempt to match Soviet weapons build-up with identical US weapons, it did attempt to gain overall superior strategic strength. Secondary sources. See also abstract 12A:877.
R. F. Kugler

447. —. LA DOCTRINE D'EMPLOI DES ARMES STRATÉGIQUES AMÉRICAINES ET LA DÉFENSE DE L'EUROPE [The doctrine of using American strategic weapons and the defense of Europe]. *Défense Natl. [France] 1974 30(10): 93-100.* Examines the motives behind Washington's redeployment of US strategic forces and evaluates the consequences for Europe's defense, 1973.

448. —. THE MILITARY BALANCE 1975-76: THE UNITED STATES AND THE SOVIET UNION. *Air Force Mag. 1975 58(12): 46-51.* Discusses the development and capabilities of the strategic forces of the United States and the USSR, along the guidelines of the summit meeting at Vladivostok in November 1974.

449. —. THE MILITARY BALANCE 1974/75: THE THEATRE BALANCE BETWEEN NATO AND THE WARSAW PACT. *Air Force Mag. 1974 57(12): 96-103.* Assesses the military balance between NATO and the Warsaw Pact, considering strengths of numbers and equipment, and weighing the advantages of geography, training, logistic support, and ideological backing; slow changes have altered the balance.

450. —. THE MILITARY BALANCE 1974/75: TABLES OF COMPARATIVE STRENGTHS. *Air Force Mag. 1974 57(12): 88-95.* Tabulates the comparative military strength of the USSR and the United States, NATO, the Warsaw Pact, and other world powers, and includes tables of defense expenditures, strength and characteristics of military forces, and military formations, 1953-75.

451. —. THE MILITARY BALANCE BETWEEN 1973/74: TABLES OF COMPARATIVE STRENGTHS. *Air Force Mag. 1973 56(12): 116-121.* Tabulates the various military strengths of world powers, particularly the United States and the USSR, including nuclear delivery vehicles, defense and national expenditures, and military manpower.

452. —. THE MILITARY BALANCE 1975-76: TABLES OF COMPARATIVE STRENGTHS. *Air Force Mag. 1975 58(12): 90-95.* Tabulates the varying military strengths of the world, 1975-76, comparing also defense spending and military manpower.

453. —. THE MILITARY BALANCE 1974/75: THE UNITED STATES AND THE SOVIET UNION. *Air Force Mag. 1974 57(12): 44-49.* Discusses the deployment of new and improved offensive and defensive missile systems in the USSR and the United States, and their continuance within the limits agreed to in 1972.

454. —. MORE NUCLEAR NATIONS? *Foreign Affairs 1974 53(1): 64-97.*

Stevenson, Adlai E., III. NUCLEAR REACTORS: AMERICA MUST ACT, *pp. 64-76.* Alarmed by India's explosion of the atomic bomb in May 1974 and the offer of US nuclear assistance to unstable Middle Eastern countries calls for a reexamination of American nuclear assistance programs. The United States should declare a conditional one-year moratorium on sales of reactors except to countries which submit their facilities to International Atomic Energy Agency (IAEA) inspection. Strict control over all materials and technology must also be insured.

Quester, George H. CAN PROLIFERATION NOW BE STOPPED?, *pp. 77-97.* Examines several countries which may follow India's example in the detonation of nuclear explosives and suggests a "patchwork quilt of hopes" whereby nuclear proliferation may yet be contained by inculcating a sense of responsibility in the signers of the Nuclear Nonproliferation Treaty (1968). 2 notes. C. W. Olson

455. —. NUCLEAR CHICKEN. *Monthly Rev. 1982 34(4): 1-11.* The sudden revival of the antinuclear movement during 1970-81 was a result of the turning-away of both sides from the posture of Mutual Assured Destruction (MAD) to counterforce measures, such as the 1979 NATO decision to reequip with cruise missiles, implying a willingness to use atomic weapons in a first-strike situation; analyzes why NATO decided to install new weapons thus upsetting the USSR and a reasonably stable balance; questions the US strong-arm role in this decision.

456. —. [THE NUCLEAR DEBATE]. *Int. Security 1977 1(4): 44-79.*

Jaipal, Rikhi. THE INDIAN NUCLEAR SITUATION, *pp. 44-51.* Defends India's underground nuclear explosion of 1974, proclaims India's peaceful intentions, and criticizes the superpowers' nuclear policies.

Long, Clarence D. NUCLEAR PROLIFERATION: CAN CONGRESS ACT IN TIME?, *pp. 52-76.* The United States should discourage nuclear proliferation by stopping subsidy and promotion of nuclear power exports, by not encouraging the use of nuclear energy, by redirecting foreign aid away from nuclear-developing nations, and by pressing other suppliers to stop proliferation. These measures, as demonstrated by the legislative efforts of Congress in response to India's "peaceful" nuclear explosion of May 1974, would be beneficial economically and environmentally.

Long, Clarence D. RESPONSES, *pp. 77-78.* Ambassador Jaipal (of India) states that India's "peaceful" explosion was not posed as a nuclear threat, but if so, how could India curb further tests? Also, Jaipal ignores the costs of nuclear power and potential effects of proliferation.

Jaipal, Rikhi. RESPONSES, *p. 79.* Long's proposed measures will not solve the problems of proliferation—the causes of international tension must be examined as well as the effects.

4

ATTEMPTS AT NUCLEAR ARMS CONTROL

457. Ackley, Richard T. WHAT'S LEFT OF SALT? *Naval War Coll. R. 1974 26(6): 43-49.* "Any effective strategic arms control measures must reduce the likelihood of nuclear war as well as reduce one's own damage if war should occur. How well SALT is able to accomplish these objectives will largely determine its future." J

458. Alger, Chadwick F. and Boulding, Elise. FROM VIETNAM TO EL SALVADOR: ELEVEN YEARS OF COPRED. *Peace and Change 1981 7(3): 35-43.* Discusses the history of the Consortium on Peace Research, Education and Development from its founding on 10 May 1970, especially its political context, organizing principles, affiliations with kindred organizations, and resources.

459. Ali, Mehrunnisa. PAKISTAN-UNITED STATES RELATIONS: THE RECENT PHASE. *Pakistan Horizon [Pakistan] 1978 31(2-3): 32-56.* Pakistan is slowly moving to a foreign policy based on nonalignment because US diplomacy has been inconsistent; one major obstacle to better relations has been American hostility to Pakistan's program for developing nuclear energy, 1970's.

460. Allen, H. C. QUIET STRENGTH? AMERICAN FOREIGN POLICY UNDER PRESIDENT CARTER. *Round Table [Great Britain] 1977 (266): 146-152.* The Inaugural Address of President Carter highlighted the need for nuclear disarmament and for morality in American foreign policy. The era of Imperial Republicanism has ended. There is a new tone in foreign and domestic policy. C. Anstey

461. Allison, Graham. COLD DAWN AND THE MIND OF KISSINGER. *Washington Monthly 1974 6(1): 38-49.* Reviews John Newhouse's *Cold Dawn: The Story of SALT* (New York: Holt, Reinhart & Winston, 1973). S

462. Allison, Graham T. and Morris, Frederic A. ARMAMENTS AND ARMS CONTROL: EXPLORING THE DETERMINANTS OF MILITARY WEAPONS. *Daedalus 1975 104(3): 99-129.* The emphasis in arms control is often on foreign relations, yet internal factors in both the United States and the Soviet Union are the predominant determinants of the weapons systems. These internal factors should be accurately analyzed and restructured so choice and

accountability can increase. Based on federal documents and secondary sources; tables, 4 charts, 37 notes. E. McCarthy

463. Alsterdal, Alvar. AVSPÄNNING OCH AVGRÄNSNING [Detente and demarcation]. *Kungliga Krigsvetenskaps Akademiens Handlingar och Tidskrift [Sweden] 1973 177(5): 151-158.* In the late 1960's and early 1970's the USSR reduced its negative official propaganda against the United States and Western Europe to an extent that suggested a relaxation of tension. However, the European Security Conference saw hostile Soviet reactions to proposals for a free exchange of ideas between the East and the West, and the author asserts that while Western Europe worked toward detente, the USSR strengthened its hold on the Eastern European countries. U. Bartels/S

464. Ambri, Mariano. BREVE STORIA DEL "SALT 2" [A brief history of SALT II]. *Civitas [Italy] 1980 31(2): 41-51.* Traces the origins and precedents for the SALT II treaty, which has been held up in the US Senate since the Soviet invasion of Afghanistan, relates it to major foreign policy themes in the Carter presidency, and considers the global significance of its provisions.

465. Arbatov, G. A. ON SOVIET-AMERICAN RELATIONS. *Survival [Great Britain] 1973 15(3): 124-129.* Discusses issues in economics and nuclear arms limitations in foreign relations and negotiations between the USSR and the United States in 1972-73, emphasizing the relaxation of Cold War attitudes.

466. Baker, Jeffrey J. W. THREE MODES OF PROTEST ACTION: THE SEARCH FOR WAYS OF MOBILIZING SCIENTISTS IN THE PUBLIC AFFAIRS ARENA. *Bull. of the Atomic Scientists 1975 31(2): 8-15.* Examines scientists' involvement in protest actions and explores three approaches: Mode 1) "Resolutions or protest statements composed by representatives or appointed or elected committees of a scientific organization and released under the name of that organization." Mode 2) "Resolutions or protest statements signed by scientists as individuals and released with no organizational name appended." Mode 3) "Resolutions or protest statements signed by scientists as individuals but released with the name of the professional organization which they represent included only for purposes of delineating the signer's primary interests or professional competence." Concludes that Mode 3 incorporates the advantage of Mode 1 (linking the prestige of an organization's name to a statement) and Mode 2 (timing) without the disadvantage of Mode 1 of associating those who do not wish to be associated with the protest action. Based on secondary sources; 10 notes. D. J. Trickey

467. Bargman, Abraham. NUCLEAR DIPLOMACY. *Pro. of the Acad. of Pol. Sci. 1977 32(4): 159-169.* Despite its apparently unilateral and self-serving behavior, the United States remains a leader in the movement for international control of atomic energy. However, little more can be achieved outside of international organizations such as the United Nations in today's changed international order. The 1978 UN Special Session on Disarmament can provide a major ideological opportunity for the United States to seek multilateral cooperation on arms control and nuclear nonproliferation. 2 notes. K. N. T. Crowther

468. Barton, John H. THE DEVELOPING NATIONS AND ARMS CONTROL. *Studies in Comparative Int. Development 1975 10(1): 67-82.* Explores the format and focus of various nuclear nonuse and nonproliferation treaties and proposals. Analyzes the attitudes of developing nations and the United States concerning the psychological, strategic, and economic aspects of arms control. There are fewer dissimilarities between developed and developing nations' arms control positions than might be expected. Policies depend on defining stable military balances and building the necessary constituencies. Solution of the arms control problem is growing more difficult. Solution is likely to be simplified if approached regionally and coupled with redesign of some international institutions and additional controls in the US-USSR nuclear arms race. Based on primary and secondary sources; 9 notes, biblio. S. A. Farmerie

469. Bates, E. A., Jr. THE SALT STANDING CONSULTATIVE COMMISSION: AN AMERICAN ANALYSIS. *Millennium: J. of Int. Studies [Great Britain] 1975 4(2): 132-145.* Examines the creation of the SALT Standing Consultative Commission (SCC), its functions, organization, and procedures, with particular reference to the Anti-Ballistic Missile System Treaty of 1972. Based on the text of the ABM Treaty of 1972 and secondary sources; 24 notes. P. J. Beck

470. Beard, Robin. THE CARTER ADMINISTRATION'S SALT-II PROPOSALS. *J. of Social and Pol. Studies 1977 2(3): 143-154.* Maintains that so many concessions have been made to the Soviets from the Carter administration's SALT-II proposals that the treaty thus far negotiated will not constrain Soviet strategic capabilities, nor diminish the Soviet threat to the US silo-based ICBM.

471. Beavers, Roy L., Jr. SALT I. *US Naval Inst. Pro. 1974 100(5): 204-219.* The 1972 Salt I "is likely to be seen in retrospect as a landmark in the panorama of affairs between the two superpowers, but as an arms limitation measure its place in history may be more uncertain." Salt I "did not end the strategic arms race" and it did not "resolve any of the fundamental political issues between the United States and the Soviet Union." 8 photos, 9 notes. A. N. Garland

472. Bennett, Roy. ALL THE KING'S HORSES MIGHT DO IT: SALT II. *Social Policy 1979 9(4): 28-32.* Discusses the prospects for Senate confirmation of the second Strategic Arms Limitation Talks treaty (SALT II); 1970's.

473. Benoit, Emile. KENNETH BOULDING AS SOCIO-POLITICAL THEORIST. *J. of Conflict Resolution 1977 21(3): 551-560.* Reviews Boulding's *Collected Papers,* vols. 4 and 5, ed. Larry O. Singell (Boulder: Colorado Associated U. Pr., 1974, 1975), focusing on his general systems approach, which relies heavily on analogies and tends to be abstract; and his fivefold classification of societal subsystems. Discusses his theory of conflict resolution, which envisages "peace research" and a "minimal world government." Other articles on peace research are contained in vol. 5. J. Tull

474. Beres, Louis Rene. THE COMING WORLD STORM. *Colorado Q. 1974 23(1): 84-90.* Urges immediate fundamental changes from competitive to cooperative foreign relations to avert atomic warfare. S

475. Bertram, Christoph. SALT II AND THE DYNAMICS OF ARMS CONTROL. *Int. Affairs [Great Britain] 1979 55(4): 565-573.* Rebuts critics' arguments against SALT II to offer an optimistic appraisal of the proposed treaty.

476. Bhatia, Shyam. BEYOND PROLIFERATION: POLITICAL FEARS AND SCIENTIFIC HOPES IN THE NUCLEAR DEBATE. *Round Table [Great Britain] 1977 (268): 325-336.* Examines the failure, 1940's-50's, to establish some measure of control over the spread of nuclear weapons, and traces the implications of the Test Ban Treaty of 1963, the Nuclear Nonproliferation Treaty of 1968, and the halting progress in the Strategic Arms Limitation Talks. The need for abundant and cheap energy in countries such as India prompted an international reactor export drive, starting with the United States Atoms for Peace Program in 1955. With the fear of a prospective oil shortage this drive has been accelerated, providing many middle and smaller states with the technology for the production of nuclear weapons. C. Anstey

477. Bonnemaison, Jacques. L'ACCORD STRATÉGIQUE DE VLADIVOSTOK, TENTATIVE DE MESURE DANS LA DÉMESURE [The Vladivostok strategic agreement, an attempt at moderation in immoderation]. *Défense Natl. [France] 1975 31(5): 27-36.* Retraces the developments that led to the Vladivostok Accord and Strategic Arms Limitation Talks in November 1974 and assesses its significance and scope.

478. Booth, Kenneth. THE STRATEGIC ARMS LIMITATION TALKS: A STOCKTAKING. *World Survey [Great Britain] 1975 (73): 1-18.* Notes the origins and results of the Strategic Arms Limitation Talks between the US and USSR, 1969-75.

479. Božić, Nemanja. PREGOVORI O OGRANIČENJU STRATEŠKOG ORUŽA: SALT-I I SALT-II [The Strategic Arms Limitation Talks: SALT I and SALT II]. *Medjunarodni Problemi [Yugoslavia] 1972 24(4): 71-79.* Discusses the course of the first round of the Strategic Arms Limitation Talks (I) from 1969 to May 1972 and considers those points likely to be carried over to SALT II.

480. Brenner, Michael. PROLIFERATION WATCH: CARTER'S BUNGLED PROMISE. *Foreign Policy 1979 (36): 89-101.* Jimmy Carter deserves credit for being the first American president to tackle forthrightly the problem of nuclear nonproliferation. His tactics and accomplishments in the matter merit less praise. Much of the difficulty derives from the complexity of the problem, but inadequate presidential leadership, leading to an undisciplined government without a definite objective, must also bear blame. Tight organization and clear purpose might enable the administration to realize its promise of containing the spread of nuclear weapons. T. L. Powers

481. Brodie, Bernard. ON THE OBJECTIVES OF ARMS CONTROL. *Int. Security 1976 1(1): 17-36.* Analyzes the present definition of and agreement upon arms control and assesses the arms race and defense budgeting.

482. Brown, Harrison. AN EARLY BRIEF ENCOUNTER. *Bull. of the Atomic Scientists 1979 35(3): 17-19.* Recounts asking (in 1947) Andrei Gromyko, Soviet delegate to the UN Security Council, to send Soviet scientists to a meeting of international scientists concerned with the future of nuclear arms.

483. Bull, Hedley. ARMS CONTROL AND WORLD ORDER. *Int. Security 1976 1(1): 3-16.* The visions of world order projected by the theory and practice of arms control that arose in the late 1950's command "little support outside the circle of the United States and the Soviet Union and their closest allies."

484. Bull, Hedley. RETHINKING NON-PROLIFERATION. *Internat. Affairs [Great Britain] 1975 51(2): 175-189.* Discusses the working of the Nuclear Nonproliferation Treaty (1968) in the light of the difficulties of administering the treaty, and in the light of the Geneva Conference to be held in 1975 to review its operation. Argues that it is important to return again to the basic issues, and to the case for and against nonproliferation. 4 notes. P. J. Beck

485. Bull, Hedley. A VIEW FROM ABROAD: CONSISTENCY UNDER PRESSURE. *Foreign Affairs 1979 57(3): 441-462.* Reviews 1978 in terms of American foreign policy developments. Despite a wide range of policies, pressures, and concerns, the Carter administration remained generally consistent with its stated foreign policy objectives. Soviet meddling in Africa and a new surge of American anti-Sovietism, for example, have not dampened President Carter's determination to finalize a SALT II treaty. M. R. Yerburgh

486. Bundy, McGeorge. EARLY THOUGHTS ON CONTROLLING THE NUCLEAR ARMS RACE: A REPORT TO THE SECRETARY OF STATE, JANUARY 1953. *Int. Security 1982 7(2): 3-27.* The now declassified report on the dangers of a nuclear war with the USSR, by the Panel of Consultants on Disarmament in 1952, that stated the need for candor with the American people, for an atomic weapons policy in harmony with our foreign policy, and for the United States to learn to communicate with the USSR.

487. Bundy, McGeorge. HIGH HOPES AND HARD REALITY: ARMS CONTROL IN 1978. *Foreign Affairs 1979 57(3): 492-502.* During 1978, the United States and the USSR moved closer to final agreement on a second strategic arms limitation treaty (SALT II). Though the US team negotiated skillfully, domestic debate increased sharply. "The year 1978 probably set the stage for the most complex test of political command, control, communications, and intelligence yet put before a president in the field of strategic policy." 4 notes.
M. R. Yerburgh

488. Burns, Richard D. ARMS CONTROL AND DISARMAMENT: TERMS AND RESOURCES. *Peace and Change 1982 8(1): 53-63.* Offers guidelines to the definition of the terms "arms control" and "disarmament" by examining general, abstract definitions and historical examples.

489. Burns, Richard Dean. DEFINING ARMS CONTROL AND DISARMAMENT: A HISTORICAL PERSPECTIVE. *Soc. for Hist. of Am. Foreign Relations. Newsletter 1978 9(4): 1-8.* Examines definitions of arms control, 18th-20th century, and uses historical examples to establish precise definitions for arms control and disarmament technique.

490. Burt, Richard. SALT II AND OFFENSIVE FORCE LEVELS. *Orbis 1974 18(2): 465-481.* Analyzes the problems of the Strategic Arms Limitations Talks (II) and differences between US and Soviet offensive nuclear arms, 1970.

491. Burt, Richard. THE SCOPE AND LIMITS OF SALT. *Foreign Affairs 1978 56(4): 751-770.* Since 1973 Soviet and US negotiators have attempted to outline the details of a new strategic arms limitation accord. It must be recognized, however, that the final terms of the SALT agreement will not altogether obviate the need for difficult US strategy questions in the years ahead. Questions over the direction of the SALT discussions include: will US strength be further eroded; will US technological initiative be dampened; and will NATO's vitality be sapped? M. R. Yerburgh

492. Byers, R. B. THE PERILS OF SUPERPOWER DIPLOMACY: DÉTENTE, DEFENSE, AND ARMS CONTROL. *Int. J. [Canada] 1980 35(3): 520-547.* In 1972 President Richard M. Nixon of the United States and Premier Leonid Brezhnev of the USSR agreed on the basic principles of détente—political coexistence, increased commercial relations, and pursuit of arms limitation—and implicitly accepted their linkage. The United States, however, failed to remind the Russians of linkage, a concept which the Russians preferred to ignore, and neither side could credit the other's striving for military parity rather than superiority. Only after the Soviet invasion of Afghanistan did President Jimmy Carter revive the concept of linkage, and his overreaction to the invasion led to the postponement of SALT II. 12 notes. E. L. Keyser

493. Caldwell, Lawrence T. THE FATE OF STRATEGIC ARMS LIMITATION AND SOVIET-AMERICAN RELATIONS. *Int. J. [Canada] 1981 36(3): 608-634.* Discusses interactions between the United States and the Soviet Union in 1981, in particular, policies since the election of Ronald Reagan. Difficulties with the Strategic Arms Limitation Talks (SALT II) stem from the attitude of the Reagan administration toward the USSR. However, there is a wide range of global issues, alliance temptations, domestic pressures, and technological dynamics which have intruded on the SALT process. Secondary sources; 45 notes. J. Powell

494. Caldwell, Lawrence T. SALT II AND THE STRATEGIC RELATIONSHIP. *Current Hist. 1979 77(450): 101-105, 130-131, 136-138.* Analyzes the provisions of the SALT II treaty and their strategic significance to the United States and the USSR; 1970's.

495. Campbell, John. EUROPEAN SECURITY AFTER HELSINKI: SOME AMERICAN VIEWS. *Government and Opposition [Great Britain] 1976 11(3): 322-336.* Examines the policy of the Ford administration toward European security as generated by the Conference on Security and Cooperation in Europe, and American public opinion on the subject, 1975.

496. Caracciolo, Roberto. I NEGOZIATI SALT NEL QUADRO DEI PROBLEMI GENERALI E DELLA POLITICA DEL DISARMO [The SALT negotiations in light of general problems and the politics of disarmament]. *Affari Esteri [Italy] 1978 10(37): 23-38.* Emphasizes the deterrent function of nuclear arms but questions the sincerity of the USSR's motives when participating in the Strategic Arms Limitation Talks.

497. Carter, Barry. WHAT NEXT IN ARMS CONTROL? *Orbis 1973 17(1): 176-196.* Discusses arms control strategies available to the United States.

498. Carter, Barry E. THE STRATEGIC DEBATE IN THE UNITED STATES. *Pro. of the Acad. of Pol. Sci. 1978 33(1): 15-29.* In the United States the public and intense debate on SALT contributes to the education of the public on nuclear issues. Chief participants in public debate are the military, the Secretary of Defense, the Secretary of State, the President and his advisers, and Congress. Each has varying amounts of influence and knowledge on the several issues which can impede as well as further debate. Care must be taken to prevent misinformation and misunderstanding while continuing to extend public participation. 10 notes. K. N. T. Crowther

499. Carter, Jimmy. SOVIET-AMERICAN RELATIONS. *State Government 1977 50(3): 141-144.* General outline of foreign policy decisions pertaining to arms limitation, human rights, and peace which affect relations between the United States and the USSR.

500. Chalfont, Lord. SALT II AND AMERICA'S EUROPEAN ALLIES. *Int. Affairs [Great Britain] 1979 55(4): 559-564.* Critical evaluation of the prospective effects of SALT II on Western Europe's security; 1970's.

501. Chayes, Abram. NUCLEAR ARMS CONTROL AFTER THE COLD WAR. *Daedalus 1975 104(3): 15-33.* As the likelihood of confrontation between the superpowers diminishes, nuclear arms control measures have declined in relative importance, but the nuclear capability of several developing countries points to the need for a strong international organization, significant limitation by the superpowers, and a comprehensive test ban. 3 notes. E. McCarthy

502. Chester, Conrad V. and Wigner, Eugene P. POPULATION VULNERABILITY: THE NEGLECTED ISSUE IN ARMS LIMITATION AND THE STRATEGIC BALANCE. *Orbis 1974 18(3): 763-770.*

503. Christopher, Robert C. THE U.S. AND JAPAN: A TIME FOR HEALING. *Foreign Affairs 1978 56(4): 857-866.* The strains in US-Japanese relations are more perceptible now than they have been for two decades. By failing to brief Japan about important foreign policy departures, by holding Japan primarily responsible for an embarrassing trade deficit, and by denying Japan the latitude to develop breeder reactors, the United States has demonstrated somewhat crude and insensitive behavior. M. R. Yerburgh

504. Clarke, Duncan L. THE ARMS CONTROL AND DISARMAMENT AGENCY: EFFECTIVE? *Foreign Service J. 1975 52(12): 12-14, 28-30.* Discusses the effectiveness of the US Arms Control and Disarmament Agency in lessening the threat of nuclear war in the 1960's and 70's.

505. Clarke, Duncan L. ARMS CONTROL AND FOREIGN POLICY UNDER REAGAN. *Bull. of the Atomic Scientists 1981 37(9): 12-19.* The Arms Control and Disarmament Agency established in 1960, has experienced traumatic shifts in leadership and support; these shifts coincide with the acceleration of the arms race, the failure to negotiate reductions, and the Reagan Administration's oversimplification, perpetually blaming the Russians.

506. Clarke, Duncan L. ROLE OF MILITARY OFFICERS IN THE US ARMS CONTROL AND DISARMAMENT AGENCY. *Military R. 1974 54(12): 47-53.* Active duty or recently retired military officers provide the civilian employees of the US Arms Control and Disarmament Agency with valuable military expertise.

507. Clarke, Duncan L. UPS AND DOWNS OF ARMS CONTROL. *Bull. of the Atomic Scientists 1974 30(7): 44-49.* Examines the status and relatively secret functioning of the Arms Control and Disarmament Agency and the General Advisory Committee on Arms Control and Disarmament, 1961-73.

508. Clemens, Walter C., Jr. ARMS CONTROL AS A WAY TO PEACE. *World Affairs 1972 135(3): 197-219.* Though perhaps not as long-lasting, arms control promises to be an easier goal than disarmament and offers aspects which disarmament does not: an ecological attitude toward foreign relations, a preventive approach to problems in world affairs, a focus for interdisciplinary studies, and sophistication of strategic studies, 1960's-72.

509. Cohen, Stuart A. SALT VERIFICATION: THE EVOLUTION OF SOVIET VIEWS AND THEIR MEANING FOR THE FUTURE. *Orbis 1980 24(3): 657-683.* Traces a clear and significant evolution of Soviet views on Strategic Arms Limitation Talks verification and concludes that "Recognition of the utility of certain intrastate agreements, especially the SALT accords, has likely provided the momentum for grudging Soviet movement in these matters. It is also likely that the utility of these same monitoring techniques for implementation of various other Soviet foreign policies, especially crisis management concerns, contributed to the evolutionary process." Based on published sources, Soviet publications, and US government documents; 55 notes, 2 appendixes.
J. W. Thacker, Jr.

510. Critchley, Julian. EAST-WEST DIPLOMACY AND THE EUROPEAN INTEREST: CSCE, MFR AND SALT II. *Round Table [Great Britain] 1974 (255): 299-306.* Examines US and Soviet concepts of detente, their interests, aims, policy objectives, and the areas open to negotiation. Discusses the achievements, progress, and success of each nation at securing its demands at the Conference on Security and Cooperation in Europe, Mutual Force Reduction, the SALT II conferences, and East-West diplomacy in Europe. C. Anstey

511. Delbourg, Denis. L'EVOLUTION STRATEGIQUE DES DEUX GRANDS ET SES CONSEQUENCES POUR LES INTERETS DES EUROPEENS [The strategic evolution of the two superpowers and its consequences for European interests]. *Défense Natl. [France] 1979 35(Jul): 43-70.* A consideration of the evolution of policy on the limitation of strategic offensive arms between the United States and the USSR, which culminated in the signing of the Strategic Arms Limitation Talks (SALT II) treaty on 18 June 1979. Based on a lecture presented to a conference organized by *Défense Nationale* on 15 March 1979 and comments made by speakers after the lecture; 23 notes. French.

512. Destler, I. M. TREATY TROUBLES: VERSAILLES IN REVERSE. *Foreign Policy 1978-79 (3): 45-65.* The defeat of the Treaty of Versailles (1920) resulted from the Wilson administration's failure to make concessions to domestic political imperatives. Experience with the SALT and Panama treaties (1978)

indicates that oversensitivity to Congress and domestic politics can complicate rather than resolve foreign policy problems. Preoccupation with domestic matters can cause an administration to raise unrealistic expectations, convey a sense of weakness to foreign governments, and lead to diplomatic failure. Diplomatic failure weakens an administration by giving an impression of incompetence, while diplomatic successes, such as the Camp David agreements (1978) can strengthen it. 4 notes. T. L. Powers

513. Doty, Paul; Carnesale, Albert; and Nacht, Michael. THE RACE TO CONTROL NUCLEAR ARMS. *Foreign Affairs 1976 55(1): 119-132.* The continuing introduction of new weapons outside the Strategic Arms Limitation Talks agreement makes a new approach to nuclear disarmament necessary. The Threshold Test Ban Treaty in July 1974 limits the yield of underground tests of nuclear weapons. Further US-USSR agreements of May 1976 to limit explosions in research on peaceful uses of nuclear science show progress, but the development of highly accurate protective armaments, which make mutual annihilation no longer certain, invalidates earlier assumptions and unbalances the basis of SALT agreements. Superpower strategic nuclear capabilities must be reduced and the spread of nuclear weapons slowed, despite the competitive impulses of nations. R. Riles

514. Doty, Paul. STRATEGIC ARMS LIMITATION AFTER SALT I. *Daedalus 1975 104(3): 63-74.* Although a balanced reduction in the number of strategic weapons has become an acceptable item of agenda since the first Strategic Arms Limitation Talks, it is in the area of qualitative reduction that most difficult negotiations will occur. Military leadership and the public must break through their psychological barriers to arms control in order to accelerate limitation. Secondary sources; 8 notes. E. McCarthy

515. Eberhard, Wallace B. FROM BALLOON BOMBS TO H-BOMBS: MASS MEDIA AND NATIONAL SECURITY. *Military Rev. 1981 61(2): 2-8.* In a democratic society such as the United States, what is the relationship between the government's right to keep secrets and the rights and responsibilities of the press in reporting such news? In the aura of solidarity that existed during World War II, the press cooperated with government efforts to censor the news about Japanese balloon bombs. In 1980, however, debate over the publication of an article in *The Progressive* magazine on H-bombs was intense. Without the unity that supported voluntary censorship in World War II, it seems likely that the press and public will break secrecy barriers when they feel like it. 2 photos, 18 notes. D. H. Cline

516. Eckhardt, William. CHANGING CONCERNS IN PEACE RESEARCH AND EDUCATION. *Bull. of Peace Proposals [Norway] 1974 5(3): 280-284.* Describes the changes in peace research in the last five years, showing how it has changed from cross-disciplinary to transdisciplinary, from crossnational to transnational, from concern about peace to emphasis on equality and freedom, and from positivism to humanism. In order to illustrate these changing patterns in peace research, a project at the Canadian Peace Research Institute is described. Secondary sources; 8 notes, biblio. R. B. Orr

517. Einhorn, Robert J. TREATY COMPLIANCE. *Foreign Policy 1981-82 (45): 29-47.* While the Soviets may have frequently cheated under past nuclear arms limitation agreements, the USSR has gained no important advantage by such violations. But US public opinion holds that US security has suffered and that the agreements are not working. To win support for arms control, the federal government should negotiate empirically verifiable treaties. By sharing information on compliance with Congress and the public, the Executive Branch can promote public confidence in compliance. M. K. Jones

518. Epstein, Edward Jay. DISINFORMATION: OR, WHY THE CIA CANNOT VERIFY AN ARMS-CONTROL AGREEMENT. *Commentary 1982 74(1): 21-28.* Discusses the success of Soviet disinformation since 1960 with reference to US estimates of their strategic strength, and argues that such disinformation renders US intelligence incapable of verifying an agreement on arms control.

519. Epstein, William. BANNING THE USE OF NUCLEAR WEAPONS. *Bull. of the Atomic Scientists 1979 35(4): 7-9.* Chronicles UN resolutions, 1961-78, seeking to ban the use of nuclear arms.

520. Espiell, Hector Gros. U.S.A. E. DUNUCLEARIZZAZIONE NELL-'AMERICA LATINA [The United States and denuclearization in Latin America]. *Riv. di Studi Pol. Int. [Italy] 1977 44(4): 565-578.* Examines the diplomatic background, 1973-77, to the signing by Jimmy Carter, 29 May 1977, of an additional protocol to the Treaty of Tlatelolco, prohibiting the development of nuclear arms in Latin America.

521. Etzold, Thomas H. SALT AND STRATEGIC REAPPRAISAL FOR THE U.S. *Marine Corps Gazette 1980 64(1): 49-53.* In the aftermath of SALT II, reviews American military and foreign policy vis-à-vis the USSR during 1960's-70's.

522. Evgen'ev, G. VENSKIE PEREGOVORY I BOR'BA ZA RAZRIADKU V EVROPE [The Vienna talks and the struggle for detente in Europe]. *Mirovaia Ekonomika i Mezhdunarodnye Otnosheniia 1981 (1): 52-61.* The representatives of 19 governments discussing Mutual Balanced Force Reduction in Central Europe have yet to prepare a draft agreement; the failure to agree on major issues is analyzed, exposing the unconstructive attitude of the NATO countries, led by the United States, to the talks. Russian.

523. Fahl, Gundolf. ESTATOS UNITOS Y LA UNIÓN SOVIÉTICA, PAÍSES ESTRATÉGICAMENTE FRONTERIZOS: ANÁLISIS E INTERPRETACIÓN DE LOS ACUERDOS SALT [United States and USSR, strategic neighbors: analysis and interpretation of the SALT agreements]. *Rev. de Política Int. [Spain] 1979 (164): 27-66.* The SALT agreements represent a definite advance in the development of international law. Using a recent sociological approach by N. Luhmann, analyzes the SALT I and SALT II basic documents from the 1972 ABM agreement to the Vladivostok declaration. Focuses on the juridical foundation of the whole SALT process. The concept of "strategic vicinity" documents the major changes which have taken place in relations between the two superpowers. The regulation of their strategic relations, by creating a complex machinery of juridical instruments, appears to delineate a partial, embryonic, system of the "universal society." Table, 200 notes. D. Ardia

524. Falk, Richard A. ARMS CONTROL, FOREIGN POLICY, AND GLOBAL REFORM. *Daedalus 1975 104(3): 35-52.* Although emphasis in arms control has been placed on the prevention of nuclear war, several other factors should be considered. Global reform should be initiated, as an educational venture at first, aimed at voluntary and nonviolent replacement of the state system, realization of the ecological unity of the planet, and a gradual renunciation of specific military options. Secondary sources; 41 notes. E. McCarthy

525. Falk, Richard A. BEYOND INTERNATIONALISM. *Foreign Policy 1976 (24): 65-113.* An adjustment of American foreign policy toward the progressive ideology associated with the Jeffersonian heritage would result in an increased awareness of global responsibilities and ethics, and would include a decreased reliance on all forms of nuclear energy together with a disavowal of "first use" of nuclear weapons; a reconciliatory diplomacy which would not rely on covert operations and the use of force; and the initiation of long-term foreign policy planning. Secondary sources; 18 notes. C. Hopkins

526. Fascell, Dante B. THE HELSINKI ACCORD: A CASE STUDY. *Ann. of the Am. Acad. of Pol. and Social Sci. 1979 (442): 69-76.* World governments have traditionally been responsive, in a variety of ways, to public pressure in the area of domestic issues; but with trends toward global interdependence and massive advances in communications technology, governmental leaders must be increasingly sensitive to their constituents' desires and concerns in the field of international relations. An important recent example of this phenomenon is embodied in the 1975 Conference on Security and Cooperation in Europe (CSCE). This article explores how nongovernmental individuals and groups have impacted this significant 35-nation agreement which is a symbol of the long-term process of East-West detente. J

527. Feld, Bernard T. THE CHARADE OF PIECEMEAL ARMS LIMITATION. *Bull. of the Atomic Scientists 1975 31(1): 8-16.* Discusses attempts at nuclear arms limitation in foreign relations and between the United States and the USSR 1959-70's, including the Partial Test-Ban Treaty of 1963 and the Non-Proliferation Treaty of 1975.

528. Feld, Bernard T. HUMAN VALUES AND THE TECHNOLOGY OF WEAPONS. *Zygon 1973 8(1): 48-58.* The use of high explosives by the United States in wartime follows an exponential curve with a six-year doubling time, the quantity of each doubling period equalling the total of all that came before from the beginning of time. Technical breakthroughs speed up this process. If we can enter into a critical reexamination of the ethics and consequences of the nuclear arms race and understand the key role scientists and engineers play in it, perhaps we can persuade such key persons to turn the race around and end it.
J. M. McCarthy

529. Feld, Bernard T. ORIGINS OF PUGWASH. *Sci. and Public Affairs 1973 29(6): 4-6.* Discusses the initiative (1957-73) of Eugene Rabinowitch (1901-73) in the international Pugwash Conferences of scientists who discuss disarmament and other nuclear issues.

530. Finan, J. S. ARMS CONTROL AND THE CENTRAL STRATEGIC BALANCE: SOME TECHNOLOGICAL ISSUES. *Int. J. [Canada] 1981 36(3): 430-459.* Examines the strategic balance between the United States and the USSR with reference to technological trends. Such technological developments as counterforce strategy, strategic manned bombers, ballistic missiles, satellites and antisatellites, verification of treaties, and aircraft with dual capabilities must all be taken into account when the superpowers sit at the negotiating table. Arms control maintains stability in a shifting and complex technological environment. 43 notes. J. Powell

531. Flanagan, Stephen J. CONGRESS, THE WHITE HOUSE AND SALT. *Bull. of the Atomic Scientists 1978 34(9): 34-40.* To combat confusion and suspicion generated by misunderstanding and lack of proper explanation of agreement terms of SALT I within Congress, suggests keeping Congress abreast of negotiation developments through creation of a standing committee on SALT and by providing necessary technical information so that alternatives may be intelligently weighed; 1972-78.

532. Floyd, David. THE PATTERN OF BLUNDERING: US NEGOTIATIONS WITH THE USSR. *Survey [Great Britain] 1980 25(2): 25-31.* A review of *Soviet Diplomacy and Negotiating Behavior: Emerging New Context for US Diplomacy,* a special study prepared for the Senate Foreign Affairs Committee by the Library of Congress, which traces the almost 50 years of US negotiating experience with the Soviet Union, revealing in the process, perhaps unwittingly, how good the Russians are at international negotiations and how bad the Americans can be at it. V. Samaraweera

533. Garn, Jake. AMERICA'S STAKE IN SALT II. *Daughters of the Am. Revolution Mag. 1979 113(6): 658-660, 751.* Examines the Salt II Treaty in a speech given by Senator Jake Garn 17 April 1979 on National Defense Night at the 88th Continental Congress.

534. Garner, William V. SALT II: CHINA'S ADVICE AND DISSENT. *Asian Survey 1979 19(12): 1224-1240.* For some 20 years the Chinese have strenuously protested US-Soviet arms control negotiations; however, the USSR is now the target of their protests rather than the United States. Recent events have increased the importance of the "China factor" in US arms control policy, increasing the difficulty with which US policymakers reconcile Sino-Soviet policy with arms control and defense policies. US policymakers must decide whether national interests would be better served by the maintenance of current Soviet military superiority over China or by China's steady acquisition of more effective military capabilities against the Soviets. 44 notes. M. A. Eide

535. Garrett, Stephen A. PROSPECTS OF PEACE OR WAR. *Virginia Q. R. 1976 52(1): 24-40.* Using Robert Heilbroner's thesis that there is little hope for man, Garrett investigates the prospects for peace. Pointing to continued violence since 1945, the increase of nuclear armament, especially in the Third World, the problems both of mutual assured destruction (MAD) and of flexible response, the inability to limit proliferation of nuclear weapons, and the decline in public interest, he concludes "that at present the prospects for permanent global peace are anything but pleasing." O. H. Zabel

536. Garthoff, Raymond L. BANNING THE BOMB IN OUTER SPACE. *Int. Security 1980-81 5(3): 25-40.* In 1963 the UN General Assembly passed Resolution 1884 banning nuclear arms in outer space, a resolution which President John F. Kennedy believed would gain more widespread American support than a treaty or executive agreement; this resolution paved the way for the 1967 Treaty on Outer Space negotiated with the USSR by President Lyndon B. Johnson and for the 1972 Anti-Ballistic Missile Treaty, but ratification of the SALT II Treaty and negotiations on an antisatellite systems ban are jeopardized by the present rift in American-Soviet relations.

537. Garthoff, Raymond L. NEGOTIATING WITH THE RUSSIANS: SOME LESSONS FROM SALT. *Int. Security 1977 1(4): 3-24.* Changing proposals on ABM levels, the 1972 Moscow negotiations, and other examples indicate that overall, the USSR stressed a political approach while the United States emphasized more concrete military concepts during the Strategic Arms Limitation Talks, 1968-76.

538. Garthoff, Raymond L. SALT I: AN EVALUATION. *World Pol. 1978 31(1): 1-25.* The SALT I Agreements, concluded in 1972, are assessed with the benefit of several years' perspective. SALT I marked a beginning to collaborative efforts at strategic arms control by the two superpowers, and in a number of respects—especially the ABM Treaty—it had a clear and favorable effect in mitigating the arms competition. The main shortcoming was a failure to reach significant restraints on strategic offensive arms, especially a ban on MIRV's at the time when that was still possible. Also, the pursuit of "bargaining chips" for arms negotiations can impede arms control and contribute to arms competition. An "oversell" of detente and SALT in 1972 prompted a swing to undervaluing both in the late 1970's. Nonetheless, on balance SALT I was a significant step forward. J/S

539. Gelb, Leslie H. WASHINGTON DATELINE: THE STORY OF A FLAP. *Foreign Policy 1974 (16): 165-181.* Based on a New York *Times* story in June 1974 which charged that Secretary of State Henry A. Kissinger had made secret agreements with the USSR regarding strategic weapons without informing Congress. Kissinger refuted this charge, although he admitted that inconsequential secret agreements had been made. In addition, it appears that during the 1972 Moscow Summit Conference, President Richard M. Nixon provided a written unilateral pledge to the Soviet Union that the United States would not build up to the allowable 710 submarine launchers—an amount provided for in the disarmament agreement. The American SALT (Strategic Arms Limitation Talks) delegation was not informed of this in detail. Kissinger apparently had also made a secret agreement with Soviet leaders permitting them to construct certain strategic weapons in excess of the disarmament agreement. This "loophole" has now been closed, perhaps as a result of media exposure. R. F. Kugler

540. Gelber, Harry G. TECHNICAL INNOVATION AND ARMS CONTROL. *World Pol. 1974 26(4): 509-541.* There are uncertainties about the meaning of innovation and ways of promoting it. To limit it is even more difficult. The two great powers seek technical advantages in research and development not only to produce more advanced weapons systems, but to improve intelligence, to exercise political pressure, and to remove uncertainties for many aspects of plan-

ning. Restraint might be approached through changes in strategy (including the requirement for assurance) and in funding, through regulating the flow of information, and through continuous negotiations. Each has drawbacks. Restraint therefore requires a continuing desire by both sides for political accommodation.
J

541. George, James L. SALT AND THE NAVY. *US Naval Inst. Pro. 1979 105(6): 28-37.* SALT II is important to the United States. It is also most important to the US Navy. Although the Navy, in general, has no major complaints with the treaty, there are two potential Navy problems: the range limits on the submarine-launched cruise missile (SLCM), and the ban on placing ballistic missiles on surface ships. Accordingly, the Navy should be more concerned than it appears to be with the SALT II treaty provisions, for they could have a greater future effect on the Navy than expected. Secondary sources; 2 photos, 5 tables, 12 notes. A. N. Garland

542. Gergorin, Jean-Louis. LES NÉGOCIATIONS SALT ET LA DÉFENSE DE L'EUROPE [SALT negotiations and European defense]. *Défense Natl. [France] 1978 34(6): 43-56.* Examines the SALT II negotiations and American concessions, particularly the limitations on cruise missiles absent in SALT I and at Vladivostok, leading to a potential disequilibrium in the USSR's favor.

543. Geyer, Alan. THE POLITICS OF DISARMAMENT IN THE THIRD NUCLEAR AGE. *Peace and Change 1982 8(1): 65-76.* Distinguishes and discusses three stages of the nuclear disarmament problem: the 1945-60 period of Cold War proliferation, the 1960-74 period of detente and negotiation, and the renewed period of crisis and build-up since 1975.

544. Ghêbali, Victor-Yves. CONSIDÉRATIONS SUR CERTAINS ASPECTS MILITAIRES DE LA DÉTENTE: LES "MESURES DE CONFIANCE" D'HELSINKI [Considerations in certain military aspects of detente: the Helsinki "measures of confidence"]. *Défense Natl. [France] 1977 33(6): 21-35.* Examines the origins of the "measures of confidence " which were brought up by the participants in the Helskini Conference on East-West relations (1975), with reference to the different countries' position toward the interpretation and application of these measures.

545. Goldschmidt, Bertrand. A HISTORICAL SURVEY OF NONPROLIFERATION POLICIES. *Int. Security 1977 2(1): 69-87.* Describes eight successive nonproliferation policies attempted during the development of atomic energy beginning in 1943, by the United States, Great Britain, USSR, and Canada.

546. Gray, Colin S. THE LIMITS OF ARMS CONTROL. *Air Force Mag. 1975 58(8): 70-72.* Considers the history of arms control agreements in the 1970's; the conflicts of interest that prompt military preparedness preclude effective arms limitation agreements.

547. Gray, Colin S. OF BARGAINING CHIPS AND BUILDING BLOCKS: ARMS CONTROL AND DEFENSE POLICY. *Internat. J. [Canada] 1973 28(2): 266-296.* Explores the impact of protracted arms control negotiations upon U. S. defense policy and finds little evidence of arms buildup abating. Table, 71 notes. R. V. Kubicek

548. Gray, Colin S. A PROBLEM GUIDE TO SALT II. *Survival [Great Britain] 1975 17(5): 230-234.* Discusses arms control and missile development issues in the 1974 SALT II negotiations in Vladivostok, emphasizing questions of military strategy and detente.

549. Gray, Colin S. RETHINKING NUCLEAR STRATEGY. *Orbis 1974 17(4): 1145-1160.* Discusses the 1972 Strategic Arms Limitation Talks. S

550. Gray, Colin S. SALT I AFTERMATH: HAVE THE SOVIETS BEEN CHEATING? *Air Force Mag. 1975 58(11): 28-33.* Discusses Soviet actions with regard to their obligations under the Strategic Arms Limitation Treaty, 1969-75.

551. Gray, Colin S. THE STRATEGIC ARMS LIMITATION TALKS: DEFENSE AND NEGOTIATION. *Air Force Mag. 1974 57(1): 32-36.* In the context of SALT II, discusses detente, the utility of military power, the lack of clarity in US nuclear strategy, and the shortcomings of SALT I.

552. Green, William C. HUMAN RIGHTS AND DETENTE. *Ukrainian Q. 1980 36(2): 138-149.* Examines the interplay between detente and human rights policy under the Carter administration. Basket Three of the Helsinki Final Act on humanitarian issues gave impetus and international prominence to Soviet dissident groups. The Carter administration's emphasis on human rights brought strong reaction from the Soviets and exposed the limitations placed on American policy by detente. Ultimately human rights policy was downplayed by the United States in the hope of gaining some success in dealing with the USSR on SALT and other more pressing international issues. 33 notes. K. N. T. Crowther

553. Grieco, Joseph M. AMERICAN FOREIGN POLICY AND NUCLE-AR-WEAPON FREE ZONES: ISSUES AND OPTIONS. *Potomac Rev. 1976 7(2): 12-23.* Nuclear-weapon free zones (NWFZs) of the type established in Latin America by the Treaty of Tlatelolco represent America's most promising means of insuring nuclear nonproliferation. Though the adherence to NWFZs might appear to hamper American military flexibility, it is in the best interest of all nations to pay serious attention to such a concept. "Either we make the small sacrifices necessary to promote nonproliferation today, or face a more nucle-arized, vastly more dangerous world tomorrow." 25 notes. M. R. Yerburgh

554. Hall, Gus. A CORRECT POLICY AT THE RIGHT TIME. *World Marxist R. [Canada] 1973 16(9): 66-71.* The 1972 Brezhnev-Nixon strategic arms agreements are a victory for the policy of peaceful coexistence and an indication of a new balance of power between capitalism and socialism. S

555. Hamilton, John A. TO LINK OR NOT TO LINK. *Foreign Policy 1981 (44): 127-144.* Linkage, or the political practice of tying a concession by one party to a concession by the other, has achieved recent prominence as an essential tool of American diplomacy during the Carter years. Although the Reagan administration has declared its intent to use linkage diplomacy, it should avoid linking other matters to the strategic arms negotiations, which are too delicate and important to benefit from this practice. L. J. Klass

556. Hartley, Anthony. THE BURDEN OF CHAOS. *Round Table [Great Britain] 1980 (280): 360-366.* American foreign policy under John F. Kennedy contained two basic axioms: to try to reduce the likelihood of a direct US-Soviet clash, and to counter the spread of Soviet influence in the Third World gained by backing "wars of national liberation." The pursuance of the first has led to the SALT I and II agreements, but the second drew the Americans into Vietnam and failed to stop the general spread of Soviet influence. Deteriorating political, social, and economic conditions in the Third World seriously threaten international stability, in turn endangering the balance of power. Secondary sources; 5 notes. E. J. Adams

557. Hoffman, Walter. ON BUILDING A NEW PEACE MOVEMENT. *Vista 1973 9(3): 24-29.* Discusses world issues for the US peace movement following the end of US involvement in the Vietnam War in 1973, emphasizing reform, disarmament, ecology, terrorism, the Panama Canal, the UN, and genocide.

558. Hopmann, P. Terrence and Smith, Theresa C. AN APPLICATION OF A RICHARDSON PROCESS MODEL: SOVIET-AMERICAN INTERACTIONS IN THE TEST BAN NEGOTIATIONS 1962-1963. *J. of Conflict Resolution 1977 21(4): 701-726.* Richardson models have often been used to describe reactive processes in arms races. This paper argues that, following the work of Otomar Bartos, negotiations may also be analyzed as a reactive process rather than as a process of discrete position changes. Four variants of the basic Richardson model were employed to determine whether the Partial Test Ban negotiations exhibited such an interactive pattern. In all four equations for the behavior of both the United States and the USSR the stimulus variable provided the greatest explanatory power, supporting the notion that these negotiations were reactive. Several important differences emerged, however, between the Test Ban negotiations and most arms races. On balance, the data gave strong support for the basic assumption of Richardson models, namely that negotiations may be treated as a highly reactive process. J

559. Howard, Michael. CONDUCTING THE CONCERT OF POWERS. *Armed Forces and Soc. 1980 6(4): 654-665.* Review article of Henry A. Kissinger's *White House Years* (1979), which presents the view that the world consists of powers pursuing their own interests according to their best perception of them and that the interest of those powers in the stability of the international system could override their mutual suspicion. Kissinger's years as negotiator for President Richard M. Nixon during 1968-72 were marked by a breakthrough in communications with China, the reconciliation of Egypt with the West, a stable settlement in Germany, and initiation of the SALT talks, but they were marred by the clandestine bombing of Hanoi and Cambodia, even though those bombings may well have helped bring North Vietnam to prefer peace by negotiation to military victory. E. L. Keyser

560. Hyland, William G. THE SOVIET UNION AND THE UNITED STATES. *Current Hist. 1981 80(468): 309-312, 343-346.* Reviews the major arguments and areas of difficulty between the United States and the USSR that have guided the Strategic Arms Limitations Treaty (SALT I and II) discussions; notes the state of readiness of the Red Army and its importance in Soviet political strategy.

561. Ignatieff, George. NEGOTIATING ARMS CONTROL. *Int. J. [Canada] 1974-75 30(1): 92-101.* Discusses negotiations, treaties, and diplomacy for disarmament among the United States, the USSR, and other nations, 1945-70's.

562. Iklé, Fred C. NUCLEAR DISARMAMENT WITHOUT SECRECY. *Freedom at Issue 1974 (28): 7-9.* Advocates the need for nuclear disarmament without government secrecy in the United States, emphasizing the potential destructive power of thermonuclear war, 1945-70's.

563. Iklé, Fred C. PREVENTION OF NUCLEAR WAR IN A WORLD OF UNCERTAINTY. *Freedom At Issue 1974 (25): 7-10.*

564. Iklé, Fred C. THE PREVENTION OF NUCLEAR WAR IN A WORLD OF UNCERTAINTY. *Foreign Service J. 1974 51(5): 10-12, 30.* Adapted from a speech before the Joint Harvard/Massachusetts Institute of Technology Arms Control Seminar. S

565. Iklé, Fred Charles. WHAT TO HOPE FOR, AND WORRY ABOUT, IN SALT. *Atlantic Community Q. 1977-78 15(4): 450-459.* Surveys the hopes and dangers in the Strategic Arms Limitation Talks and proposes four tests by which we can judge an agreement: 1) deterring nuclear attack, 2) balancing USSR and US nuclear strength, 3) reducing nuclear arsenals, and 4) finding an enforceable agreement. J/S

566. Jackson, Henry M. CREDIBLE DETERRENCE IN A SALT II ENVIRONMENT. *Air Force Mag. 1973 56(1): 44-46.* The US delegation which will continue the Strategic Arms Limitation Talks with the USSR must not lose sight of America's need to maintain a strong military to guarantee peace and mutual respect between the two superpowers, 1973.

567. Jackson, William D. POLICY ASSESSMENT AT THE CROSSROADS: THE SOVIETS AND SALT. *Bull. of the Atomic Scientists 1979 35(4): 10-14.* In light of what US policy in SALT should be, assesses the growth of Soviet strategic power, 1969-78, and the dichotomy of the USSR's public support for a nuclear arms ban while maintaining a strong (and growing) nùclear arsenal.

568. Jain, J. P. THE SALT AGREEMENTS. *India Q.: J. of Int. Affairs [India] 1976 32(1): 68-75.* Analyzes the bilateral discussions between the United States and the USSR concerning the Strategic Arms Limitation Talks (SALT), tracing the proposed limitations on intercontinental ballistic missiles, submarine-launched missiles, multiple reentry vehicles, and nuclear weapons of all sorts. These efforts sought to prevent nuclear war and to effect savings on military expenditures. Since the Russians remained unwilling to allow on-site inspections and the Americans were determined to have such guarantees, there was little expectation of any significant accord. Secondary sources. S. H. Frank

569. Johnston, Whittle. THE NEW DIPLOMACY OF PRESIDENT CARTER. *Australian J. of Pol. and Hist. [Australia] 1978 24(2): 159-173.* Comments on the significance of elections when the Presidency changes party hands. Analyzes the main tenets of Jimmy Carter's approach: moral principles, emphasis on arms control and limitation of the nuclear arms race, the reinforcement of bonds with democracies, and improved relations with the Third World.

Concludes with examples of the constraints placed on foreign policy execution and a comparison with Woodrow Wilson. Documented from newspapers, journals, and addresses; 40 notes. W. D. McIntyre

570. Juda, Lawrence. NEGOTIATING A TREATY ON ENVIRONMENTAL MODIFICATION WARFARE: THE CONVENTION ON ENVIRONMENTAL WELFARE AND ITS IMPACT UPON ARMS CONTROL NEGOTIATIONS. *Int. Organization 1978 32(4): 975-991.* Discusses the negotiations leading to the signing of the Convention on the Prohibition of Military or Any Other Hostile Use of Environmental Modification Techniques on behalf of the United States by Secretary of State Vance in 1977; discusses the significance of advances in arms controls and disarmament on the part of the USSR and the United States in this context.

571. Kaiser, Karl. THE GREAT NUCLEAR DEBATE: GERMAN-AMERICAN DISAGREEMENTS. *Foreign Policy 1978 30: 83-110.* The oil crisis of 1973-74 convinced many nations that a reliable supply of energy could come only from atomic breeder reactors, while the Indian nuclear detonation of 1974 demonstrated that the purely peaceful use of such technology could no longer be assured. Fearful of the further spread of atomic weapons, the United States tightened its interpretation of the 1968 Nonproliferation Treaty and opposed an already concluded sale of breeder technology from Germany to Brazil. This precipitated the greatest diplomatic crisis between the NATO allies since World War II, and demonstrated the breakdown of the international nuclear nonproliferation consensus. 7 notes. T. L. Powers

572. Kaiser, Karl. À LA RECHERCHE D'UN ORDRE NUCLÉAIRE MONDIAL. RÉFLEXIONS SUR LES DIVERGENCES GERMANO-AMÉRICAINES EN MATIÈRE D'ÉNERGIE NUCLÉAIRE [The search for a world nuclear order: remarks on divergences between America and Germany regarding nuclear energy]. *Pol. Étrangère [France] 1978 43(2): 145-171.* Analyzes the problems of power resources and nuclear arms; stresses the divergences between US and Germany nuclear policy since 1968, and dwells on the contradiction between a universal policy of nonproliferation and the world powers' arms races.

573. Karenin, A. THE SOVIET UNION IN THE STRUGGLE FOR DISARMAMENT. *Int. Affairs [USSR] 1975 (9): 13-22.* Discusses the role of the USSR in negotiations with the United States for nuclear disarmament in the 1970's, emphasizing detente.

574. Kennedy, Edward M. ARMS CONTROL IS CRUCIAL. *Center Mag. 1976 9(2): 12-20.* Focuses on foreign relations between the US and the USSR, detente, and the need for arms control, 1961-75.

575. Kincade, William H. THINKING ABOUT ARMS CONTROL AND STRATEGIC WEAPONS. *World Affairs 1974 136(4): 364-376.* Critical review of three works on theories and national policies on strategic arms control. S

576. Kiselyak, Charles. ROUND THE PRICKLY PEAR: SALT AND SURVIVAL. *Orbis 1979 22(4): 815-844.* Discusses arguments about SALT II, from a desire for a treaty on any terms to a desire for unequivocal US strategic superiority. Studies made to date "erroneously assume both sides have hard and soft targets equal in number and vulnerability, and they ignore several Soviet capabilities which greatly increase the USSR's strategic potency." Quotes President Carter's Paris pledge: "Our central security problem today is maintaining our will to keep the military strength we need, while seeking every opportunity to build a better peace." Table, 3 fig., 29 notes. E. P. Stickney

577. Kistiakowsky, G. B. THE GOOD AND THE BAD OF NUCLEAR ARMS CONTROL NEGOTIATIONS. *Bull. of the Atomic Scientists 1979 35(5): 7-10.* Though seemingly ineffective, SALT is preferable to its alternative —unlimited nuclear arms stockpiling.

578. Kistiakowsky, George B. and Stone, Jeremy J. THE NUCLEAR BUTTON. *Bull. of the Atomic Scientists 1976 32(3): 53-58.* Presents different views toward the Federation of American Scientists' 1975 proposal for an amendment to the War Powers Resolution requiring the US president to consult Congress before employing nuclear arms.

579. Kohl, Wilfrid L. THE NIXON-KISSINGER FOREIGN POLICY SYSTEM AND U.S.-EUROPEAN RELATIONS: PATTERNS OF POLICY MAKING. *World Pol. 1975 28(1): 1-43.* No single model adequately explains the American foreign policymaking process. At least six models are required, singly or in some combination, to understand recent American foreign policy formation under the Nixon Administration. The six models are: democratic politics, organizational process/bureaucratic politics, the royal-court model, multiple advocacy, groupthink, and shared images or mind-sets. After a review of the rules of the foreign policy game in Washington and the main elements of the Nixon-Kissinger National Security Council system, the article seeks to apply the models to a number of cases in recent American policy making toward Europe. U.S.-Soviet relations, the "Year of Europe," and Nixon's New Economic Policy of August 1971 are examined as cases of royal-court decision making. A second category of cases exhibits mixed patterns of decision making: SALT, the Berlin negotiations, U.S. troops in Europe, MBFR, and US trade policy. Bureaucratic variables alone explained policy outcomes in international economic policy-making in the autumn of 1971, and an organizational process model was found to be dominant generally in the formation of recent international monetary policy, led by the Treasury Department. The conclusion considers the relationships between the models and certain kinds of policies. J

580. Korb, Lawrence J. THE ARMS CONTROL IMPLICATIONS OF THE CARTER DEFENSE BUDGET. *Naval War College Rev. 1978 31(2): 3-16.* Historical evidence suggests that the basic policies of the full term of a President can be deduced from an examination and analysis of his initial budget. The arms control philosophy of the present administration is examined here. J

581. Kriesberg, Louis. NONCOERCIVE INDUCEMENTS IN U.S.-SOVIET CONFLICTS: ENDING THE OCCUPATION OF AUSTRIA AND NUCLEAR WEAPONS TESTS. *J. of Pol. and Military Sociol. 1981 9(1): 1-17.* Applies a typology of inducements in conflict to the conflict resulting in the Austrian State Treaty in 1955 and the partial nuclear test ban agreement in 1963. Focuses on US and Soviet leaders' several objectives and methods. The agreements reached were based on a convergence of positions shaped by domestic considerations and a number of interrelated actors. Within this context, persuasion and reward made limited, but not trivial, contributions. J/S

582. Krohn, Richard W. SALT: ARMS LIMITATION AND DEFENSE MAXIMIZATION. *Towson State J. of Int. Affairs 1979 13(2): 117-130.* Assesses the Strategic Arms Limitation Talks as true attempts to limit nuclear arms but also as ceilings under which both the USSR and the United States attempt to maximize their strategic forces.

583. Kruzel, Joseph. SALT II: THE SEARCH FOR A FOLLOW-ON AGREEMENT. *Orbis 1973 17(2): 334-365.* Concerns problems of the Strategic Arms Limitations Talks II of 1972, focusing on foreign relations and military strategy. S

584. Kruzel, Joseph J. ARMS CONTROL AND AMERICAN DEFENSE POLICY: NEW ALTERNATIVES AND OLD REALITIES. *Daedalus 1981 110(1): 137-158.* Examines serious superpower arms control negotiations during 1960-80 to tentatively evaluate costs and benefits and to help to determine what role, if any, arms control should play in future relations between the United States and the USSR.

585. Lamb, John and Mandell, Brian. HOW ARMS CONTROL BEGINS AT HOME: THE AMERICAN AND SOVIET CASES. *Int. J. [Canada] 1981 36(3): 575-607.* Discusses militarist and arms control tendencies within the major institutions concerned with arms control policymaking in the United States and the USSR. These domestic organizations for arms control policymaking have contributed to competition between the great powers by institutionalizing the competition without providing effective ways of reconciling competing demands. 36 notes. J. Powell

586. Legault, Albert. MBFR: THE LONG-AWAITED TEST OF THE SUPERPOWERS' GOOD FAITH. *Int. Perspectives [Canada] 1973 (9-10): 14-17.* US troop reduction in Europe is the proper outcome of Mutual and Balanced Force Reductions conferences, because it will act as a test of faith between East and West, make NATO countries improve their defense systems, and generally allow Europe to regain a world power status.

587. Legault, Albert. NUCLEAR POLICY SHOULD BE MORE OPEN AND LESS AMBIGUOUS. *Internat. Perspectives [Canada] 1976 (1): 8-13.* Discusses the need for less government secrecy about Canada's supplying nuclear technology to India and other nations, 1968-70's, emphasizing the implications of the Treaty on the Non-Proliferation of Nuclear Weapons (1968).

588. Lehman, Christopher M. and Hughes, Peter C. "EQUIVALENCE" AND SALT II. *Orbis 1977 20(4): 1045-1054.* Since the 1950's the US strategic nuclear arms policy has been based largely on two objectives: deterrence and stability. Each administration has come up with its own particular program to meet those objectives. The programs have included massive retaliation, controlled response, assured destruction, and essential equivalence. Because it does not seem those objectives will be changed by the Jimmy Carter administration, it remains to be seen how that administration will define equivalence, particularly because of the various Strategic Arms Limitation Talks and agreements. Because of the recent Soviet military buildup, the US approach to equivalence is of critical importance to this country. Secondary sources; 7 notes. A. N. Garland

589. Lehman, John F., Jr. SALT: ASKING THE RIGHT QUESTION. *Air Force Mag. 1977 60(5): 28-32.* Discusses the Strategic Arms Limitations Talks (SALT II), contending that they should be signed, strategic equality should be maintained, and regional imbalances (particularly in Europe and the Middle East) should be remedied; examines the Carter administration's policy, 1977.

590. Lellouche, Pierre. BREAKING THE RULES WITHOUT QUITE STOPPING THE BOMB: EUROPEAN VIEWS. *Int. Organization 1981 35(1): 39-58.* Beginning with the Ford administration policy and continuing with the Carter nonproliferation policy, a major nuclear controversy opposed the American and European nuclear suppliers throughout the 1970's. The first area of controversy was the question of conditions for technological transfers to the Third World, and the area of the plutonium economy. Most of the problems raised in the 1970's are still open. Major points of disagreement remain: full scope safeguards, the question of breeder reactors, and plutonium economy. The major uncertainty for the future will be whether nuclear energy as a whole will remain in depression or grow again. J/S

591. Listerud, Gunnar. UTVIKLINGEN I DET BILATERIALE AVTALEVERKET MELLOM USA OG SOVJET-UNIONEN 1958-1973 [Developments in the system of bilateral agreements between the United States and the USSR]. *Internasjonal Politikk [Norway] 1976 (4): 843-857.* The continuation of the development of a community of interests between the United States and the USSR, begun in 1958, required the cessation of the arms race, normal commercial relations, mutual restraint, negotiations, and consultation.

592. Lodal, Jan M. SALT II AND AMERICAN SECURITY. *Foreign Affairs 1978-79 57(2): 245-268.* The United States and the USSR have virtually completed construction of a new strategic arms limitation treaty (SALT II). Its implications for international stability are enormous. Though ratification will not compromise American security, "... rejection would lead the rest of the world to see the United States as hesitant, insecure, and incapable of mounting an effective political and military effort to establish a stable peace in the nuclear age." 19 notes. M. R. Yerburgh

593. Lodal, Jan M. VERIFYING SALT. *Foreign Policy 1976 (24): 40-64.* Charges that the Soviets cheated on the SALT I agreements can be largely discounted, nor is there much fear that present and near-future technical means of verification will fail to detect Soviet SALT II violations. Future agreements

such as a Comprehensive Test Ban, European force reductions, reduced defense budgets, reduced force levels, and qualitative limitations in SALT require more difficult verification. The lack of air-tight verification methods, however, should not completely determine US disarmament policy. 6 notes. C. Hopkins

594. Long, F. A. ARMS CONTROL FROM THE PERSPECTIVE OF THE NINETEEN-SEVENTIES. *Daedalus 1975 104(3): 1-13.* Since 1960 certain international problems remain: nuclear war and nuclear deterrence, verification of arms control and disarmament, and negotiation of arms control agreements. What has changed is an increase in the political complexity of control and a decrease in technical problems. Current issues include the difficulties of negotiation processes, the need for more effective international organizations, attention to unilateral programs of control and disarmament, and realization of the impact of the qualitative arms race. Based on secondary sources; 12 notes.
E. McCarthy

595. Luck, Edward C. THE ARMS TRADE. *Pro. of the Acad. of Pol. Sci. 1977 32(4): 170-183.* Increasing international arms transfers have in the past prompted the UN to undertake regional arms limitations and embargoes and to propose registration and publication of arms transfer information. These policies have not been successful; however, the United States has apparently reviewed its policy and is renewing efforts toward multilateral restraints. While nuclear weapons growth must be controlled, conventional arms growth also needs restraint. Major arms suppliers must be convinced that cooperation rather than competition is more beneficial for all. K. N. T. Crowther

596. Luttwak, Edward N. WHY ARMS CONTROL HAS FAILED. *Commentary 1978 65(1): 19-28.* Discusses why US and Soviet strategic policy in the last 10 years has resulted in an increase in nuclear arms in both countries even in the face of the SALT negotiations.

597. Mack, Charles. FROM AUSTERLITZ TO MOSCOW: AN AMERICAN SUCCESS STORY. *Am. Heritage 1978 29(1): 48-57.* From Napoleon's victory at Austerlitz (1805) to the signing of the SALT agreements (1972), mankind endured the seemingly unlimited scourge of mass war. Today's nuclear balance, illustrated by the early SALT agreements, has produced greater security because of the annihilating character of the new weapons. 6 illus.
J. F. Paul

598. Makins, Christopher J. SALT AND BEYOND: BRINGING IN THE ALLIES. *Foreign Policy 1979 (35): 91-108.* America's European allies must be closely involved in future arms limitation negotiations with the USSR. Divergent European and American attitudes toward the limitation process, sensitivity to the shifting balance of power within the alliance from America to Europe, and the usual West European difficulty in attaining a common perspective on anything, all will work against effective multilateral negotiations; but failure to undertake that approach could create strains and corrosion in the alliance. Note.
T. L. Powers

599. Maland, Charles. *DR. STRANGELOVE* (1964): NIGHTMARE COMEDY AND THE IDEOLOGY OF LIBERAL CONSENSUS. *Am. Q. 1979 31(5): 697-717.* Discusses the film *Dr. Strangelove* as a response to the

ideology of liberal consensus, which deemed American society sound and Communism a threat to the United States. This ideology contributed to the Cold War mentality and the nuclear arms race. Stanley Kubrick resolved to make a black comedy about such thinking. The film frequently links the sex drive with the war drive and lambastes anti-Communism paranoia, society's inability to realize the enormity of the nuclear threat, nuclear strategies, and the blind faith often put in technological progress. The film's importance comes from its uncharacteristic portrayal of the Cold War, since most films of the 40's and 50's were ardently supportive of the patriotic ideology. 4 photos, 31 notes. S

600. Mandelbaum, Michael. IN DEFENSE OF SALT. *Bull. of the Atomic Scientists 1979 35(1): 15-21.* Assesses criticisms against SALT I and II, comments on the insistence of the criticism from Congress, and speculates about what may be gained from SALT.

601. Marshall, Charles Burton. CHANGING CONCEPTS OF PEACE: AN AMERICAN FOCUS. *Social Res. 1975 42(1): 52-63.* Discusses expectations for peace prevailing in US foreign policy from World War II to the present. S

602. Menchén Benítez, Pedro. LAS CONVERSACIONES SALT [The SALT Conversations]. *R. General de Marina [Spain] 1975 188(5): 499-518.* Describes the evolution of the Strategic Arms Limitation Talks (SALT) between the United States and the Soviet Union. The destructive power of strategic nuclear arms in both states has created a true balance of terror. As a result, the 1972 SALT I accords limited the number of missiles and launch sites of each state to the number adequate to achieve the mutual destruction of both superpowers. US opinion has been surprised and critical of these accords as SALT I allows an official policy of equality in strategic arms, gives the Soviet Union a first-strike capability, and yet provides no defense against nuclear attack. The United States had for decades enjoyed a military superiority, but now the Soviet Union is negotiating from a position of strength. The best that the United States can hope for in a future SALT II agreement is an exact parity in nuclear arms. Details American and Soviet missile systems. 3 photos. W. C. Frank

603. Mets, David R. ARMS CONTROL SINCE HIROSHIMA. *U.S. Naval Inst. Pro. 1973 99(12): 18-26.* After many false starts, some modest but significant progress in arms control has been made. However, general and complete disarmament is a long way in the future and probably will come only if based on far-reaching political changes that are of sufficient scope to satisfy the security information needs of the United States without the intrusive inspection so abhorrent to the Soviet Union. Illus. J. K. Ohl

604. Meyer, Stephen M. POLITICS AMONG DATA: A REVIEW OF THE *SIPRI YEARBOOK 1979.* *Naval War Coll. Rev. 1980 33(2): 108-112.* Review article on arms control and national security as presented in the annual yearbook published by the Stockholm International Peace Research Institute (SIPRI).

605. Migolatyev, A. CURBING THE ARMS RACE: MAJOR TASK OF OUR TIME. *Int. Affairs [USSR] 1977 (3): 85-93.* The watchword of US militarism is now "realistic deterrence," which expresses a reluctant admission of the new balance of global forces. Nonetheless, reactionaries still hope to

frustrate detente by escalating weapons rivalry. The United States and its NATO allies are increasing spending on weaponry, and US arms exports have risen dramatically since 1970. Disarmament agreements are urgently needed to prevent nuclear proliferation, chemical weapons, and entirely new weapons of mass destruction. Moves toward a comprehensive nuclear test ban have been frustrated by the unjustified emphasis certain states have laid on the question of inspection.
L. W. Van Wyk

606. Minter, Charles S. NATO FORCES—PROSPECTS FOR MBFR. *J. of the Royal United Services Inst. for Defence Studies [Great Britain] 1974 119(3): 3-7.* Describes NATO consultative proceedings for developing policy on Mutual Balanced Force Reductions and the aims and principles of MDFR from the NATO viewpoint. Reviews the current state of negotiations with the Warsaw Pact in Vienna. Concludes that "the ultimate success of the enterprise depends equally importantly upon the political resolve of our leaders in the West" who may, in the event of East-West tension, be required to mobilize forces, including if necessary "forces removed by MBFR back in to the area of reductions." Text of a lecture given at the Royal United Services Inst., 27 February 1974.
D. H. Murdoch

607. Molineu, Harold. NEGOTIATING HUMAN RIGHTS: THE HELSINKI AGREEMENT. *World Affairs 1978 141(1): 24-39.* Focuses on the negotiations at the European Security Conference leading up to the Helsinki Agreement of 1975.

608. Moore, David W. SALT AND BEYOND: THE PUBLIC IS UNCERTAIN. *Foreign Policy 1979 (35): 68-73.* Public support for the Strategic Arms Limitation agreements (SALT II) is not as great as public opinion polls have indicated. Polling techniques have disguised the low level of public awareness of the issue. Such support as exists is highly unstable, and is contingent upon Soviet behavior. In the end, public support for SALT II will reflect people's confidence in the president.
T. L. Powers

609. Muravchik, Joshua. EXPECTATIONS OF SALT I: LESSONS FOR SALT III. *World Affairs 1981 143(3): 278-297.* "Reviews the record of public discussion of the Strategic Arms Limitations Treaty (SALT I) at the time when it was signed, presented to the nation, and ratified by the Congress" in order "to recall what those who negotiated, endorsed, and approved SALT I thought it would accomplish, and to begin to judge the extent to which their goals were fulfilled"; 1972-80.

610. Nacht, Michael. ABM ABCS. *Foreign Policy 1982 (46): 155-174.* From the perspective of 10 years after the negotiations of the antiballistic Strategic Arms Limitation Treaty, neither the USSR nor America would benefit from renegotiation or abrogation of the treaty. Technologically feasible and reliable possibilities for credible ballistic missile defense systems do not exist.
M. K. Jones

611. Nacht, Michael. THE VLADIVOSTOK ACCORD AND AMERICAN TECHNOLOGICAL OPTIONS. *Survival [Great Britain] 1975 17(3): 106-113.* Discusses military strategy and the role of technology in the Vladivostok Accord (1974) and in nuclear arms limitation talks between the United States and the USSR in the 1970's, emphasizing the strategic capabilities of ICBM's.

612. Naumov, P. THE US "HAWKS" AND THEIR WORRIES. *Int. Affairs [USSR] 1978 (2): 95-98.* Senators Robert Dole and Robert Byrd, journalists Evans and Novak, Committee on the Present Danger members Dean Rusk, Eugene Rostow, Paul Nitze, and David Packard, et al., have been trying to sabotage SALT II since 1977 in order to embarrass Jimmy Carter and the Democratic Party (at least this is the motivation of the Republican "hawks") and to try to foster US weapons superiority.

613. Navrozov, Lev. SALT: A MAGINOT LINE ON PAPER. *Midstream 1979 25(6): 5-19.* Based on what has been known (or not known) by the United States since 1950 about actual Soviet armed strength and the USSR's ability to camouflage its military installations and production, Congressman Les Aspin (see *Scientific American,* February 1979) and Jan M. Lodal (see abstract 592), are incorrect in their assessment that the SALT II treaty is verifiable.

614. Neff, Thomas L. and Jacoby, Henry D. NONPROLIFERATION STRATEGY IN A CHANGING NUCLEAR FUEL MARKET. *Foreign Affairs 1979 57(5): 1123-1143.* The Nuclear Nonproliferation Act (US, 1978) places severe restriction on the technological activities of countries wishing to acquire US nuclear goods and services. Unilateral solutions of this nature are increasingly irrelevant at a time when the nations of the world are discovering alternative markets. "Dealing with proliferation is increasingly an international task requiring the cooperation of a number of key nations." M. R. Yerburgh

615. Neubroch, H. CONTROLLING ARMS CONTROL. *Round Table [Great Britain] 1980 (278): 210-213.* A review article of Duncan L. Clarke's *Politics of Arms Control: The Role and Effectiveness of the U.S. Arms Control and Disarmament Agency* (New York and London: Free Pr., 1979), which traces the antecedents and development of ACDA since 1961.

616. Newcombe, Alan G. DOLLARS AND SENSE OF PEACE. *Peace Res. Rev. [Canada] 1977 7(3): 1-14.* Military spending neither benefits the international economy nor produces desired national security; peace research, especially in the United States, is a necessary alternative.

617. Nikolayev, Y. THE VLADIVOSTOK MEETING: IMPORTANT PROGRESS. *Internat. Affairs [USSR] 1975 (2): 3-9.* Views the Vladivostok Accord between Leonid Brezhnev and Gerald R. Ford as an important step of detente, especially in the area of strategic arms limitation. The success of the 1974 meeting has made detente irreversible and as such a triumph of the Soviet policy of peaceful coexistence. Despite these gains, however, there are still major areas of concern: the continuing opposition within the United States toward detente and the still unresolved Middle East question. D. K. McQuilkin

618. Nitze, Paul H. THE VLADIVOSTOK ACCORD AND SALT II. *R. of Pol. 1975 37(2): 147-160.* The Vladivostok Accord may lead to future nuclear instability if missile throw-weights and "heavy" bombers are not more precisely defined. Without definition, "any ceiling on the number of missiles and heavy bombers is meaningless." L. E. Ziewacz

619. Nye, Joseph S. MAINTAINING A NONPROLIFERATION REGIME. *Int. Organization 1981 35(1): 15-38.* The world has seen a surprising nuclear stability thus far. Only a few nations have chosen to develop nuclear weaponry. An international nonproliferation regime—a set of rules, norms, and institutions—haltingly and imperfectly has discouraged the proliferation of nuclear weapons capability. Political wisdom begins with efforts to maintain the existing regime with its presumption against proliferation. J/S

620. Nye, Joseph S. NONPROLIFERATION: A LONG-TERM STRATEGY. *Foreign Affairs 1978 56(3): 601-623.* Reviews American nuclear nonproliferation efforts during the 1950's-70's. We must "work with the international community in minimizing the weapons potential of civilian technology, and building institutions to manage the remaining areas of overlap. The development and diffusion of nuclear technology requires of all involved nations a commitment to caution and to shared benefits: that which cannot be effectively safeguarded or postponed must be jointly controlled. The alternative would make us all poorer." 13 notes. M. R. Yerburgh

621. Nye, Joseph S., Jr. PROLIFERATION WATCH: WE TRIED HARDER (AND DID MORE). *Foreign Policy 1979 (36): 101-104.* President Jimmy Carter's policies have succeeded in slowing the rate of nuclear nonproliferation. His style of leadership, while often criticized for being messy and messianic, has been effective as well. "His habit of asking for a whole loaf may produce half a loaf of global reform, while asking cautiously for an incremental slice would have produced half a slice of reform." T. L. Powers

622. Nye, Joseph S., Jr. RESTARTING ARMS CONTROL. *Foreign Policy 1982 (47): 98-113.* The visibility and political vulnerability of arms limitation by treaty, the deteriorated status of USSR-American relations, and the strong inherent linkage of arms control and other aspects of US-Soviet relations makes it imperative that the strategic arms control process be adjusted to the post detente period. A promising approach is the pursuit of as many policy instruments as possible, such as Strategic Arms Limitation Talks (SALT), Strategic Arms Reduction Talks (START), and Transparency and Communication Talks (TAC) in order to create a nuclear stability framework. M. K. Jones

623. Panofsky, Wolfgang K. H. THE MUTUAL-HOSTAGE RELATIONSHIP BETWEEN AMERICA AND RUSSIA. *Foreign Affairs 1973 52(1): 109-118.* For nearly two decades the nuclear armaments of the United States and the Soviet Union have been powerful enough to hold each other's civilian population hostage against nuclear attack. Critics point out that the Strategic Arms Limitation Talks (SALT I) treaty prohibits extensive antiballistic missile deployments, and leaves mass slaughter as the only option if deterrence fails. 4 notes. R. Riles

624. Pastusiak, Longin. OBJECTIVE AND SUBJECTIVE PREMISES OF DETENTE. *J. of Peace Res. [Norway] 1977 14(2): 185-193.* Pastusiak attempts to identify the objective and subjective factors that have made the United States enter the road of coexistence and détente with socialist countries. The main objective factors are: changes in the balance of power between East and West; changes in the balance of power within the West; transformations in the configu-

ration of the contemporary world; nuclear parity between USSR and USA; fiasco of previous US policy toward socialist countries; the rising role of domestic priorities in the United States; growing economic interdependence of the American economy; and last but not least the détente policy of the socialist countries which constantly offers the capitalist world the idea of cooperation and coexistence. The factors of subjective nature are: changes in attitudes of American public opinion toward socialist countries, and the traditional pragmatism of the American establishment. Pastusiak also discusses various reservations raised by the critics of détente in the United States. Pastusiak comes to the conclusion that objective factors play the most important role in shaping US policy toward détente and thus the process of relaxation of tension between East and West tends to be more durable. J

625. Paul, Ron. SALT-FREE DEFENSE. *Reason 1980 11(11): 36-38, 40-41.* The Salt II treaty and its continuation of the policy of Mutual Assured Destruction must be abandoned; US national security policy should embrace noninterventionism.

626. Petrovski, V. RAZORUZHENIE I NESOSTOIATEL'NOST' UKHRISHCHRENII EGO PROTIVNIKOV [Disarmament and the failure of the schemes of its adversaries]. *Mirovaia Ekonomika i Mezhdunarodnye Otnosheniia [USSR] 1982 (6): 3-11.* Despite the attempts of the United States and its NATO allies to use the myth of superior USSR military strength as an excuse for their own arms build-up, the USSR and its Warsaw Pact cosigners have led the international struggle for disarmament since 1932. Russian.

627. Petrovski, V. F. PROBLEMY MIRA I RAZORUZHENIIA I SOVREMENNAIA AMERIKANSKAIA POLITOLOGIIA [Problems of peace and disarmament in US politology]. *Novaia i Noveishaia Istoriia [USSR] 1979 (1): 31-48.* The author criticizes the concepts of *peace through force* and *balance of fear,* and exposes the maneuvers by the opponents of peace and disarmament proposing the so-called code of detente. He shows that a genuinely scientific approach to problems of war and peace requires a resolute dissociation from the apologetics of force, a consistent struggle for detente and real disarmament, and stronger international legal guarantees of security. J

628. Petrovski, V. F. ROL' I MESTO SOVETSKO-AMERIKANSKIKH OTNOSHENII V SOVREMENNOM MIRE [The role and place of Soviet-American relations in the present-day world]. *Voprosy Istorii [USSR] 1978 (10): 79-96.* Highlights the significance of Soviet-American relations for the maintenance of international peace and security. The chief lesson of history consists in the fact that in our epoch the only realistic basis for the maintenance and development of relations between the USSR and the United States is provided by the principle of peaceful coexistence. To the extent to which Soviet-American relations correspond to this basic principle, they benefited the peoples of both countries and promoted the cause of international peace. Shows the consistent implementation by the CPSU and the Soviet government of Lenin's directives concerning the development of relations with the United States and reveals the chief reason of the inconsistency and contradictory approach to Soviet-American relations manifested by the ruling element of the United States. The author makes a comprehensive analysis of the agreements and other documents signed in

1972-74 as a result of four top-level meetings, attaching particular importance to those of them which are intended to lessen the war danger and to limit the armaments race. The improvement of Soviet-American relations dictated by the interests of universal peace requires that these relations should not be based on any transient considerations, but must be predominantly guided by long-term considerations dictated by the real concern for peace. J

629. Pierre, Andrew J. THE DIPLOMACY OF SALT: A REVIEW ESSAY. *Int. Security 1980 5(1): 178-197.* Reviews Strobe Talbott's *Endgame: The Inside Story of SALT II* (New York: Harper and Row, 1979).

630. Pogodin, V. RAZRIADKA I NERASPROSTRANENIE IADERNOGO ORUZIIA [Détente and nonproliferation of nuclear weapons]. *Mirovaia Ekonomika i Mezhdunarodnye Otnosheniia [USSR] 1981 (3): 141-143.* Reviews three American reports on the world's nuclear arms, the nuclear industry, and nuclear exports issued in 1978. Russian.

631. Pomerance, Jo. WHY DO WE CONTINUE NUCLEAR TESTING? *Vista 1973 9(3): 20-23, 52.* Discusses treaties for the discontinuance of nuclear testing by the United States and the USSR, 1963-73.

632. Prince, Howard T., II. SALT, NATIONAL SECURITY, AND INTERNATIONAL POLITICS. *J. of Int. and Comparative Studies 1971 4(1): 14-27.* Outlines the "low-profile" Nixon Doctrine of 1970, reviews American Strategic Arms Limitation Talks (SALT) objectives, and concludes that despite weapon asymmetry and the technological weapons race of both sides, the shift from confrontation to negotiation offers some hope that armed conflict can be avoided.

633. Ramberg, Bennett. TACTICAL ADVANTAGES OF OPENING POSITIONING STRATEGIES: LESSONS FROM THE SEABED ARMS CONTROL TALKS 1967-1970. *J. of Conflict Resolution 1977 21(4): 685-700.* This study assesses Soviet and American positioning strategy during the negotiation of the 1971 treaty banning nuclear weapons and other weapons of mass destruction from the seabed. It finds the superpowers employed both maximalist and equitable postures in the initial draft treaties they submitted. Evidence suggests that the negotiation would have been concluded more expeditiously, the outcome being the same, had the United States and the Soviet Union at the outset adopted equitable positioning strategies on the three focal points of the negotiation—the comprehensiveness of the prohibition, its geography, and verification—rather than opting for preferences commonly recognized as anathema to the other. J

634. Ranger, Robin. MBFR: POLITICAL OR TECHNICAL ARMS CONTROL? *World Today [Great Britain] 1974 30(10): 411-418.* A limited agreement on mutual force reductions, avoiding deadlock over the technical requirements of MBFR, may well prove acceptable to both super-powers as a measure symbolizing their political détente; it would, however, do little to curb their use of military technology. J

635. Ranger, Robin. WEIGHING CHANCES FOR PROGRESS IN THE FIELD OF ARMS CONTROL. *Internat. Perspectives [Canada] 1974 (5): 27-31.* Discusses the negotiations between NATO and the Warsaw Pact on Mutual Balanced Force Reductions, 1973-74.

636. Rathjens, G. W. CHANGING PERSPECTIVES ON ARMS CONTROL. *Daedalus 1975 104(3): 201-214.* The 1960's consensus on the military threat to national security and the arms-control effort needed to reduce it no longer exists. Four current policy issues should be analyzed: the interest in flexible response with strategic weapons, the desirability of damage limitation, attitudes to strategic arms limitations, and mutual balanced force reductions in Europe. 4 notes. E. McCarthy

637. Redick, John R. REGIONAL RESTRAINT: U.S. NUCLEAR POLICY AND LATIN AMERICA. *Orbis 1978 22(1): 161-200.* The Treaty of Tlatelolco (1967), currently in force for 21 Latin American states, established Latin America as the first inhabited continent free of nuclear arms. Four militarily significant countries—Argentina, Brazil, Chile, and Cuba—are not full parties to the treaty. Both signatory and nonsignatory states are among those in Latin America with active nuclear energy programs. Worries in the United States about the possibility of these states developing nuclear weapons has led to American efforts to block export of nuclear technology and equipment to them, most notably in the case of Brazil's acquisition of hardware and technology from West Germany. This led to bitter recriminations between the United States and Brazil. An answer may lie in directly involving nations with accelerating nuclear programs, like Argentina and Brazil, in the search for a solution to the problem of nuclear proliferation. 73 notes. J. C. Billigmeier

638. Rowny, Edward L. THE SOVIETS ARE STILL RUSSIANS. *Survey [Great Britain] 1980 25(2): 1-9.* A reflection on the author's six years of experience as the military representative on the United States SALT negotiating team. V. Samaraweera

639. Ruehl, Lothar. LA NÉGOTIATION SALT II OU LA DIFFICULTÉ DE COMPTER [The SALT II negotiations, or the difficulties of counting]. *Défense Natl. [France] 1977 33(3): 45-70.* The problems surrounding the number of nuclear arms held by the USSR and the United States which were central to the SALT II negotiations, must be seen in the context of the political framework which led to the formulation of the SALT I agreement.

640. Sakamoto, Yoshikazu. NEW DIMENSIONS OF DISARMAMENT PROCESSES. *Japan Q. [Japan] 1982 29(2): 169-175.* A basic failure of the disarmament discussion is the inadequate manner of conceptualizing the goals and processes of disarmament by decisionmakers, and the definition of such terms as "balance" and "deterrence." Notes the need for disarmament proposals. F. W. Iklé

641. Schelling, Thomas C. A FRAMEWORK FOR THE EVALUATION OF ARMS CONTROL PROPOSALS. *Daedalus 1975 104(3): 187-200.* In simple binary decisions on whether or not to possess a given weapon, there are 16 possible pairs of bargaining positions for two negotiating nations, ranging from absolute yes to no, and the position chosen may be nonnegotiable or negotiable.

Since proposals rarely are simple binary decisions, the alternative-preference configurations increase according to the factors involved. 4 notes.

E. McCarthy

642. Schlafly, Phyllis. FAULTY INTELLIGENCE IS DANGEROUS TO OUR DEFENSE. *Daughters of the Am. Revolution Mag. 1979 113(5): 474-477.* The USSR is using the SALT negotiations with the United States to deceive US intelligence about the number and sophistication of Soviet nuclear arms.

643. Schlafly, Phyllis. THE SALT-SELLERS TRY AGAIN. *Daughters of the Am. Revolution Mag. 1978 112(8): 802-806.* Reviews national defense technology and its relation to Strategic Arms Limitation Talks, 1972-78, advising that rather than ask the USSR to freeze production of nuclear weapons and missile systems, the United States should proceed with all planned defense systems.

644. Schorr, Brian L. TESTING STATUTORY CRITERIA FOR FOREIGN POLICY: THE NUCLEAR NON-PROLIFERATION ACT OF 1978 AND THE EXPORT OF NUCLEAR FUEL TO INDIA. *New York U. J. of Int. Law and Pol. 1982 14(2): 419-466.*

645. Scoville, Herbert, Jr. MIRV CONTROL IS STILL POSSIBLE. *Survival [Great Britain] 1974 16(2): 54-59.* Discusses the prospects for slowing proliferation of MIRVs (multiple warhead missiles) in the United States and the USSR in upcoming SALT II negotiations.

646. Shapely, Deborah. ARMS CONTROL AS A REGULATOR OF MILITARY TECHNOLOGY. *Daedalus 1980 109(1): 145-157.* Examines why arms control has failed to regulate military technology during the 1970's.

647. Sharp, Jane M. O. RESTRUCTURING THE SALT DIALOGUE. *Int. Security 1981 6(3): 144-176.* Distinguishes between the positive and negative components of traditional arms control diplomacy as exemplified by the Strategic Arms Limitation Talks (SALT).

648. Shrivastava, B. K. AMERICAN PERSPECTIVES ON *DETENTE. Int. Studies [India] 1974 13(4): 577-607.* Discusses the role of strategic arms limitation negotiations and economic relations in the Nixon administration's policy of detente toward the USSR 1969-73.

649. Shulman, Marshall D. ARMS CONTROL AND DISARMAMENT: A VIEW FROM THE USA. *Ann. of the Am. Acad. of Pol. and Social Sci. 1974 (414): 64-72.* While the interests of the United States and the Soviet Union are in many respects competitive, they do overlap in one important respect: namely, the avoidance of nuclear war. Yet, the fact is that despite the strategic arms limitation talks (SALT), the competition in strategic weapons continues and may even be exacerbated by these negotiations. For numerous and varied reasons—which are discussed—we have reached the point at which both countries are acquiring war-fighting capabilities. The crisis can be offset only if people take the trouble to study, to learn and to seek to understand the problems and, then, to make their voices heard. J

650. Shulman, Marshall D. ARMS CONTROL IN AN INTERNATIONAL CONTEXT. *Daedalus 1975 104(3): 53-61.* Political considerations are so intertwined with arms control problems today that technical approaches alone cannot solve them. Several sources of potential conflict are activities of multinational corporations, domestic social and political instabilities in industrialized nations, the effect of advancing industrialization on environment and resources, and the interpenetration of societies by transportation and communication. Economic and political power must be balanced with military power to achieve international security. E. McCarthy

651. Simon, Anne L. BAN THE BOMBERS: A PROPOSAL FOR THE MUTUAL ELIMINATION OF STRATEGIC BOMBERS. *Int. Problems [Israel] 1977 16(3-4): 59-70.* The long-range bombers are an obsolete part of the US nuclear deterrent forces and should be eliminated as a strategy in the arms limitations negotiations with the USSR.

652. Smart, Ian. JANUS: THE NUCLEAR GOD. *World Today [Great Britain] 1978 34(4): 118-127.* Examines US policy on nuclear nonproliferation, 1968-77, especially in light of recent increased exports of nuclear matter from France and West Germany; though President Jimmy Carter appears outwardly to discourage proliferation, if restrictions are too tight, multiplication may occur.

653. Smith, A. Robert. THE NEW SCIENTIST-ADVOCATES: BEYOND SILENT PARTNERSHIP IN PUBLIC AFFAIRS. *Bull. of the Atomic Scientists 1975 31(2): 16-18.* Traces the trend toward increased political activity by scientists beginning in 1946 with establishment of the Federation of American Scientists and predicts an even more influential role for scientists in the future, perhaps as scientist-statesmen. Based on secondary works. D. J. Trickey

654. Smith, Gerald C. SALT AFTER VLADIVOSTOK. *J. of Internat. Affairs 1975 29(1): 7-18.* The Vladivostok Accord, together with the ABM Treaty, irreversibly stabilized the power balance by insuring the survivability of ICBM's, thereby precluding nuclear blackmail. Problems of increased US throw weight, definitions of heavy bombers, and further nuclear testing remain.
R. D. Frederick

655. Smith, Gerard and Rathjens, George. REASSESSING NUCLEAR NONPROLIFERATION POLICY. *Foreign Affairs 1981 59(4): 875-894.* The Carter Administration's nuclear nonproliferation policy had limited success because it saw in the spread of nuclear power facilities an increased weapons production capability. This logic led to controls on exports of nuclear power technology which proved ineffective in preventing foreign nations from achieving nuclear capability. Future policy should emphasize international cooperation in setting guidelines for nuclear technology and materials exports, and in achieving greater international security and energy self-sufficiency. 10 notes.
A. A. Englard

656. Smith, Gerard C. NEGOTIATING AT SALT. *Survival [Great Britain] 1977 20(3): 117-120.* Americans should not expect quick results from the Strategic Arms Limitation Talks, because the USSR has some advantages over the United States, and also because Soviet participants are only beginning to understand the processes and complexities of nuclear strategy.

657. Soapes, Thomas F. A COLD WARRIOR SEEKS PEACE: EISENHOWER'S STRATEGY FOR NUCLEAR DISARMAMENT. *Diplomatic Hist. 1980 4(1): 57-71.* "[President Dwight D.] Eisenhower's efforts to reduce nuclear armaments, the potential for surprise attack, and East-West tensions illustrate his earnest desire for a cessation of the nuclear arms race, his ability to make decisions independently of his advisers, and his ability to formulate a plan for achieving his objectives. The historian's estimate of Eisenhower's effort to achieve these objectives must be tempered, however, by the fact that Eisenhower's conventional Cold War attitude toward the Soviet Union placed a major obstacle in his path. It was an obstacle he did not overcome." 41 notes.
T. L. Powers

658. Soutou, Georges-Henri. DE QUELQUES ASPECTS POLITIQUES DES NEGOCIATIONS SALT [Some political aspects of the SALT negotiations]. *Défense Natl. [France] 1981 37(Jul): 85-110.* Compares American and Soviet attitudes during Strategic Arms Limitation Talks. French.

659. Stashevski, S. and Stakh, G. KOSMOS DOLZHEN BYT' MIRNYM [Outer space is to be peaceful]. *Mirovaia Ekonomika i Mezhdunarodnye Otnosheniia [USSR] 1982 (2): 15-24.* Focuses attention on the problem of the demilitarization of outer space, as one aspect of the more general problem of disarmament. The authors analyze the new Soviet initiative—the draft treaty prohibiting the deployment of any weapons in outer space submitted to the UN. The authors consider the attempts of the United States and its allies to divert the world community from a discussion of the vital problems of outer space demilitarization and pay special attention to the latest American plans for the militarization of outer space. Russian. J/S

660. Steiner, Barry H. ON CONTROLLING THE SOVIET-AMERICAN NUCLEAR ARMS COMPETITION. *Armed Forces and Soc. 1978 5(1): 53-71.* "Preventing an 'uncontrolled arms race' has now become important enough to be the key issue in Soviet American Detente." Each superpower has budgeted its resources for competition rather than for a showdown; they have not competed to the limit of their economic capacities. The first major development in controlling super power competition came in 1959 when the American government accepted the standard of likely instead of possible Soviet force adjustments. The resulting acceleration of modernization of weaponry has been spectacular in long-range ballistic missiles. Argues that "aspiration to control arms competition requires the super powers to reevaluate the way they have pursued the rivalry thus far, and greater acknowledgement that their commitment to competition is reducible." 45 notes. E. P. Stickney

661. Stubbs, Richard and Ranger, Robin. MECHANISTIC ASSUMPTIONS AND UNITED STATES STRATEGY. *Int. J. [Canada] 1978 33(3): 557-572.* Mechanistic images fostered by the social and intellectual climate and internalized by American theorists have pushed arms control and military strategy along questionable lines. 35 notes. R. V. Kubicek

662. Sweet, William. NON-PROLIFERATION TREATY AND THE THIRD WORLD. *Alternatives: J. of World Policy [Netherlands] 1976 2(4): 405-420.* The effectiveness of the Nuclear Non-Proliferation Treaty has remained

limited to the major alliance systems of the two superpowers, and its future now depends on the reconciliation of sharply differing views concerning why and how the spread of nuclear weapons should be prevented. India has always been the most articulate critic of the Treaty, and the position adopted by New Delhi ever since the test of a peaceful nuclear explosive in May 1974 simultaneously highlights the weaknesses in the line traditionally taken on proliferation by the superpowers, and suggests how the Treaty could be revised to meet Third World demands. India might now take an initiative aiming at revision of the Non-Proliferation Treaty according to a scheme in which the so-called 'near-nuclear-weapon states' would acquire, in the act of ratifying the Treaty, the right to participate in an international agency that would be responsible for the development and use of peaceful nuclear explosives. J

663. Talbott, Strobe. SCRAMBLING AND SPYING IN SALT II. *Int. Security 1979 4(2): 3-21.* Discusses encryption of telemetry during the flight testing of missiles and its role in the issue of verification during the SALT negotiations of 1977-79.

664. Thee, Marek. ARMS CONTROL: THE RETREAT FROM DISARMAMENT: THE RECORD TO DATE AND THE SEARCH FOR ALTERNATIVES. *J. of Peace Res. [Norway] 1977 14(2): 95-114.* The main thesis in this paper is that arms control as implemented in recent years and reflected in a number of multilateral and bilateral US Soviet accords has not halted the arms race, but rather impelled its course. An effort must be made to change direction and initiate the process of real disarmament. Arms control has meant a retreat from disarmament. It has come to symbolize a practice of building security not on less but on more arms. Deterrence has become the main theme of arms control, and has meant the establishment of a threat system which requires a constant augmentation of armaments, so as to enhance the retaliatory power of the adversary. The paper analyses the flaws and drawbacks of arms control and reviews the achievements and failures of the arms control agreements. It then discusses the armaments dynamics and prospects of disarmament. Two basic mainstays of armaments dynamics today lie in the domain of politics and technology. To undercut armaments dynamics the *modus operandi* in these two fields must be changed. In the political domain, a most important precondition for change is greater openness in questions of security. One promising strategy to follow would be unilateral reciprocated initiatives to gradually reduce armaments. In the field of technology, military research and development must be brought under control. Disarmament can only come through a radical departure from dominant arms control concepts. There is a need to return to the idea of general and complete disarmament. J

665. Toinet, Marie-France. LA "NOUVELLE" STRATÉGIE AMÉRICAINE ET SALT II [The "new" American strategy and SALT II]. *Défense Natl. [France] 1974 30(6): 45-57.* Examines the real motivations behind the official announcement of a more ambitious US nuclear arms policy in 1973, tracing its link with Strategic Arms Limitation Talks (II) and with President Richard M. Nixon's growing weakness in the United States.

666. Toinet, Marie-France. LE PRÉSIDENT CARTER ET SALT II [President Carter and SALT II]. *Défense Natl. [France] 1977 33(12): 9-24.* Sees continuing interaction of US internal policy with the negotiations for a SALT II agreement, and considers the conflicting demands President Jimmy Carter has to deal with, especially the demand for the maintenance of a diplomatic status quo with the USSR.

667. Tomilin, Y. PROBLEM OF ARMED FORCES REDUCTIONS IN EUROPE. *Int. Affairs [USSR] 1973 (4): 37-42.* Discusses procedural problems in the opening of negotiations for armed forces reductions and disarmament in Europe in 1973; the USSR has had a positive interest in European disarmament since 1955, while the United States and NATO until recently have attempted to block such initiatives.

668. Tonelson, Alan. SALT AND BEYOND: NITZE'S WORLD. *Foreign Policy 1979 (35): 74-90.* Paul Nitze's opposition is one of the greatest obstacles to Senate ratification and public approval of the Strategic Arms Limitation agreements (SALT II). This opposition derives from Nitze's background as a post-World War II cold warrior and from his view of the world as an arena for a clear-cut struggle between good and evil. Here, only military power is of any consequence, nuclear weapons do not differ qualitatively from others, and genuine compromise is impossible. This simplistic outlook ignores the complexities of reality. 9 notes. T. L. Powers

669. Tumkovskii, R. G. SOVETSKO-AMERIKANSKIE PEREGOVORY OB OGRANICHENII STRATEGICHESKIKH VOORUZHENII [Soviet-American talks on the limitation of strategic arms]. *Voprosy Istorii [USSR] 1979 (3): 70-86.* The author examines the most important aspect of the present stage in Soviet-American relations: the talks aimed at drafting the second agreement on the limitation of strategic offensive arms (SALT-2). The article shows the Soviet Union's tireless and consistent efforts to achieve a just solution of this problem on the basis of the principle of equality and equal security, which precludes the possibility of gaining unilateral military advantages. Secondary sources; 55 notes. J

670. Ulsamer, Edgar. SALT II'S GRAY-AREA WEAPON SYSTEMS. *Air Force Mag. 1976 59(7): 80-85.* Discusses the Air Force Association's symposium "Tomorrow's Strategic Options," 28-29 April 1976.

671. Urban, Josef. SOVĚTSKO-AMERICKÉ JEDNÁNÍ O OMEZENÍ STRATEGICKÝCH ZBRANÍ [Soviet-American negotiations on strategic arms limitation]. *Hist. a Vojenství [Czechoslovakia] 1979 28(3): 98-118.* Reflects on the story of attempts to limit strategic arms since 1963 and cites some Soviet sources reproaching the United States for lack of interest in a final agreement with the USSR. A US military buildup inside the NATO system, US refusal to accept a military strategic balance between the two superpowers, and the hesitation of the Carter administration to present the results of the negotiations to Congress have stalled action. 24 notes. G. E. Pergl

672. Vallaux, François. SALT II: AMÉRICAINS ET SOVIÉTIQUES AU PIED DU MUR [SALT II: Americans and Russians with their backs to the wall]. *Défense Natl. [France] 1975 31(12): 57-68.* Reviews the steps toward

disarmament agreements since 1963, and considers the results and major problems resulting from the Vladivostok Accord in 1974, stressing the importance and urgency of the Strategic Arms Limitation Talks.

673. Vernant, Jacques. LES DEUX GRANDS FIN 1977 [The two great powers at the end of 1977]. *Défense Natl. [France] 1977 33(12): 115-121.* Reassesses relations between the United States and the USSR at the end of 1977, with reference to positions adopted on the SALT agreements, the Middle East problem, and international policy.

674. Vernant, Jacques. LES ENTRETIENS DE VLADIVOSTOK [The Vladivostok talks]. *Défense Natl. [France] 1975 31(1): 105-111.* Discusses the significance of the arms limitation talks between Gerald R. Ford, Leonid Brezhnev, and Henry A. Kissinger on 23 and 24 November 1974 in Vladivostok.

675. Viktorov, V. TEN YEARS OF THE MOSCOW TREATY. *Int. Affairs [USSR] 1973 (9): 32-37.* Discusses the effectiveness of the Nuclear Test Ban Treaty (1963) drawn up by the USSR, the United States, and Great Britain; says the USSR has struggled for disarmament and an end to nuclear tests, 1954-70's.

676. Vorontsov, G. A. EVOLIUTSIIA VNESHNEI POLITIKI SSHA V USLOVIIAKH RAZRIADKI [Evolution of US foreign policy in the conditions of detente]. *Voprosy Istorii [USSR] 1976 (9): 45-63.* The article examines the evolution of US foreign policy course in the conditions marked by a relaxation of international tension. U.S. recognition of the need for peaceful co-existence took place under the impact of the altered alignment of forces on the international arena and the growing influence exerted by the socialist world community, by the forces of peace and progress. Highlighting the various aspects of Soviet-American relations, the author analyzes the sources, root causes and concrete forms of manifestation of the new tendencies in U.S. foreign policy. The policy of peaceful co-existence, of better relations with the U.S.S.R. is meeting with increasing support among the vast majority of Americans, because it constitutes the only reasonable alternative to the cold war and to the heightening of tensions. At the same time the evolution of America's foreign policy towards strengthening the positions of realistically-minded politicians prompts the opponents of detente in the United States to step up their activity. J

677. Vukadinović, Radovan. AMERIČKO-SOVJETSKI STRATEGIJSKI DIJALOG [The American-Soviet strategic dialog]. *Politička Misao [Yugoslavia] 1977 14(3): 361-372.* Scrutinizes the SALT I agreement (1973) and the preparations for SALT II and asserts that despite several setbacks, the talks between the USSR and the United States serve the purpose of detente.

S. Košak

678. Warnecke, Steven J. NON-PROLIFERATION AND INFCE: AN INTERIM ASSESSMENT. *Survival [Great Britain] 1979 21(3): 116-124.* Assesses the ability of the International Nuclear Fuel Cycle Evaluation (INFCE), in which 50 nations are participating from mid-1978 to mid-1980, to deal with nuclear safeguards, nuclear proliferation, and new technology.

679. Welch, Richard E., Jr. DETENTE AND ITS ANALYSTS. *Peace and Change 1977 4(3): 3-7.* Discusses detente between the United States and the USSR, giving synopses of foreign policy critics and analysts and noting the effects of the Strategic Arms Limitation Talks.

680. Werner, Roy A. VLADIVOSTOK AND SALT II: IS THE WORLD SAFER? *J. of the Royal United Services Inst. for Defence Studies [Great Britain] 1975 120(1): 56-58.* Presents a critique of the Strategic Arms Limitation Talks (II) and the Vladivostok Accord (1974). "SALT is political gamesmanship" in which the initiative has passed to the Soviet Union. It disguises a continued arms race with escalating costs resulting from technological innovation. SALT II may also affect European stability as US and allied security policy goals diverge. Concludes by proposing achievement of arms control by mutual phased reduction of both nuclear and conventional weapons, a ban on mobile ICBMs, explicit limitation on testing and use of new weapons, and verification of these steps by third parties via satellites jointly financed by the United States and the USSR. 8 notes. D. H. Murdoch

681. Wessell, Nils H. ISSUES IN SOVIET-AMERICAN RELATIONS: CONFERENCE REPORT. *Orbis 1981 25(1): 209-222.* Reports on a December 1980 meeting of Soviet and American scholars on American-Soviet security issues. The major topics covered at the conference were the Strategic Arms Limitation Talks and Soviet-American relations, the problems of international security in Asia, the instability of the Third World, and the enhancement of security in Europe. A list of the participants. J. W. Thacker, Jr.

682. Westervelt, Donald R. THE ESSENCE OF ARMED FUTILITY. *Orbis 1974 18(3): 689-705.* Discusses the future of the Strategic Arms Limitation Talks and problems of nuclear arms. S

683. Winne, Clinton H., Jr. SALT AND THE BLUE-WATER STRATEGY. *Air U. R. 1974 25(6): 25-35.* Examines the possible impact of the Strategic Arms Limitation Talks agreements on the US defense establishment with emphasis on the so-called "Blue Water Strategy," which advocates the movement of US nuclear deterrents to sea. After a discussion of possible advantages and disadvantages of the "Blue Water Strategy," concludes that the multiple delivery system (ICBM's, submarines, and manned bombers) would be more advantageous. Based on government reports; 5 illus., 15 notes. J. W. Thacker, Jr.

684. Wolfe, Bertram. COULD AMERICA'S NUCLEAR POLICIES BE COUNTERPRODUCTIVE? *Bull. of the Atomic Scientists 1980 36(1): 43-48.* In the effort to control world-wide proliferation, US policies are no longer so influential in the development of nuclear technology of other countries; covers 1946-80.

685. Woolsey, R. James. CHIPPING AWAY AT THE BARGAINS. *Daedalus 1975 104(3): 175-185.* A review of recent congressional debates on arms control issues, especially the classic ABM controversy of 1969, reveals five classic positions. They can be described as the Assured Destroyer, Damage Limiter, Compleat Controller, Cost Cutter, and Political Container. Their arguments are predictable, as is the fact that the goal of negotiated strategic arms control is often a pawn in a domestic political debate. E. McCarthy

686. Wyle, Frederick. EUROPEAN SECURITY: BEATING THE NUMBERS GAME. *Foreign Policy 1973 (10): 41-54.* Critiques the present negotiations for Mutual Balanced Force Reductions. S

687. Yankelovich, Daniel. FAREWELL TO "PRESIDENT KNOWS BEST." *Foreign Affairs 1979 57(3): 670-693.* Since the Vietnam debacle, an enlarged public role in the formulation of public policy has materialized. Though some observers fear that the United States may merely be giving up the abuses of an imperial presidency for the more volatile abuses of an "imperial public," public attitudes toward the Panama Canal, strategic arms limitation talks (SALT), the Middle East, and South Africa represent a healthy "public mind." Under our system, public involvement is the only real alternative to an imperial presidency. 40 notes. M. R. Yerburgh

688. Yochelson, John. MBFR: THE SEARCH FOR AN AMERICAN APPROACH. *Orbis 1973 17(1): 155-175.* Discusses America's foreign policy position on MBFR (Mutual Balanced Force Reductions) in Europe.

689. Yochelson, John. THE SEARCH FOR AN AMERICAN APPROACH. *Survival [Great Britain] 1973 15(6): 275-283.* Discusses the prospects for a possible lessening of US influence in NATO and the defense of Western Europe against the USSR in the 1970's, emphasizing foreign policy and military strategy, including mutual balanced force reductions.

690. Young, Wayland. DISARMAMENT: THIRTY YEARS OF FAILURE. *Int. Security 1978 2(3): 33-50.* The quest for general and comprehensive disarmament down to the level required to maintain order within nation states has been going on at least since the publication of Émeric Crucé's *La Nouvelle Cynée* in 1623. The attempt has been particularly important in the last 30 years, but has failed because there is no comprehensive plan, no overall framework for discussion, and the two greatest superpowers do not trust each other, nor are they trusted by many of the other nations.

691. —. [ARMAMENTS-ARMS CONTROL-DISARMAMENT]. *Bull. of Peace Proposals [Norway] 1975 6(2): 131-155.* Sixteen summaries of reports and articles, 1974-75, discussing problems such as disarmament and arms control, the Vladivostok Agreement (1974), military expenditures and the arms trade, force reduction in Europe, a Nordic nuclear free zone, U.N. peacekeeping, and a proposed world disarmament conference. R. B. Orr

692. —. DOCUMENTATION: US POLICIES ON ARMS TRANSFERS AND NON-PROLIFERATION. *Survival [Great Britain] 1981 23(5): 231-232.*
—. PRESIDENTIAL DIRECTIVE ON ARMS TRANSFER POLICY 8 JULY 1981, *pp. 231-232* Reproduces the Reagan administration's announcement of new policies on arms transfers.
—. PRESIDENTIAL STATEMENT ON NON-PROLIFERATION 16 JULY 1981, *pp. 232-233.* Reprints the Reagan administration's policy statement on nuclear nonproliferation.

693. —. OUTLOOK FOR DETENTE AND DISARMAMENT: TWO INTERVIEWS. *World Marxist Rev. [Canada] 1977 20(4): 80-87.*
Chandra, Romesh. NEW INITIATIVES INSPIRE OPTIMISM, *pp. 80-83.*

Kade, Gerhard. THE SEARCH FOR MUTUALLY ACCEPTABLE MEASURES, *pp. 83-87.* The authors discuss international attitudes toward and possibilities for USSR-US detente and mutual disarmament.

694. —. QUESTIONS OF POWER, MATTERS OF RIGHT. *Round Table [Great Britain] 1977 (267): 207-214.* Examines the implications of Jimmy Carter's election to the Presidency for detente, human rights, and nuclear disarmament. Analyzes the effects of Henry A. Kissinger's policies on American attitudes toward foreign policy, particularly his support for undemocratic regimes in Indochina, the Middle East, and Latin America, his lack of foresight over the South African issue, and his treatment of the 1973 oil crisis. C. Anstey

695. —. SALT AND MBFR: THE NEXT PHASE—REPORT OF A TRILATERAL CONFERENCE. *Survival [Great Britain] 1975 17(1): 14-24.* Discusses topics covered at a 1974 conference on Strategic Arms Limitation Talks between the United States and the USSR, including the military objectives of Western Europe and Japan.

696. —. THE SALT PROCESS: WHY AND HOW. *Am. Soc. of Int. Law. Pro. 1978 72: 50-56.*

Warnke, Paul C. THE SALT PROCESS: WHY AND HOW, *pp. 50-54.* The basic premise of the Strategic Arms Limitations Talks with the USSR is the establishment of greater national security for both sides and the maintenance of international nuclear stability; 1972-78.

Walker, George K. DISCUSSION, *pp. 54-56.* Summarizes Warnke's responses to questions from the audience pertaining to the use of the neutron bomb, suspected trangressions of the treaty by the USSR, weapons systems, and the role of the UN.

5

NUCLEAR REACTORS AND PUBLIC REACTION

697. Abbotts, John. RADIOACTIVE WASTE: A TECHNICAL SOLUTION? *Bull. of the Atomic Scientists 1979 35(8): 12-18.* Discusses the political problems of radioactive waste disposal focusing on case histories in Kansas in 1970 and Michigan in 1975.

698. A.G. LA NOUVELLE POLITIQUE ÉNERGÉTIQUE DES ÉTATS-UNIS [The new energy policy of the United States]. *Défense Natl. [France] 1978 34(2): 105-116.* Examines President Jimmy Carter's energy policy of conservation and conversion and considers the ramifications of nuclear alternatives for world power supply.

699. Anthony, Robert. ACCOUNTABILITY AND CREDIBILITY IN THE MANAGEMENT OF COMPLEX HAZARDOUS TECHNOLOGY. *Policy Studies Rev. 1982 1(4): 705-715.* Shows how the conflicts over different safety actions arise; presents three policy options based on information exchange that could reduce the level of conflict, with the general points illustrated primarily by reference to a leak of liquid high-level nuclear wastes from a storage tank at Hanford, Washington.

700. Austin, Nancy. DIARY OF THREE MILE ISLAND INCIDENT, 3/28/79-4/4/79. *Social Educ. 1979 43(6): 458-459.* Day by day, first person impressions of the Three Mile Island incident in 1979.

701. Bamière, Christine. FUSION THERMONUCLÉAIRE PAR LASER: PROGRÈS AMÉRICAINES ET SOVIÉTIQUE [Thermonuclear fusion by laser: US and USSR progress]. *Défense Natl. [France] 1975 31(5): 127-134.* Traces and compares Soviet and US progress in research and development of thermonuclear fusion over the past few years.

702. Barkan, Steven E. STRATEGIC, TACTICAL AND ORGANIZATIONAL DILEMMAS OF THE PROTEST MOVEMENT AGAINST NUCLEAR POWER. *Social Problems 1979 27(1): 19-37.* Describes how the contemporary antinuclear protest movement has resorted to a variety of protest tactics, particularly illegal occupation of plant sites, to make its grievances known, and focuses on differences within the movement over strategic, tactical, and organizational matters, identifying some of the tactical problems atomic energy protesters have encountered in the courts.

703. Benedict, Robert; Bone, Hugh; Leavel, Willard; and Rice, Ross. THE VOTERS AND ATTITUDES TOWARD NUCLEAR POWER: A COMPARATIVE STUDY OF "NUCLEAR MORATORIUM" INITIATIVES. *Western Pol. Q. 1980 33(1): 7-23.* The study focuses upon recent ballot propositions in four western states which proposed additional state regulation for nuclear power plants. Three frameworks are developed which contain differing perspectives upon the variables of general attitudes toward the role of technology in society; economic growth and energy development; the mix of citizens, experts and politicians in decisionmaking for nuclear plants; and cost-benefit analysis of such plants. A very high correlation is found between the variable of cost-benefit analysis and the vote, with a substantial relationship discerned between the vote and the variables of attitudes toward economic growth, and the mix of decisionmakers. In contrast, among the political variables of party identification, liberalism-conservatism, and degree of voter confusion about the issue, a mixed pattern is found, with only the party identification variable displaying a moderate relationship to the vote. Finally the implications of the findings for the substance of energy policy and the methods of decisionmaking are discussed. 8 tables, 42 notes. J

704. Blair, Bruce G. and Brewer, Garry D. THE TERRORIST THREAT TO WORLD NUCLEAR PROGRAMS. *J. of Conflict Resolution 1977 21(3): 379-403.* Terrorism in the global setting has become the predominant form of confrontation between differing subcategories of societies that seek to overcome each other, regardless of size. In the case of nuclear terrorism, the consequences of failure are potentially catastrophic. While the logic of our strategic nuclear policy is clear, the same clarity does not hold for policies directed at nuclear terrorism. In the former case, a prevailing view is that the risk of nuclear war is low because the United States responds vigilantly to nuclear threats posed by other nations. In the latter case, there is no terrorist prevention doctrine, nor is there an institutional focus for preventing terrorism that is even remotely commensurate with that which exists for deterring nuclear war. We here consider the dimensions of the nuclear terrorism problem, discuss these with respect to the Minuteman Intercontinental Ballastic Missile system, consider the capabilities and objectives of potential terrorist groups, and formulate some basic recommendations for improving the current state of affairs. J

705. Bonetti, Alberto. NOTIZIE DA HARRISBURG PER L'ITALIA [News from Harrisburg for Italy]. *Ponte [Italy] 1979 35(5): 550-560.* Examines the development of Italian policy on nuclear energy, 1970's, in light of the lessons of the accident at the Three Mile Island nuclear plant in Harrisburg, Pennsylvania, March 1979.

706. Brady, David and Althoff, Phillip. THE POLITICS OF REGULATION: THE CASE OF THE ATOMIC ENERGY COMMISSION AND THE NUCLEAR INDUSTRY. *Am. Pol. Q. 1973 1(3): 361-384.* Examines relations between politico-economic and scientific-technological variables in policymaking as they pertain to the nuclear industry under federal regulations through the Atomic Energy Commission, 1946-70.

707. Bromberg, F. and Kurenkov, Iu. ATOMNAIA ENERGETIKA—NOVYI UZEL IMPERIALISTICHESKIKH PROTIVORECHII [Atomic energetics in a new tangle of imperialist contradictions]. *Mirovaia Ekonomika i Mezhdunarodnye Otnosheniia [USSR] 1981 (9): 59-68.* Analyzes the socioeconomic impact of the intensive development of nuclear power on the world energy situation, noting that in the early 70's, US monopolies became firmly established in reactor engineering due to greater unit capacities and more advanced power reactors, but that difficulties developed in connection with terms of construction, engineering, and safety problems; the nuclear fuel supplies until recently controlled by the United States are now also becoming a sphere of competitive struggle. Russian.

708. Bromet, Evelyn and Dunn, Leslie. MENTAL HEALTH OF MOTHERS NINE MONTHS AFTER THE THREE MILE ISLAND ACCIDENT. *Urban & Social Change Rev. 1981 14(2): 12-15.* Presents the results of a study on the mental health of mothers with children living near the Three Mile Island power plant in Pennsylvania during the nuclear accident of 1979; the accident "had both acute and chronic mental health effects among mothers of small children living near the plant" due to stress.

709. Brucer, Marshall. THE TUCSON TRITIUM TRIALS. *Reason 1980 11(11): 30-35.* The public and regulatory furor in 1979 over the use of tritium by the American Atomics Corporation in Tucson, Arizona, was unfounded, because tritium is one of the safest radioactive substances.

710. Bupp, Irvin C. THE ACTUAL GROWTH AND PROBABLE FUTURE OF THE WORLDWIDE NUCLEAR INDUSTRY. *Int. Organization 1981 35(1): 59-76.* The stalemate over the future of nuclear power is particularly deep in the United States. Administrative and personnel problems in the Nuclear Regulatory Commission, slow progress on radioactive waste disposal by the Department of Energy, severe financial problems for most electric utilities, and drastic reductions in the rate of electricity demand growth combine to make continuation of the five-year-old moratorium on reactor orders inevitable. Many of the 90 plants under construction may never operate and some of the 70 in operation may shut down before the end of their economic life. Covers 1973-80.
J/S

711. Burness, H. Stuart; Montgomery, W. David; and Quirk, James P. THE TURNKEY ERA IN NUCLEAR POWER. *Land Econ. 1980 56(2): 188-202.* Examines the rationale behind General Electric Co.'s and Westinghouse's turnkey contracts, assuming responsibility for the design, construction, and testing of 13 nuclear reactors, 1962-66. Concludes that the losses suffered by the plant builders were learning costs and that the post-turnkey surge in orders for atomic power plants was a response to the utilities' regulated ratemaking process. 5 tables, 18 notes, 8 ref. E. S. Johnson

712. Camilleri, J. A. THE MYTH OF THE PEACEFUL ATOM. *Millennium: J. of Int. Studies [Great Britain] 1977 6(2): 111-127.* In spite of an appreciation of the dangers of nuclear proliferation, early efforts in the post-1945 period to control the peaceful use of atomic energy through the UN Atomic Energy Commission (1946) and the USA's Baruch Plan proved abortive, mainly because

of superpower disagreements. Subsequent agreements on safeguards, such as the 1970 Nonproliferation Treaty, have failed to provide meaningful barriers, especially with the increased application of nuclear technology in the world. A further problem is the potential political instability of many new members of the nuclear club. The United States is a key influence, but President Carter's new policy initiative of 1977—the seven-point plan—is unlikely to solve the misuse problem. 43 notes. P. J. Beck

713. Cohen, Bernard L. PERSPECTIVES ON NUCLEAR DEBATE. *Bull. of the Atomic Scientists 1974 30(8): 35-39.* Refutes criticism of possible proliferation of atomic power plants: discharge of radioactive material, threat to life in the event of a nuclear accident, poisonous potential of plutonium, and the concurrent proliferation of nuclear weapons, asserting that public controversy over nuclear power plants is irrelevant.

714. Cohen, Bernard L. RADIATION FANTASIES. *Reason 1980 11(11): 24-29, 35.* Fears about the dangers of low-level radiation, particularly as voiced by scientist John W. Gofman in recent years, are unfounded; many prestigious scientific evaluation committees worldwide have found that low-level radiation is not dangerous.

715. Comey, David Dinsmore. WILL IDLE CAPACITY KILL NUCLEAR POWER? *Bull. of the Atomic Scientists 1974 30(9): 23-28.* Assesses the cost to taxpayers of unused capacity in atomic power plants and the concomitant cost in public utilities, 1973-74.

716. Connor, James E. PROSPECTS FOR NUCLEAR POWER. *Pro. of the Acad. of Pol. Sci. 1973 31(2): 63-73.* Examines the relevance of America's nuclear power industry to the country's energy problem and possible changes the industry might undergo in the future. S

717. Cook, Earl. THE ROLE OF HISTORY IN ACCEPTANCE OF NUCLEAR POWER. *Social Sci. Q. 1982 63(1): 3-15.* Nuclear power is a viable political option in Canada but not in the United States, due to experience, confidence in government, and energy economies.

718. Daneke, Gregory A. THE POLITICAL ECONOMY OF NUCLEAR DEVELOPMENT. *Policy Studies J. 1978 7(1): 84-89.* Examines the nuclear industry as a model for the ramifications of federal government subsidization both as it promotes the development of energy sources and as it supplies a sagging industry with artificial incentives and buttresses which protect it from the reality of economic conditions, 1954-78.

719. DelSesto, Steven L. NUCLEAR REACTOR SAFETY AND THE ROLE OF THE CONGRESSMAN: A CONTENT ANALYSIS OF CONGRESSIONAL HEARINGS. *J. of Pol. 1980 42(1): 227-241.* Throughout its existence, the Joint Committee on Atomic Energy favored nuclear development. In a selected sample of Joint Committee hearings, *The Status of Nuclear Reactor Safety, 1973-74,* members assumed role behaviors consistent with a pronuclear stance. They cooperated with witnesses for the Atomic Energy Commission, the atomic power industry, and the entire nuclear community while deprecating the testimony from environmentalists, concerned citizens, and others opposed to

nuclear development. The Committee essentially fulfilled its statutory mandate to develop civilian nuclear power, with its decline due to narrow overspecialization and the changing political climate induced by the environmental movement. 5 tables, 12 notes. A. W. Novitsky

720. DeVolpi, A. ENERGY POLICY DECISION-MAKING: THE NEED FOR BALANCED INPUT. *Bull. of the Atomic Scientists 1974 30(10): 29-33.* Examines Atomic Energy Commission interconnections various power companies and assesses the clout which environmental groups and the general public have in determining energy policy; discusses the separation of military use of nuclear power and nonmilitary nuclear activity, asserting that environmentalists need to press for more responsible public disclosure of nuclear proliferation.

721. Farrell, Thomas B. and Goodnight, G. Thomas. ACCIDENTAL RHETORIC: THE ROOT METAPHORS OF THREE MILE ISLAND. *Communication Monographs 1981 48(4): 271-300.* Examines communication practices and the rhetoric involving the Three Mile Island nuclear incident in 1979 and their presuppositions, discussing three world views that have gradually advanced the power of technical reason over social concerns in general and nuclear power in particular.

722. Faulkner, Peter. WHISTLE BLOWER: PETER FAULKNER, NUCLEAR ENGINEER. *Civil Liberties Rev. 1978 5(3): 41-49.* Peter Faulkner, systems application engineer for the Nuclear Services Corp., submitted confidential documents and a criticism of the engineering deficiencies in atomic power systems to Senate Subcommittee's Energy Research and Development Administration hearing in 1974.This resulted in his dismissal by his employers.

723. Flood, Michael. NUCLEAR SABOTAGE. *Bull. of the Atomic Scientists 1976 32(8): 29-36.* Discusses the vulnerability of nuclear installations and facilities to sabotage.

724. Friedman, Sharon M. BLUEPRINT FOR BREAKDOWN: THREE MILE ISLAND AND THE MEDIA BEFORE THE ACCIDENT. *J. of Communication 1981 31(2): 116-128.* Based on the report of a presidential task force to investigate the breakdown of communication during the Three Mile Island accident in 1979, finds that both the local utility and the media share the blame for overemphasizing the safety, cleanliness, and economy of atomic energy and underplaying its potential problems in the public information available before the accident.

725. García Ferrando, Manuel. EL DEBATE PUBLICO SOBRE EL USO DE LA ENERGIA NUCLEAR [The public debate over the use of nuclear energy]. *Rev. Española de Investigaciones Sociol. [Spain] 1981 (16): 57-90.* Traces the origins of both pronuclear and antinuclear sentiments in the United States and Spain; presents results of a poll, and sociodemographic characteristics of antinuclear proponents. Spanish.

726. Gilinsky, Victor. PLUTONIUM, PROLIFERATION AND THE PRICE OF REPROCESSING. *Foreign Affairs 1978-79 57(2): 374-386.* The United States and Europe have developed divergent views on the propriety of reprocessing spent nuclear fuels, e.g., plutonium, for export. Over US protest,

France and Great Britain are prepared to provide such materials to countries lacking destructive nuclear capabilities. The American position is an appropriate one; "strict, uniform, and universal rules" should be constructed before such ventures. M. R. Yerburgh

727. Gordon, Suzanne. THE ULTIMATE SINGLE ISSUE. *Working Papers Mag. 1982 9(3): 20-25.* Discusses the new peace movement against the threat of nuclear war, which is supported by the middle classes rather than the fringe groups that spearheaded the 1960's antiwar movement.

728. Gravel, Mike. PLUTONIUM RECYCLE: THE CIVIL LIBERTIES VIEW. *Civil Liberties R. 1976 3(1): 38-42.* The important point is not that we must watch for infringements of civil liberties under the guise of plutonium security. Rather, it is that the horrific consequences of plutonium terrorism could indeed justify the curtailment of such liberties. J

729. Green, Harold P. "BORN CLASSIFIED" IN THE AEC: A LEGAL PERSPECTIVE. *Bull. of the Atomic Scientists 1981 37(10): 28-30.* A former Atomic Energy Commission lawyer covers the development of atomic energy law and the statutory definition of restricted data and asks for a reexamination of the implications of the restricted data concept, since the information control provisions are dangerous to a democratic society, 1945-75.

730. Greene, Mark R. A REVIEW AND EVALUATION OF SELECTED GOVERNMENT PROGRAMS TO HANDLE RISK. *Ann. of the Am. Acad. of Pol. and Social Sci. 1979 (443): 129-144.* Six government insurance programs are described and evaluated against three criteria: 1) size and significance of the economic burden imposed by the risks covered, 2) public acceptance of the program, and 3) necessity of government initiative to accept the risk if private insurers could not do so. It is concluded that government handling of flood and swine flu liability risks are justified, but the government programs in riot reinsurance, crime insurance, and nuclear energy liability should be terminated. Federal crop insurance is of questionable necessity in view of the basic insurability of this risk by private insurers. J

731. Hamilton, Mary A. ENERGY POLICY AND CHANGING PUBLIC-PRIVATE SECTOR RELATIONSHIPS. *Policy Studies J. 1978 7(1): 90-95.* Discusses joint public-private activity in energy policy and the development and commercialization of new energy technologies in order to assess changes in government-business relations which must occur in order to achieve current energy policy objectives, 1978.

732. Harnik, Peter. THE ETHICS OF ENERGY-PRODUCTION AND USE: DEBATE WITHIN THE NATIONAL COUNCIL OF CHURCHES. *Bull. of the Atomic Scientists 1979 35(2): 5-9.* Preparation by the National Council of Churches of a "Policy Statement on the Ethical Implications of Energy Production and Use" emphasized debate on the ethics of the energy problem and resulted in the realization that the energy industry neither encouraged nor welcomed debate on ethics, and that the information gap between so-called experts and the general public remains quite wide, 1976-78.

733. Hewlett, Richard G. "BORN CLASSIFIED" IN THE AEC: A HISTORIAN'S VIEW. *Bull. of the Atomic Scientists 1981 37(10): 20-27.* Describes the genesis and evolution of the "born classified" concept regarding atomic energy during 1946-75, when the Atomic Energy Commission consistently relied upon this concept in administering its authority to control dissemination of classified information.

734. Hill, Gladwin. INITIATIVES: A SCORECARD. *Working Papers for a New Soc. 1977 4(4): 33-37.* Discusses the usefulness of nuclear safety initiatives in the 1976 elections, even though they lost in most states; considers the nature of the political referendum.

735. Holdren, John P. THE NUCLEAR CONTROVERSY AND THE LIMITATIONS OF DECISION-MAKING BY EXPERTS. *Bull. of the Atomic Scientists 1976 32(3): 20-22.* Discusses disagreements among scientific experts about the danger of nuclear reactor accidents and the 1976 California Nuclear Safeguards Initiative, including the problems of sabotage and radioactive wastes.

736. Ichord, Robert F., Jr. PACIFIC BASIN ENERGY DEVELOPMENT AND U.S. FOREIGN POLICY. *Orbis 1977 20(4): 1025-1043.* Changes in the balance of power have been taking place in the Pacific Basin since 1973. Energy has become the key variable in the region. There are new nuclear power programs, major petroleum and coal development activities, and the ongoing development of other energy resources. The United States believes in the development of expanded energy systems throughout the Basin and will work closely with the governments of Japan, Indonesia, and Australia to reduce dependence on the petroleum-producing Middle East countries as well as to assist poor countries in the region. Secondary sources; 3 charts, 27 notes. A. N. Garland

737. Ingram, Timothy H. NUCLEAR HIJACKING: NOW WITHIN THE GRASP OF ANY BRIGHT LUNATIC. *Washington Monthly 1973 4(11): 20-28.*

738. Ioisysh, A. I. PRAVOVYE PROBLEMY FIZICHESKOI ZASHCHITY IADERNYKH MATERIALOV [Legal problems of the physical protection of nuclear materials]. *Sovetskoe Gosudarstvo i Pravo [USSR] 1980 (8): 83-87.* Examines international programs to guarantee physical protection of nuclear materials used for atomic energy in Europe and the United States. Russian.

739. Janke, Peter. NUCLEAR DENIAL: THE POLITICS OF DIRECT DEMOCRACY. *Contemporary Rev. [Great Britain] 1979 234(1359): 174-180.* Examines the hostility surrounding the nuclear issue in Europe, Japan, and the United States and assesses the extent to which it has been manipulated by politically orientated groups hostile to the survival of liberal democracy, suggesting that only the USSR has anything to gain by the slowing down of western nuclear technology.

740. Jezer, Marty. THE SOCIALIST POTENTIAL OF THE NO-NUKE MOVEMENT. *Radical Am. 1977 11(5): 63-71.* The antinuclear power movement is the most visible example of US radical political protest. The Clamshell

Alliance is a single-issue mass movement based on acts of personal moral courage which might be converted into a vehicle for massive left-wing political action. Discussions revolve around such points as collectivity versus individual action, confrontation tactics as opposed to building an organizational base, environmentalism versus economic issues, and the question of antinuclear power agitation as a single cause in relation to a multi-issue approach to social and economic problems. The political left might look at the Clamshell Alliance as a possible way to draw environmentalism into radical politics and to effect a working relationship with the working class. Based mainly on personal participation in the Clamshell Alliance. N. Lederer

741. Katz, Neil H. and List, David C. SEABROOK: A PROFILE OF ANTINUCLEAR ACTIVISTS, JUNE 1978. *Peace and Change 1981 7(3): 59-69.* Discusses the protesters of the June 1978 antinuclear rally at the site of a partly built atomic power plant at Seabrook, New Hampshire, especially their backgrounds, ideologies, and perceptions of effectiveness of the rally and toward the Clamshell Alliance, as studied by a research team from Syracuse University.

742. Keating, William Thomas. POLITICS, ENERGY, AND THE ENVIRONMENT: THE ROLE OF TECHNOLOGY ASSESSMENT. *Am. Behavioral Scientist 1975 19(1): 37-74.* Discusses public participation in the Atomic Energy Commission's assessment of the costs of nuclear power 1946-73, in a special issue entitled "Policy Content and the Regulatory Process." S

743. Keller, Edward B. THE FLAP OVER "PLUTONIUM: AN ELEMENT OF RISK." *J. of Communication 1979 29(3): 54-61.* Describes the controversy surrounding the television documentary *Plutonium: An Element of Risk,* aired by only 11 out of 270 PBS member stations in 1977, and discusses earlier documentaries dating to 1971 on atomic energy by producer Don Widener.

744. Kemeny, John G. POLITICAL FALLOUT. *Society 1981 18(5): 5-9.* The President's Commission on the Accident at Three Mile Island, which convened on 11 April 1979, found that the nuclear industry suffered from the belief that its equipment and systems were foolproof, that industry operators were not trained adequately for their work, and that the Nuclear Regulatory Commission was a "total disaster."

745. Khalilzad, Zalmay and Benard, Cheryl. ENERGY: NO QUICK FIX FOR A PERMANENT CRISIS. *Bull. of the Atomic Scientists 1980 36(10): 15-20.* Despite their faith in modern technology, proponents of atomic power and those favoring alternative energy sources will not find the answers based on technical solutions because the industrial world's energy problems are too complex; 1940's-80.

746. Klein, Jeffrey S. THE NUCLEAR REGULATORY BUREAUCRACY. *Society 1981 18(5): 50-56.* The accident at Three Mile Island raised questions concerning the effectiveness of the Nuclear Regulatory Commission.

747. Kopkind, Andrew. WHAT TO DO TILL THE MOVEMENT ARRIVES. *Working Papers for a New Soc. 1978 6(1): 43-49.* Discusses various social and political projects available to political activists from the 1960's who cannot find a cause to support; discusses local action groups, community movements, and national anti-nuclear groups, 1970's.

748. Kowarski, L. CERN'S FIRST DIRECTOR-GENERAL, 1954-1955. *Rice U. Studies 1980 66(3): 123-131.* Describes Felix Bloch's role in shaping the early operations of the Conseil Européen pour la Recherche Nucléaire (CERN) in Geneva as its first director general; prior to his appointment he had lived in the United States as a research nuclear physicist.

749. Krieger, David M. TERRORISTS AND NUCLEAR TECHNOLOGY. *Bull. of the Atomic Scientists 1975 31(6): 28-34.* Warns that terrorist groups will achieve nuclear weapon capability unless steps are taken "almost immediately to halt both nuclear weapon and nuclear power plant proliferation." Explores ways in which terrorist groups "may gain possession of nuclear materials, including weapons; the way in which they may use nuclear weapons and other nuclear technologies to their benefit; and various courses of action designed to minimize the possibilities of terrorists utilizing nuclear technology to their benefit and society's detriment." Based on primary and secondary sources; map, 21 notes.
D. J. Trickey

750. Lanouette, William J. "NO LONGER CAN THE NRC SAY. . . ." *Bull. of the Atomic Scientists 1979 35(6): 6-8.* Reviews the events of Three Mile Island, Pennsylvania, 1979, assessing the relative roles of human error, mechanical failure, and design errors which led to the release of radioactivity from the atomic power plant; examines confusion of federal regulation from the Atomic Energy Commission and the Nuclear Regulatory Commission.

751. Lapp, Ralph H. TOWARD NUCLEAR EDUCATION. *Sci. and Public Affairs 1973 29(6): 6-8.* As editor of *Science and Public Affairs (Bulletin of the Atomic Scientists),* Eugene Rabinowitch (1901-73) promoted interdisciplinary and public knowledge of nuclear science and technology, 1940's-70's.

752. Lee, Kai N. NUCLEAR POWER AND ELECTRICAL ENERGY. *Policy Studies J. 1981 9(7): 1087-1092.* Reviews Peter de Leon's *Development and Diffusion of the Nuclear Power Reactor* (1979), William Ramsay's *Unpaid Costs of Electrical Energy* (1979), and Elizabeth S. Rolph's *Nuclear Power and Public Safety* (1979) which discuss the history, problems, and future of nuclear power.

753. Lellouche, Pierre. INTERNATIONAL NUCLEAR POLITICS. *Foreign Affairs 1979-80 58(2): 336-350.* Examines American interest in establishing the International Nuclear Fuel Cycle Evaluation, 1970's, to study the implications of the growth of atomic energy in response to world hostility to American desires to prevent nuclear proliferation.
S

754. Leopold, Richard W. HISTORICAL ADVISORY COMMITTEES: STATE, DEFENSE, AND THE ATOMIC ENERGY COMMISSION. *Pacific Hist. R. 1975 44(3): 373-385.* The historical advisory committees for the Defense Department (consisting of academics and military personnel), the State

Department, and the Atomic Energy Commission (consisting entirely of academics) advise on the handling of manuscript collections, on publication projects, and on the issuing of grants. The committees have been a great help to civil servants, but problems remain. Committee members tend not to be sufficiently critical of agencies, and absenteeism and insufficient preparation have been obstacles. Based on personal correspondence and published government documents; 22 notes.
W. K. Hobson

755. Lester, Richard. SECRECY, PATENTS AND NON-PROLIFERATION. *Bull. of the Atomic Scientists 1981 37(5): 35-38.* Using the development of the molecular laser isotope in the 1970's as an example, demonstrates that though US law prohibits the filing of classified patents overseas, fundamental scientific discoveries cannot be kept secret.

756. Light, Alfred R. THE CARTER ADMINISTRATION'S NATIONAL ENERGY PLAN: PRESSURE GROUPS, AND ORGANIZATIONAL POLITICS IN THE CONGRESS. *Policy Studies J. 1978 7(1): 68-75.* Relates the Carter administration's record on energy policy, emphasizing policy formation and relations with Congress, 1977-78.

757. Light, Alfred R. THE HIDDEN DIMENSION OF THE NATIONAL ENERGY PLAN: EXECUTIVE POLICY DIRECTION IN NUCLEAR WASTE MANAGEMENT. *Publius 1979 9(1): 169-187.* The problems in getting the National Energy Acts through Congress have necessitated not even including the more difficult aspects such as radioactive waste disposal. Analyzes intergovernmental bickering, the powers of states to veto locations, political trade-offs, government-private industry conflicts, complications in the proposed Carlsbad (New Mexico) disposal site, and attempts through the Executive Planning Council to avoid a preemption/state veto showdown. The process illustrates the power of even relatively "weak" actors to slow things down in the federal system.
R. V. Ritter

758. Light, Alfred R. THE 1979 ENERGY CRISIS SYMPOSIUM: INTRODUCTION. *Publius 1980 10(1): 43-45.* Summarizes "centrally-directed" federalism evident in symposium articles on energy regulation in this issue. Evaluates 1979 trends toward a Department of Energy monopoly on such regulation in the case of nuclear waste management, in particular through burial of nuclear wastes at the Waste Isolation Pilot Plant in Carlsbad, New Mexico.
C. B. Schulz

759. Lovins, Amory B.; Lovins, L. Hunter; and Ross, Leonard. NUCLEAR POWER AND NUCLEAR BOMBS. *Foreign Affairs 1980 58(5): 1137-1177.* Examines the questions concerning the uses and effectiveness of nuclear power, and concludes that atomic energy is the least effective substitute for oil, is not economical, and is the major force behind nuclear arms proliferation. More effective means of energy production are available, including energy conservation and utilization of existing non-nuclear technology, which would displace oil and meet energy needs. Based on documents of various US government agencies, articles in business and scientific journals, and books dealing with energy topics; 45 notes.
A. A. Englard

760. Mabbutt, Fred R. BROKEN ARROW: THE LEGACY OF HIROSHIMA. *Mankind 1978 6(4): 30-32, 38-40.* Danger of nuclear pollution through warfare, individual terroristic employment of atomic weapons, and accidents in the peaceful use of nuclear power is prevalent in contemporary society. The building of atomic power plants and their maintenance has been accompanied by increasing numbers of "broken arrows," namely accidents at such installations. The number and range of such accidents at the Oklahoma Kerr-McGee plant alone have been enough to generate justifiable alarm and concern among the public. International agencies established to control the use of nuclear energy are impotent. Accidents involving military transport of nuclear arms bode ill for the future security of humanity from nuclear contamination.

N. Lederer

761. Marrett, Cora Bagley. ACCIDENT ANALYSIS. *Society 1981 18(5): 66-72.* Discusses the membership of the President's Commission on the Accident at Three Mile Island, 5 April 1979, their procedures, their opposition to the Nuclear Regulatory Commission, and their influence.

762. Marx, Leo. REFLECTIONS ON THE NEO-ROMANTIC CRITIQUE OF SCIENCE. *Daedalus 1978 107(2): 61-74.* Discusses the widespread criticism of science in the United States which has taken place largely since World War II. Anxiety about the nuclear arms race, pollution, the harmful effects of chemicals, and moral revulsion over the uses of scientific technology during the Vietnam War are key factors in this dissatisfaction. Discusses 19th century attitudes about the interrelation of science and society by examining the writings of Emerson, Wordsworth, Coleridge, Carlyle, and Shelley, and compares them with 20th century attitudes expressed by C. P. Snow, I. A. Richards, and Theodore Roszak. Argues that current criticism of science has a long history in Western thought; it is a "new phase of the 'romantic reaction' that began some two centuries ago."

S

763. Mazuzan, George T. CONFLICT OF INTEREST: PROMOTING AND REGULATING THE INFANT NUCLEAR POWER INDUSTRY, 1954-1956. *Historian 1981 44(1): 1-14.* The Atomic Energy Act of 1954 assigned the US Atomic Energy Commission (AEC) both promotional and regulatory responsibilities that made inevitable conflicts of interest and hindered the objective analysis of safety issues for the next 20 years. How to answer the complex safety questions in order to protect the public without stymieing reactor development was the central problem facing the regulators. A lack of safety criteria and the need for efficiency necessitated ignoring safety standards. Traces the history, operation, regulations and research procedures of the AEC which demonstrate the predominance of promotion over safety; nonetheless that inclination reflected the assumptions, concerns, and priorities of the times in which the atomic industry came into existence and set the general tone of regulation for the next 20 years. Based on primary sources; 36 notes. R. S. Sliwoski

764. McCracken, Samuel. THE HARRISBURG SYNDROME. *Commentary 1979 67(6): 27-39.* Examines recent events related to the atomic power industry, such as the Three Mile Island incident and the film *The China Syndrome,* to study varying attitudes.

765. McCracken, Samuel. THE WAR AGAINST THE ATOM. *Commentary 1977 64(3): 33-47.* Most nuclear power discussions are grossly misinformed. The anti-nuclear power critique cites radiation danger, meltdown danger, risk of explosion, lack of economy, thermal pollution, incitement to terrorism, and nuclear weapons proliferation as reasons to curtail the use of nuclear power. These charges are not true. D. W. Johnson

766. Mitchell, Robert Cameron. FROM ELITE QUARREL TO MASS MOVEMENT. *Society 1981 18(5): 76-84.* Opposition to nuclear energy in the United States has gone through three phases since the early 1950's, including legal protest and direct action, and is now a social movement with goals of decentralization, egalitarianism, and participatory democracy.

767. Moylan, Maurice P. EMPLOYMENT IN THE ATOMIC ENERGY FIELD, 1973. *Monthly Labor Rev. 1974 97(9): 23-27.* Surveys employment in atomic energy in the United States between 1963 and 1973, showing the impact of the gradual replacement of government by private industry in peaceful atomic activities.

768. Murphy, Dervla. THE REALITY OF NUCLEAR POWER. *Blackwood's Mag. [Great Britain] 1979 326(1967): 193-211.* Details the history, politics, process, dangers, and costs of nuclear weapons, reactors, and waste throughout the world, 1950's-79.

769. Nelkin, Dorothy. ANTI-NUCLEAR CONNECTIONS: POWER AND WEAPONS. *Bull. of the Atomic Scientists 1981 37(4): 36-40.* Examines the evolution of the antiatomic power movement and the peace movement, their overlapping interests, and their organizational links, 1960-80.

770. Nelkin, Dorothy. THE ROLE OF EXPERTS IN A NUCLEAR SITING CONTROVERSY. *Bull. of the Atomic Scientists 1974 30(9): 29-36.* Examines the impact of organization and expertise in environmental pressure groups in 1973 in New York, where local citizens' groups and the academic community halted the installation of an atomic power plant at Cayuga Lake.

771. Nelson, Jon P. THREE MILE ISLAND AND RESIDENTIAL PROPERTY VALUES: EMPIRICAL ANALYSIS AND POLICY IMPLICATIONS. *Land Econ. 1981 57(3): 363-372.* Examines the prices of houses sold in two residential areas within four miles of the Three Mile Island (TMI) nuclear plant accident. Regression analysis of sales between the fourth quarter of 1977 and the fourth quarter of 1979 found no significant drop in value or retardation of appreciation. Comparison with home sales in a larger area centered on TMI likewise showed no negative effect. Either the effects of TMI were viewed as short term, or the housing market could not react within the short time span of the study. 5 tables, 14 notes, biblio. E. S. Johnson

772. Newcomb, Richard. THE AMERICAN COAL INDUSTRY. *Current Hist. 1978 74(437): 206-209, 228.* Discusses the 50-year decline of the American coal industry, and its recent revival prospects since the mid-1970's, in spite of environmental problems, because of increased doubt that a workable nuclear technology could be developed soon.

773. Nimmo, Dan and Combs, James E. FANTASIES AND MELODRAMAS IN TELEVISION NETWORK NEWS: THE CASE OF THREE MILE ISLAND. *Western J. of Speech Communication 1982 46(1): 45-55.* Examines television coverage of the Three Mile Island nuclear accident in Pennsylvania during 1979, comparing the four major rhetorical stances adopted by the three networks during the course of the incident; focuses on melodramatic elements in the networks' reports.

774. Noorani, A G. INDO-U.S. NUCLEAR RELATIONS. *Asian Survey 1981 21(4): 399-416.* Discusses the differences between India and the United States regarding the continued supply of low-enriched uranium fuel to the Tarapur Atomic Power Station under the US-Indian agreement of 1963, which have brought into question the continuance of the accord. Based on government documents and newspapers; 30 notes. M. A. Eide

775. Overton, Jim. NUCLEAR BAILOUT OR GRASSROOT ALTERNATIVE. *Southern Exposure 1979 7(4): 116-120.* Studies the most feasible long-term sources of energy for municipalities. There are several ways in which private utilities have fended off the growth of government-owned or -subsidized power developments, sometimes themselves turning to government-owned power facilities or to available low-interest government funds on whatever grounds. This applies especially in the case of the financing of nuclear reactors and ownership division with municipalities. The local development of alternative energy plans gives coops and municipalities "a sounder defense against the continuing assaults of the private utilities." Photo. R. V. Ritter

776. Perrow, Charles. NORMAL ACCIDENT AT THREE MILE ISLAND. *Society 1981 18(5): 17-26.* Examines the nuclear accident at the Three Mile Island power plant.

777. Perry, Harry. DEVELOPING ALTERNATIVE ENERGY SOURCES. *Current Hist. 1975 69(407): 32-36, 52-53.* Examines possible use of renewable energy resources such as the nuclear breeder reactor, fusion, bituminous sands and oil shale, and solar, geothermal, and tidal energies.

778. Primack, Joel and Von Hippel, Frank. NUCLEAR REACTOR SAFETY. *Bull. of the Atomic Scientists 1974 30(8): 5-12.* Discusses federal policy concerning nuclear reactor safety, concluding to guarantee plant safety, electric energy consumption needs to be slowed, the Atomic Energy Commission needs a strong regulatory agency, and independent studies need to be made.

779. Rasmussen, Norman; Crandell, Michael, interviewer. MEASURING NUCLEAR RISKS. *Center Mag. 1981 14(3): 15-21.* Interviews Norman Rasmussen, Professor of Nuclear Engineering at the Massachusetts Institute of Technology and the director for WASH-1400, the study of the safety of nuclear reactors, discussing how risks associated with the operation of nuclear reactors are quantified and how they compare with the risks of alternative energy sources.

780. Reed, Adam V. WHO CAUSED THREE MILE ISLAND? *Reason 1980 12(4): 16-23.* Examines the chain of events behind the Three Mile Island nuclear plant accident in Pennsylvania and places primary blame on inadequate safety regulations established by the Nuclear Regulatory Commission.

781. Reynolds, William. DEATH TRIPS: TRANSPORTATION OF NUCLEAR WASTE. *Southern Exposure 1979 7(4): 56-60.* The South is the major crossroads for the US transportation network for nuclear materials and fuel. In addition this network also carries the wastes from every nuclear powered facility, government or private. Regulations for such shipments are inadequate, and some are being formulated, both at local and national levels. The Nuclear Regulatory Commission, the Department of Energy, and the Department of Transportation are all deeply involved, but sometimes at cross-purposes. Citizens at local and state levels have sometimes been able to bring pressure successfully to improve safety regulations. Photo, map. R. V. Ritter

782. Reynolds, William. THE SOUTH: GLOBAL DUMPING GROUND. *Southern Exposure 1979 7(4): 49-56.* Surveys former and current facilities for atomic waste disposal in the South. Covers the Oak Ridge National Laboratory, Tennessee; the Savannah River Plant near Aiken, South Carolina; Maxey Flats near Morehead, Kentucky; and the Chem-Nuclear Systems waste site at Barnwell, South Carolina. The operation of these disposal facilities illustrates the many unresolved problems remaining in the disposal of all types of nuclear wastes —problems of a governmental, political, economic, or technical nature. The possible locations of future dumps, largely in the South, can be charted, but precise location, usually in areas of least resistance, remains an unknown. No one seems to be able to answer the basic problems of safety in a convincing manner. Based on government waste management documents and other sources; 6 photos. R. V. Ritter

783. Rhodes, Richard. A DEMONSTRATION AT SHIPPINGPORT. *Am. Heritage 1981 32(4): 66-73.* Traces the development of atomic power from its beginnings in the United States through the establishment of "the world's first full-scale atomic electric plant devoted exclusively to peacetime uses" at Shippingport, Pennsylvania, in 1957. Discusses the roles of government and private industry, of Presidents Truman and Eisenhower, and of Admiral Hyman Rickover. 4 illus. J. F. Paul

784. Rhodes, Suzanne. BARNWELL: ACHILLES HEEL OF NUCLEAR POWER. *Southern Exposure 1979 7(4): 44-48.* Asks what responsibility does the nuclear industry have for its wastes? Where does federal responsibility begin? What role do taxpayers have in policy decisions regarding wastes and energy? The Barnwell Nuclear Fuel Plant of Allied General Nuclear Services in South Carolina illustrates typical inherent problems. Describes its history, the plant operations, its financial difficulties, unresolved feasibility questions, and the federal bailout. Opposition still continues, not only from South Carolina residents because of the recurrent accidents and lack of a waste disposal storage program, but also from the national antinuclear efforts of such organizations as the Palmetto Alliance. 2 photos. R. V. Ritter

785. Richardson, Robert A. THE SELLING OF THE ATOM: HOW OAK RIDGE ASSOCIATED UNIVERSITIES INITIATED A NEW KIND OF INFORMATIONAL ENDEAVOR. *Bull. of the Atomic Scientists 1974 30(8): 28-35.* Discusses workshops designed by Oak Ridge Associated Universities, Inc., a nonprofit corporation of more than 40 colleges and universities with operating contracts from the Atomic Energy Commission, to dispel misconceptions about

atomic energy and to aid in educating citizenry in populated areas adjacent to proposed nuclear plant sites, 1970-74.

786. Rochlin, Gene I.; Held, Margery; Kaplan, Barbara G.; and Kruger, Lewis. WEST VALLEY: REMNANT OF THE AEC. *Bull. of the Atomic Scientists 1978 34(1): 17-26.* Traces the history of nuclear reprocessing plant at West Valley, New York, 1954-78 which is the only nonfederal radioactive wastes management outlet in America.

787. Rycroft, Robert W. U.S. ENERGY DEMAND AND SUPPLY. *Current Hist. 1978 74(435): 100-103, 130-131.* Sees a need to develop alternative energy sources including wind, solar, and atomic power.

788. Schleimer, Joseph D. THE DAY THEY BLEW UP SAN ONOFRE. *Bull. of the Atomic Scientists 1974 30(8): 24-27.* A scenario for sabotaging atomic power plants, including possible sources of nuclear materials, entrance to plant facilities, and results of destruction of the plant for surrounding urban areas; examines the threat of nuclear power in the hands of terrorist groups.

789. Schneider, Steven A. [COMMON SENSE ABOUT ENERGY].

COMMON SENSE ABOUT ENERGY—PART ONE: WHERE HAS ALL THE OIL GONE? *Working Papers for a New Soc. 1978 6(1): 30-42.* Surveys present use of nuclear energy, coal, natural gas, and oil, and speculates on the international reserves of gas and oil as well as the unexplored potential in these areas, 1975-78.

COMMON SENSE ABOUT ENERGY, PART TWO: LESS IS MORE: CONSERVATION AND RENEWABLE ENERGY. *Working Papers for a New Soc. 1978 6(2): 49-59.* Examines the need for research in alternative forms of energy, asserting that though it is obvious that other forms of energy generation must be developed, environmentally sound and economically viable methods will not be necessarily simple to adjust to.

790. Seryogin, I. USSR-USA: SCIENTIFIC AND TECHNICAL COOPERATION. *Int. Affairs [USSR] 1973 (5): 81-83.* Discusses US-USSR cooperation in atomic energy research, ecology, space exploration, medicine, and agriculture in 1972.

791. Shields, Mitchell J. and Brooks, Bill. THE LEAVINGS OF POWER. *Southern Exposure 1979 7(1): 108-114.* Cleanup efforts on the site of the Georgia Nuclear Laboratory in Dawson County, Georgia, began in 1970 and remain an issue in 1979 because radioactive waste was improperly stored and the cleanup was insufficient; the lab was built by Lockheed in the 1950's and closed in 1970 after the site was sold to the city of Atlanta.

792. Simpson, John A. SOME PERSONAL NOTES. *Bull. of the Atomic Scientists 1981 37(1): 26-32.* In 1945, the author was the first chairman of the Atomic Scientists of Chicago, which grew out of his evening seminars on the implications of atomic power; here he considers present problems and ramifications of atomic energy and nuclear arms.

793. Skinner, Scott and Burlingham, Bo. NUCLEAR REACTION. *Working Papers for a New Soc. 1976 3(4): 34-39.* Discusses problems of nuclear reactors.

794. Skogmar, Gunnar. NUCLEAR ENERGY AND DOMINANCE: SOME INTERRELATIONSHIPS BETWEEN MILITARY AND CIVIL ASPECTS OF NUCLEAR ENERGY IN US FOREIGN POLICY SINCE 1945. *Cooperation and Conflict [Norway] 1980 15(4): 217-235.* A basic point of departure is that the military and civil aspects of atomic energy cannot be separated. American nuclear policy is seen as passing through stages with fundamental changes occurring around 1953, in the early sixties, and after the energy crisis in the seventies. The themes most fully discussed are the American deployment of nuclear weapons in Western Europe, the nonproliferation policy, and the control of the nuclear fuel market. One main conclusion is that the dimensions of nonproliferation and dominance/dependence in the energy field are closely intertwined. J/S

795. Spector, Judith A. WALTER MILLER'S *A CANTICLE FOR LEIBOWITZ:* A PARABLE FOR OUR TIME? *Midwest Q. 1981 22(4): 337-345.* A discussion of the relation of science fiction to science, with a reference to the applicability of Walter Miller's *A Canticle for Leibowitz* (1959) to the 1979 Three Mile Island accident. Most people prefer science fiction to science, for science fiction "offers solutions which incorporate religious, ethical, mythical, and imaginative elements." Faith, engendering belief in the seemingly impossible, is an essential component of science fiction and is "central to *A Canticle* and to the anti-nuclear movement." 16 notes, biblio. M. E. Quinlivan

796. Speth, Gus. THE NUCLEAR RECESSION. *Bull. of the Atomic Scientists 1978 34(4): 24-27.* Rising costs in nuclear plant construction and fuel, and the sharp drop in the growth of electricity demand as a result of the Arab oil embargo have caused a drop in the need for atomic power plants, 1973-78.

797. Speth, J. Gustave; Tamplin, Arthur R.; and Cochran, Thomas B. PLUTONIUM RECYCLE: THE FATEFUL STEP. *Bull. of the Atomic Scientists 1974 30(9): 15-22.* Plutonium's dangerous toxicity, basically unknown nature, and improper safeguards for storage and transport make extremely important the Atomic Energy Commission's 1974 decision to allow the commercial use of plutonium in nuclear reactors.

798. Stewart, Larry R. CANADA'S ROLE IN THE INTERNATIONAL URANIUM CARTEL. *Int. Organization 1981 35(4): 657-689.* In early 1972 Canada participated in an international uranium cartel designed to control the world price and supply of uranium. This study focuses on Canada's role in the formation and operation of that cartel, the domestic political reaction when its existence was discovered, and the implications of this for Canadian-American relations. Domestic economic considerations were a major factor that led to a break with traditional Canadian foreign policy. Related to this are the close corporate connections between the Canadian and American uranian industry and the enormous impact of American domestic policies on Canada. The uranium case also offers support to the theory that transnational relations and other multinational processes threaten democratic control of foreign policy. J

799. Stockman, David A. THE WRONG WAR? THE CASE AGAINST A NATIONAL ENERGY POLICY. *Public Interest 1978 (53): 3-44.* The three axioms that underlie Jimmy's Carter's National Energy Plan lack validity: 1) that the exhaustion of conventional fossil fuels is imminent and that "business as usual" in the marketplace is inadequate to cope with the problem, 2) that the world energy marketplace was irremediably impaired by the events of October 1973, and now is dependent on a politically motivated cartel that represents a major national economic and security peril, and 3) that homegrown energy is better for the domestic economy. The creation of a giant bureaucracy to regulate in an area where the supposed problems cannot be validated is a major mistake.
R. V. Ritter

800. Swain, Bruce M. *THE PROGRESSIVE,* THE BOMB AND THE PAPERS. *Journalism Monographs 1982 (76): 1-45.* Describes how the *Progressive* magazine article "How a Hydrogen Bomb Works" not only sparked a court case, *US* v. *The Progressive* (US, 1979), but created an uproar in the press that forced the government to drop its effort to block publication under circumstances that showed that publication of the article did not divulge "secrets" not already available to the public; the *Progressive* case raised the question of the prohibition of publication of secrets relating to the construction of nuclear arms, even in a society affording freedom of the press.

801. Sweet, William. UNRESOLVED: THE FRONT END OF NUCLEAR WASTE DISPOSAL. *Bull. of the Atomic Scientists 1979 35(5): 44-48.* Surveys work by the Nuclear Regulatory Commission and associated interest groups in forcing federal action on proper atomic waste disposal, especially the Uranium Mill Tailings Radiation Control Act (US, 1978).

802. Sylves, Richard T. CARTER'S NUCLEAR LICENSING REFORM VS. THREE MILE ISLAND. *Publius 1980 10(1): 69-79.* The March 1979 nuclear accident at the Three Mile Island power plant of the Pennsylvania Public Utility Commission prevented passage of the Nuclear Siting and Licensing Act, which would have increased state participation in locating and approving construction of nuclear power plants. The bill was sponsored in part to reduce the lead-time in Nuclear Regulatory Commission (NRC) processes of approval from the current 10-12 years. Environmental, nuclear utility, and state government responses to the accident all tend to prolong such licensing and increase separate federal and state regulatory activities, thus increasing potential costs of nuclear power and reducing its potential as a power alternative. Based on US government documents and press reports; fig., 40 notes.
C. B. Schulz

803. Temples, James R. THE POLITICS OF NUCLEAR POWER: A SUBGOVERNMENT IN TRANSITION. *Pol. Sci. Q. 1980 95(2): 239-260.* Analyzes the federal government's role in the development of the nuclear power industry and the licensing of commercial nuclear power plants, and finds evidence of a gradual long-term shift from a "distributive" to a "regulatory" posture toward the nuclear industry.
J

804. Tiezzi, Enzo. RISCHI BIOLOGICI E COSTI SOCIALI DELLA SCELTA NUCLEARE: *SHUT DOWN:* UN PROCESSO PER ASSASSINIO [Biological risks and social costs of the nuclear choice. *Shut Down:* a trial for

murder]. *Ponte [Italy] 1980 36(5): 426-436.* Reviews *Shut Down: Nuclear Power on Trial,* Albert Bates, ed. (1979), which is a dossier on the risks to human life represented by the nuclear plants and experiments connected with them.

805. Ulsamer, Edgar. THE US CAN'T TURN BACK THE NUCLEAR CLOCK. *Air Force Mag. 1977 60(5): 33-39.* Discusses methods of nuclear energy and technology, including current reactor systems and breeder reactors, and their effect on the uranium supply; examines present treaties and controls on proliferation and testing as well as extant treaties, 1946-77.

806. Vig, Norman J. and Bruer, Patrick J. THE COURTS AND RISK ASSESSMENT. *Policy Studies Rev. 1982 1(4): 716-727.* Comments on judicial review of technical regulations; outlines the major choices facing judges; details the reaction of the Supreme Court in recent cases involving nuclear waste disposal and occupational health protection, which illustrates the current schisms within the judiciary.

807. von Hippel, Frank. LOOKING BACK ON THE RASMUSSEN REPORT. *Bull. of the Atomic Scientists 1977 33(2): 42-47.* The 2,400-page report by the Nuclear Regulatory Commission (NRC) has assembled a great deal of useful material. However, the 12-page summary's conclusions are deceptive. It has failed to put the risks associated with nuclear energy into perspective. Therefore, the report's usefulness in policymaking is limited. Based on a study of the Nuclear Regulatory Commission's report, Reactor Safety Study, published government documents and secondary works; 2 tables, fig., 9 notes.
D. J. Trickey

808. Walker, J. Samuel. NUCLEAR SAFETY, THE ATOMIC ENERGY COMMISSION, AND THE STATES. *Wisconsin Mag. of Hist. 1982 65(3): 158-175.* Briefly discusses the background and provisions of the Atomic Energy Act (US, 1954) and then examines in detail the evolution of support for a more active role for states in atomic regulation that resulted in the 1959 amendment. Seeks to explain the basis for the states' claims, their limited success, and the ambiguity of the amendment in order to shed light on the continuing debate of the 1970's and 80's. 43 notes, 4 illus. N. C. Burckel

809. Walker, J. Samuel. THE SOUTH AND NUCLEAR ENERGY, 1954-62. *Prologue 1981 13(3): 175-191.* Discusses the South's interest in and activities directed toward harnessing atomic energy in terms of the South's two primary postwar concerns: industrial growth and protection of states' rights. The South reacted quickly to establish the administrative mechanism to promote the use of nuclear technology. The Southern Governors Conference in its October 1955 meeting was asked to study and make recommendations on the appropriate regional action. The governors moved to coordinate the growth of nuclear technology by establishing a Regional Advisory Council on Nuclear Energy in February 1957. The South in its concern for states' rights fought for and won recognition for independent state functions in nuclear regulatory matters. Based on the Joint Committee on Atomic Energy File, Records of the Joint Committee of Congress; 10 photos, 37 notes. M. A. Kascus

810. Walsh, Edward J. RESOURCE MOBILIZATION AND CITIZEN PROTEST IN COMMUNITIES AROUND THREE MILE ISLAND. *Social Problems 1981 29(1): 1-21.* Based on data gathered from the growth and development of grassroots citizen protest organizations after the Three Mile Island nuclear accident in 1979, assesses and refines existing social movements theories, focusing on the importance of grievances in the emergence and continuation of organized protest.

811. Weinberg, Alvin M. SALVAGING THE ATOMIC AGE. *Wilson Q. 1979 3(3): 88-112.* Reviews the history of atomic power and suggests necessary changes.

812. Wenner, Lettie McSpadden and Wenner, Manfred W. NUCLEAR POLICY AND PUBLIC PARTICIPATION. *Am. Behavioral Scientist 1978 22(2): 277-310.* Describes the historical and political context of the 1976 atomic power plants moratoriuim initiatives in California, Arizona, Ohio, Colorado, Montana, Oregon, and Washington; and analyzes the voting patterns, reporting on the effects of demography and situational variables on election results.

813. Wheeler, Keith. [THE NATIONAL BUREAU OF STANDARDS].

THE VERY INQUISITIVE AGENCY THAT TAKES AMERICA'S MEASURE. *Smithsonian 1978 9(6): 42-51.* Discusses the National Bureau of Standards, in the Commerce Department, from 1901-78. Located in Gaithersburg, Maryland, and near Boulder, Colorado, over 3,000 scientists, engineers, and technicians work at researching almost any problem that needs a solution. Conceived basically to take custody of standards by which every quantity and quality applicable to science, engineering, manufacturing, and commerce could be measured with uniformity, it has in addition evolved into an arm of industry, and friend to the consumer. 12 illus.

NEUTRON BEAMS, A THINKING ROBOT AND ULTRASOUND. *Smithsonian 1978 9(7): 88-96.* Mentions projects including standards for cancer therapy, nuclear power plant design, ultrasonic devices for medical evaluations, defining and tracing environmental contaminants, standards for computers, upgrading automation, and MHD electric power generation.

S. R. Quéripel

814. Wilkes, Owen and Mann, Robert. THE STORY OF NUKEY POO. *Bull. of the Atomic Scientists 1978 34(8): 32-36.* Examines the environmental and monetary costs, particularly radioactive waste disposal, incurred by the Navy in decommissioning its nuclear reactor located at McMurdo Sound, Antarctica, 1960-71, which failed to supply adequate electricity for the base there.

815. Willrich, Mason and Marston, Philip M. PROSPECTS FOR A URANIUM CARTEL. *Orbis 1975 19(1): 166-184.* Discusses the possibilities for a uranium cartel among countries with nuclear power industries in the 1970's, emphasizing the implications for US trade and foreign policy.

816. Willrich, Mason. TERRORISTS KEEP OUT! *Bull. of the Atomic Scientists 1975 31(5): 12-16.* Assesses possible nuclear materials theft and recommends increased safeguards to prevent and detect theft, adjacent location of nuclear sites and support facilities to avoid nuclear material transport, "spiking" of nuclear waste to endanger any improper handling of materials, and international cooperation on nuclear safeguard policies.

817. Wilson, Carroll L. NUCLEAR ENERGY: WHAT WENT WRONG. *Bull. of the Atomic Scientists 1979 35(6): 13-17.* Reviews atomic reactors development, 1947-78, offering suggestions on future plant construction, environmental responsibility, and security in transfer of plutonium and radioactive waste, to assuage public fears and introduce responsibility into the atomic power industry.

818. Winters, Francis X. DES REVOLUTIONNAIRES MALGRE EUX. LES EVEQUES AMERICAINS S'OPPOSENT A LA COURSE AUX ARMEMENTS [Reluctant revolutionaries: American bishops oppose the arms race]. *Etudes [France] 1982 357(1): 5-19.* Describes the vociferous, extensive moral condemnation of the nuclear arms race by the hierarchy of the Catholic Church in America, to illustrate the church's increasing sense of civic responsibility in the face of crucial public issues; 1979-81. French.

819. Worthley, John A. and Torkelson, Richard. MANAGING THE TOXIC WASTE PROBLEM: LESSONS FROM THE LOVE CANAL. *Administration & Soc. 1981 13(2): 145-160.* Presents findings on the nature of the problem of toxic waste disposal drawn from a study of the first major toxic waste disaster at Love Canal, New York, and also identifies "legislative, organizational, and fiscal dimensions that require policy research and development" in order to generate the collaborative effort between public- and private-sector interests necessary to deal with the challenge.

820. Yanarella, Ernest. THE POLITICS OF THE "PEACEFUL ATOM." *Peace and Change 1981 7(3): 45-58.* Examines the history and politics of the civilian use of atomic power in the United States from the Manhattan Project antecedents in 1939 to Three Mile Island in 1979, with special attention to the organization of the Atomic Energy Commission and the Joint Committee on Atomic Energy.

821. Zinberg, Dorothy. THE PUBLIC AND NUCLEAR WASTE MANAGEMENT. *Bull. of the Atomic Scientists 1979 35(1): 34-39.* Surveys current controversy over atomic waste management, measures confidence expressed in public opinion polls, and discusses uranium and plutonium theft as they affect public policy formation, 1970's.

822. —. [NUKES, OIL, AND POLITICS]. *Dissent 1979 26(3): 271-286.*
Bromwich, David. A HIGH STANDARD OF DYING, *pp. 271-274.*
Connolly, Peter. IDEOLOGY AND THE POLITICS OF ENERGY, *pp. 274-278.*
Harrington, Michael. NUCLEAR POWER AND CORPORATE PRIORITIES, *pp. 278-280.*
Orr, David W. PERSPECTIVES ON ENERGY, *pp. 280-284.*
Howe, Irving. NUCLEAR POWER: *HOW* DO WE KNOW?, *pp. 285-286.*
The incident at Three Mile Island, Pennsylvania, 28 March 1979, throws into relief many of the dangers in atomic energy, US dependence on enormous quantities of energy from a dwindling supply of increasingly expensive oil and incompletely tamed nuclear energy, the vested interests of many entrepreneurs, business and labor leaders, and politicians in "conventional" energy sources, and the growing influence of interest groups who deplore large-scale nuclear and petroleum power industries and advocate alternative, small-scale energy systems. C. Moody

823. —. [SOME SOCIAL AND POLITICAL DIMENSIONS OF NUCLEAR POWER]. *Am. Pol. Sci. Rev. 1981 75(1): 132-145.*

Nelkin, Dorothy. SOME SOCIAL AND POLITICAL DIMENSIONS OF NUCLEAR POWER: EXAMPLES FROM THREE MILE ISLAND, *pp. 132-142.* Draws examples from the Three Mile Island accident to review several characteristics of nuclear technology, its scale and costs, its complexity, its uncertain and unpredictable physical effects, and its indirect risks. Explores the implications for social, political and administrative institutions as they grope for ways to manage the risks of nuclear power in the context of critical public scrutiny.

Sills, David L. A COMMENT ON DOROTHY NELKIN'S "SOME SOCIAL AND POLITICAL DIMENSIONS OF NUCLEAR POWER: EXAMPLES FROM THREE MILE ISLAND," *pp. 143-145.* Agrees with Nelkin's conclusions and views her article as the climax of a concern for the social science aspects of atomic energy and as "the harbinger of a time when these aspects will constitute the core . . . of nuclear energy research." J/S

824. —. [SPECIAL REPORT (THREE MILE ISLAND). PART I]. *Bull. of the Atomic Scientists 1980 36(1): 17-31.*

—. NUCLEAR POWER IN 1980, *p. 17.* Introduces the following section of articles.

Gilinsky, Victor. THE IMPACT OF THREE MILE ISLAND, *pp. 18-20.* Outlines the history of atomic energy development in the United States since the 1950's, focusing on events that led to the Three Mile Island accident in Pennsylvania; safety will be more important than ever before.

Lanouette, William J. THE KEMENY COMMISSION REPORT, *pp. 20-24.* Discusses the Kemeny Report on the accident at Three Mile Island, pointing out a variety of interpretations; summarizes the most important recommendations for improved safety, 1979.

—. EXCERPTS FROM: REPORT OF THE PRESIDENT'S COMMISSION ON THE ACCIDENT AT THREE MILE ISLAND, *pp. 24-31.* Excerpts sections of the Kemeny Report on the accident at Three Mile Island in Pennsylvania, 1979, on subjects such as causes and the right to information.

SUBJECT INDEX

Subject Profile Index (ABC-SPIndex) carries both generic and specific index terms. Begin a search at the general term but also look under more specific or related terms.

Each string of index descriptors is intended to present a profile of a given article; however, no particular relationship between any two terms in the profile is implied. Terms within the profile are listed alphabetically after the leading term. The variety of punctuation and capitalization reflects production methods and has no intrinsic meaning; e.g., there is no difference in meaning between "History, study of" and "History (study of)."

Cities, towns, and counties are listed following their respective states or provinces; e.g., "Ohio (Columbus)." Terms beginning with an arabic numeral are listed after the letter Z. The chronology of the bibliographic entry follows the subject index descriptors. In the chronology, "c" stands for "century"; e.g., "19c" means "19th century."

Note that "United States" is not used as a leading index term; if no country is mentioned, the index entry refers to the United States alone. When an entry refers to both Canada and the United States, both "Canada" and "USA" appear in the string of index descriptors, but "USA" is not a leading term. When an entry refers to any other country and the United States, only the other country is indexed.

The last number in the index string, in italics, refers to the bibliographic entry number.

A

Abelson, Philip H. (reminiscences). California, University of, Berkeley. Lawrence Radiation Laboratory. Nuclear physics. 1935-40. *1*

ABM. Defense Policy. Strategic Arms Limitation Treaty. USSR. 1982. *610*

—. Disarmament. ICBM's. MIRV. Strategic Arms Limitation Talks. 1969-73. *100*

ABM controversy. Arms control. Congress. 1969-75. *685*

—. Defense Policy. Military Strategy. 1965-80. *114*

ABM Treaty. Arms control. Vladivostok Accord. 1974. *654*

Accidents. Diaries. Nuclear power plants. Pennsylvania (Three Mile Island). 1979. *700*

—. Federal Regulation. Nuclear power plants. Nuclear Regulatory Commission. Pennsylvania (Three Mile Island). Safety. 1979. *780*

—. Kemeny Report. Nuclear power plants. Pennsylvania (Three Mile Island). 1950-79. *824*

Accountability. Nuclear waste. Safety. Washington (Hanford). 1970's. *699*

Acquisition Cost Evaluation project. Air Forces. Brown, George S. (interview). Inflation. Military Finance. 1970's. *181*

Action-reaction theory. Arms race. International Relations (discipline). Models. USSR. 1970's. *331*

AEC. *See* Atomic Energy Commission.

Aeronautics, Military. Bombers, long-range. Strategic Arms Limitation Talks. USSR. 1950-77. *651*

Aerospace systems. Defense Policy. Military Strategy. 1977. *128*

Afghanistan. Arms race. Cold War. Foreign Relations. USSR. 1960-80. *355*

—. Foreign Policy. 1979-80. *367*

Air Force Association (symposium). Balance of Power. Military Strategy. National Security. 1964-76. *407*

—. Military Strategy. Strategic Arms Limitation Talks (II). Weapons. 1976. *670*

Air Forces. Acquisition Cost Evaluation project. Brown, George S. (interview). Inflation. Military Finance. 1970's. *181*

—. Armament Development and Test Center. Eglin Air Force Base. Florida. 1960's-70's. *179*

—. Defense Policy. Military Strategy. North American Air Defense Command. 1970's. *415*

—. Lemay, Curtis. Military Capability. Nuclear Arms. Strategic Air Command. 1945-48. *88*

—. Military Capability. Weapons. 1970's. *183*

—. Missiles. Space Vehicles. 1954-74. *118*

Air Forces (Air Targets Division). Bureaucracies. Military intelligence. Morality. Nash, Henry T. (account). Nuclear War. Planning. 1950's-70's. *147*

Air Warfare. Armaments. Cruise Missiles. 1914-80. *185*

—. Atomic bomb. *Enola Gay* (aircraft). Wendover Field. World War II (transportation). 1944-45. *60*

Aircraft carriers. Naval Strategy. 1945-74. *177*

Airplanes, Military. B-1 (aircraft). Lobbyists. Rockwell International Corp. Unemployment. 1975. *149*

—. B-47 (aircraft). Peck, Earl G. (account). Strategic Air Command Wing. ca 1960. *153*

—. B-52 bomber. Technology. Vietnam. 1952-75. *124*

Airplanes, Military (bombers). B-58 (aircraft). 1946-79. *119*

Allied General Nuclear Services. Barnwell Nuclear Fuel Plant. Nuclear power plants. South Carolina. ca 1963-79. *784*

Allies. Arms control. Balance of power. Europe. USSR. 1979. *598*

Allison, Graham *(Essence of Decision)*. Committees. Decisionmaking. Foreign policy. Nuclear War. World War II. 1944-45. *54*

Alperovitz, Gar. Diplomacy. Historiography. Nuclear Arms. World War II. 1943-46. 1965-73. *323*

—. Foreign Relations. Nuclear War. World War II (review article). 1945. *27*

Alperovitz, Gar (review article). Decisionmaking. Historiography. Japan (Hiroshima). Nuclear War. World War II. 1945-65. *52*
Alvarez, Luis W. (reminiscences). California, University of, Berkeley. Lawrence Radiation Laboratory. Nuclear Physics. 1934. *4*
American Aeronautics and Astronautics Associations. ICBM's. USSR. 1970's. *406*
American Atomics Corp. Arizona (Tucson). Manufactures. Public Health. Radiation. Tritium. Working Conditions. 1979. *709*
Anderson, Herbert L. (reminiscences). Chicago, University of. Columbia University. Fermi, Enrico. Nuclear Science and Technology. Szilard, Leo. 1933-45. *6*
—. Chicago, University of. Illinois. Nuclear chain reaction. World War II. 1942. *5*
Angola. Foreign policy. Mozambique. Nuclear potential. South Africa. USA. 1970-76. *427*
Annexation. Intergovernmental Relations. New Mexico (Albuquerque). Nuclear Science and Technology. Weapons. 1940-78. *157*
Antarctic (McMurdo Sound). Navies. Nuclear reactors. Radioactive waste. 1960-71. *814*
Antiballistic Missiles. *See* ABM.
Anti-Communist Movements. *Dr. Strangelove* (film). Films. Films. Humor. Kubrick, Stanley. Liberalism. Nuclear Arms. 1950-64. *599*
Anti-nuclear power movement. Clamshell Alliance. Leftism. Political protest. 1970's. *740*
—. Community movements. Political activism. 1970's. *747*
—. Nuclear Arms. Peace Movements. 1960-80. *769*
Antisatellite weapons. Armaments. Space. USSR. 1960-80. *145*
Antiwar Sentiment. Cruise missiles. Great Britain. Neutron Bomb. 1978-81. *441*
Arizona. Grand Canyon. Mines. Uranium. 1540-1969. *156*
Arizona (Tucson). American Atomics Corp. Manufactures. Public Health. Radiation. Tritium. Working Conditions. 1979. *709*
Armament Development and Test Center. Air Forces. Eglin Air Force Base. Florida. 1960's-70's. *179*
Armaments. Air Warfare. Cruise Missiles. 1914-80. *185*
—. Antisatellite weapons. Space. USSR. 1960-80. *145*
—. Defense Policy. USSR. 1960's-70's. *396*
—. Disarmament. Nuclear proliferation. Wars, significance of. 1945-75. *205*
—. Documents. Federal Policy. Nuclear nonproliferation. 1981. *692*
—. Electric Power. Nuclear Science and Technology. 1950-75. *221*
—. Military strategy. 1945-78. *335*
Armaments industry. 1970's. *310*
Armies. Defense Policy. Nuclear Arms (tactical). Politics. Rose, John P. (review article). 1950-81. *239*
—. Foreign Relations. NATO. 1950-80. *365*
—. Military strategy. 1970's. *284*
Arms control. ABM controversy. Congress. 1969-75. *685*
—. ABM Treaty. Vladivostok Accord. 1974. *654*
—. Allies. Balance of power. Europe. USSR. 1979. *598*
—. Austria (Vienna). Detente. Mutual Balanced Force Reductions. 1973-80. *522*
—. Balance of Power. Technology. USSR. 1960-80. *530*
—. Balance of Power. Technology. Weapons. 1970's. *135*
—. Brezhnev, Leonid. Ford, Gerald R. Foreign Relations. Kissinger, Henry A. USSR. Vladivostok Accord. 1974. *674*
—. Canada. China. India. Nuclear Arms. Nuclear power. Regionalism. USA. 1951-78. *387*
—. Competition. Foreign Policy. Institutions. USSR. 1970-80. *585*
—. Defense policy. 1973. *547*
—. Defense policy. Foreign Relations. USSR. 1960-80. *584*
—. Defense Policy. National security. 1960's-75. *636*
—. Detente. Foreign relations. USA. USSR. 1961-75. *574*
—. Detente. Mutual Balanced Force Reductions. USA. USSR. 1974. *634*
—. Deterrence. Foreign Relations. International organizations. 1960-75. *594*
—. Developing nations. Foreign Policy. Nuclear arms. USA. 1975. *468*
—. Developing nations. International organization. Nuclear capability. USA. 1950-75. *501*
—. Disarmament. Foreign Policy. Military strategy. USSR. 1945-70's. *214*
—. Disarmament. Foreign Relations. USA. USSR. 1970's. *664*
—. Disarmament. Seabed talks. USSR. 1967-70. *633*
—. Disarmament. Treaties. USSR. 1977. *570*
—. Disarmament. USA. USSR. 1974. *649*
—. Domestic Policy. USA. USSR. Weapons. 1963-75. *462*
—. Environment. Industrialization. International security. Multinational corporations. 1975. *650*
—. Europe. USA. 1958-73. *497*
—. Executive Branch. Public opinion. Treaties. USSR. 1972-82. *517*
—. Foreign Policy. Military Strategy. Nuclear arms. Strategic Arms Limitation Talks. 1960's-78. *596*
—. Foreign Policy. Nuclear energy. 1945-80. *292*
—. Foreign Policy. USSR. 1979-82. *622*
—. Foreign Relations. Great Powers. 1960's-72. *508*
—. International Security. World government. 1975. *524*
—. Mechanism (philosophy). Military strategy. Social Conditions. 1950's-60's. *661*
—. Military Intelligence. USSR. 1960-82. *518*
—. Mutual Balanced Force Reductions. NATO. Warsaw Pact. 1973-74. *635*
—. National Security (review article). Stockholm International Peace Research Institute. 1979. *604*
—. Political change. USA. USSR. 1945-73. *603*
—. Preparedness. 1970's. *546*
—. Strategic Arms Limitation Talks (II). USSR. 1970's. *475*
—. Technical innovation. USA. USSR. 1945-73. *540*
—. Technology, military. 1970's. *646*
—. USA. USSR. World order. 1950's-75. *483*
Arms control agreements. China. Detente. Nuclear strategy. USA. USSR. 1970's. *401*
Arms Control and Disarmament Agency. Carter, Jimmy (administration). Civil defense. Evacuations. Nuclear War. Public opinion. 1978. *95*
—. Clarke, Duncan L. (review article). 1961-80. *615*
—. Disarmament. Federal Government. General Advisory Committee on Arms Control and Disarmament. 1961-73. *507*
—. Foreign policy. Reagan, Ronald (administration). 1960-81. *505*

—. Military officers, role of. 1974. *506*
—. Nuclear War (threat of). 1960's-70's. *504*
Arms control (definitions). Disarmament. 18c-20c. *489*
Arms control (myths of). International Security. Strategic balance. USA. USSR. 1975. *286*
Arms control objectives. Defense budget. 1975. *481*
Arms control proposals. Decisionmaking. 1975. *641*
Arms control (qualitative). International Security. Strategic Arms Limitation Talks. 1974-75. *514*
Arms control (term). Disarmament (term). 1970-82. *488*
Arms race *See also* Nuclear Arms Race.
—. Action-reaction theory. International Relations (discipline). Models. USSR. 1970's. *331*
—. Afghanistan. Cold War. Foreign Relations. USSR. 1960-80. *355*
—. Balance of Power. Deterrence. Foreign Relations. USSR. 1970's. *312*
—. Beam energy. Grechko, Andrei. ICBM's. USSR. 1960's-70's. *414*
—. Bishops. Catholic Church. 1979-81. *818*
—. Carter, Jimmy (administration). Cruise missiles. Tomahawk (missile). World War II. 1977-80. *160*
—. Cold War. Economic Conditions. International Trade. Truman, Harry S. (administration). 1948-49. *347*
—. Detente. Disarmament. NATO. 1950's-76. *605*
—. Detente. Nuclear strategy. USA. USSR. 1972-74. *319*
—. Diplomacy. Negotiations. Nuclear test ban. Richardson models. USSR. 1962-63. *558*
—. Disarmament. Political Theory. 1946-79. *317*
—. Disarmament. USA. USSR. 1945-70's. *277*
—. Federal Policy. Germany, West. Nuclear arms. Power resources. 1968-70's. *572*
—. Foreign Relations. Great Powers. USSR. 1945-78. *660*
—. Foreign relations. Technologies. USA. USSR. 1973. *343*
—. International relations theory. Models. USSR. 1948-70. *289*
—. Management. Military. Scientific Experiments and Research. Technology. 1945-75. *190*
—. Military Capability. Nuclear Arms. USSR. 1960-80. *404*
—. Military potential. USA. USSR. 1940-74. *445*
—. Missiles. Navies. Nuclear submarines. Trident (missile). 1960-80. *158*
—. Nuclear Arms. Research. 1946-74. *267*
—. Nuclear Arms. Technology. Values, human. 1973. *528*
—. Strategic Arms Limitation Talks (II). 1816-1965. *426*
—. USSR. 1960's-70's. *349*
—. War. 1816-1965. *425*
Arms race model. USA. USSR. 1945-73. *315*
Arms race, qualitative. International security. Military innovation. Nuclear test ban. Strategic Arms Limitation Treaty (1972). 1972-75. *220*
Arms races, rationales for. International security. 1974. *273*
Arms Trade. Europe, Western. Middle East. Nuclear equipment sales. USA. 1973-75. *306*
—. UN. 1945-77. *595*
Army Corps of Engineers. Journalism. Nuclear energy. Oak Ridge *Journal*. Tennessee. 1943-49. *2*
Atomic bomb. Air Warfare. *Enola Gay* (aircraft). Wendover Field. World War II (transportation). 1944-45. *60*
—. Franck, James. Nuclear arms race. Political Attitudes. Truman, Harry S. (administration). World War II. 1944-45. *62*
—. Franck, James. Policymaking. Scientists. Stimson, Henry L. World War II. 1945. *58*
—. New Mexico (Los Alamos). World War II. 1943-46. *41*
Atomic Bomb Casualty Commission. Civilian survivors. Japan (Hiroshima, Nagasaki). Medical Research. Radiation Effects Research Foundation. 1945-75. *63*
Atomic bomb development. Diplomacy. Nuclear Science and Technology. World War II. 1942-45. *46*
Atomic bombs. Japan. Public opinion. USA. World War II. 1945-49. *68*
Atomic energy. *See* Nuclear energy.
Atomic Energy Commission. Decisionmaking. Energy. Federal policy. Military. Pressure Groups. 1974. *720*
—. Defense Department. Historical advisory committees. State Department. 1947-74. *754*
—. Deterrence. Gillet, Edward B. (interview). Military Strategy. Nuclear Arms (warheads). Research. 1970's. *411*
—. Engineers. Manhattan Project. Nuclear technology. Scientists. 1938-50. *28*
—. Environment. Nuclear power. Public participation. Technology assessment. 1946-73. *742*
—. Espionage. Federal Bureau of Investigation. Greenglass, David. Nuclear Arms. Rosenberg case. Trials. 1942-74. *142*
—. Executive Order 10450. Loyalty. Oppenheimer, J. Robert. Strauss, Lewis. 1946-54. *117*
—. Federal regulation. Politics. Public Policy. Science and Government. 1946-70. *706*
—. Federal Regulation. Public Relations. Safety. 1954-56. *763*
—. Greenglass, David. Nuclear Science and Technology. Rosenberg case. Secrecy. Trials. 1945-51. *75*
—. Hijacking. Nuclear Arms. 1970-74. *737*
—. Joint Committee on Atomic Energy. Politics. 1939-79. *820*
—. Law. Security classification. 1945-75. *729*
—. Legislation. Safety. State Government. 1954-70. *808*
—. Marshall Islands. Nuclear Science and Technology. Radiation exposure. 1942-72. *81*
—. New York (West Valley). Radioactive waste. 1954-78. *786*
—. Nuclear power plants. Nuclear Regulatory Commission. Pennsylvania (Three Mile Island). 1979. *750*
—. Security classification. 1944-75. *733*
Atomic power. *See* Nuclear power.
Atomic Scientists of Chicago. Nuclear arms. Nuclear energy. Simpson, John A. (views). 1945-80. *792*
Atomic structure, theory of. Magnets, floating. Mayer, Alfred Marshall. Physics. Scientific Experiments and Research. 1878-1907. *57*
Attitudes. Canada. Nuclear power. USA. 1960's-70's. *717*
—. *China Syndrome* (film). Nuclear Energy. Pennsylvania (Three Mile Island). 1978-79. *764*
—. Cruise missiles. Defense Department. Politics and the Military. Weapons. 1964-76. *140*
—. Decisionmaking. Nuclear power plants. State Government. Voting and Voting Behavior. Western States. 1963-78. *703*

—. Democracy. Europe. Interest Groups. Japan. Nuclear Science and Technology. USSR. 1970's. *739*
—. Economic development. Military Strategy. Nuclear arms. Technology. War. 1914-68. *351*
—. Economics. Foreign relations. Nuclear arms limitation. USA. USSR. 1972-73. *465*
—. Exports. France. Great Britain. Nuclear proliferation. Plutonium, reprocessed. 1946-78. *726*
—. Fallout. Nuclear arms. Ozone layer, destruction. 1945-75. *126*
—. Generations, political. Nuclear war. 1974. *295*
—. Hydrogen bomb. Scientists. 1950-55. *105*
—. Institute of Defense Analysis (Jason group). Scientific Experiments and Research. Vietnam War. 1966. *223*
—. Science and society. 19c-20c. *762*
Austria. Foreign Relations. Nuclear Test Ban Treaty. USSR. 1955. 1963. *581*
Austria (Vienna). Arms control. Detente. Mutual Balanced Force Reductions. 1973-80. *522*

B

Bainbridge, Kenneth T. (reminiscences). Manhattan Project. New Mexico (Los Alamos). Nuclear Arms. World War II. 1945. *8*
Balance of Power. Air Force Association (symposium). Military Strategy. National Security. 1964-76. *407*
—. Allies. Arms control. Europe. USSR. 1979. *598*
—. Arms control. Technology. USSR. 1960-80. *530*
—. Arms control. Technology. Weapons. 1970's. *135*
—. Arms race. Deterrence. Foreign Relations. USSR. 1970's. *312*
—. China. Detente. Foreign policy. USSR. 1945-79. *218*
—. Communism. Foreign Policy. USSR. 1965-78. *336*
—. Cruise missiles. Defense Department. Military Capability. Strategic Arms Limitation Talks (II). ca 1950-79. *125*
—. Cuban Missile Crisis. Kennedy, John F. Military Strategy. USA. USSR. 1962. *236*
—. Defense Policy. Nuclear Arms. Schlesinger, James R. USSR. 1960's-70's. *380*
—. Defense Policy. Nuclear Arms. USA. USSR. 1974-75. *241*
—. Defense policy. Sweden. USSR (Kola Peninsula). 1945-74. *322*
—. Defense spending. Manpower. Military Capability. 1975-76. *452*
—. Developing Nations. Foreign policy. USSR. 1961-79. *556*
—. Diplomacy. Nuclear arms. Politics. 1945-79. *217*
—. Energy development. Foreign policy. Pacific Basin. USA. 1973-77. *736*
—. Europe. Strategic Arms Limitation Talks (II). USSR. 1972-78. *542*
—. Foreign Relations. USSR. 1960-69. *363*
—. France. Great Britain. Newspapers. USSR. 1948-73. *262*
—. ICBM's. USSR. 1972-75. *453*
—. Metaphysics. Nuclear Arms. USSR. 1970's. *222*
—. Meyer, John C. (interview). Military Strategy. USSR. 1970's. *412*
—. Military. 1953-75. *450*
—. Military. NATO. Warsaw Pact. 1962-75. *449*
—. Military Capability. USSR. 1980. *238*
—. Military Ground Forces. Nuclear Arms. USSR. 1969-77. *193*
—. Strategic Arms Limitation Talks. USA. USSR. 1972. *554*
Baldwin, Paul H. (account). Fighter squadrons, night. Nuclear War. Philippines. World War II (aerial operations). 1942-45. *9*
Barnwell Nuclear Fuel Plant. Allied General Nuclear Services. Nuclear power plants. South Carolina. ca 1963-79. *784*
Baruch, Bernard. Cold War. USSR. 1946. *110*
Baruch Plan. Foreign Relations. Nuclear Energy. Nuclear proliferation. UN. 1945-77. *712*
Bates, Albert (review article). Nuclear Energy. 1970-79. *804*
Battles. Deterrence. Morale. Nuclear War. World War II. 1944-81. *130*
Beam energy. Arms race. Grechko, Andrei. ICBM's. USSR. 1960's-70's. *414*
The Beginning or The End (film). Films. Manhattan Project. Nuclear Arms. Public Opinion. Scientists. 1947. *69*
Berlin crisis. Civil defense. Fallout shelters. 1961-62. *129*
Bibliographies. National security. 1960's-70's. *274*
Bikini Atoll. Hydrogen bomb. Japanese. *Lucky Dragon* (vessel). Pacific Dependencies (US). 1954-75. *164*
—. Navies. Nuclear testing. Operation Crossroads. 1946. *116*
Bishops. Arms race. Catholic Church. 1979-81. *818*
Bismarck, Otto von. Cuban missile crisis. Diplomacy. Escalation theory. Franco-Prussian War. Kennedy, John F. McNamara, Robert S. 1870. 1962. *391*
Bloch, Felix. Conseil Européen pour la Recherche Nucléaire. Nuclear Science and Technology. 1954-55. *748*
Bohr, Niels. Nuclear energy. World War II. 1943. *51*
Bombers, long-range. Aeronautics, Military. Strategic Arms Limitation Talks. USSR. 1950-77. *651*
Bombing. Japan. World War II. 1937-45. *18*
Boston Study Group (report). Brookings Institution (report). Budgets. Defense spending. 1979. *321*
Boulding, Kenneth (review article). Peace research. Sociopolitical theory. 1974-75. 1977. *473*
Brazil. Diplomacy. Germany. NATO. Nuclear nonproliferation. 1968-77. *571*
—. Germany, West. Nuclear technology, transfer of. USA. 1945-76. *257*
Brezhnev, Leonid. Arms control. Ford, Gerald R. Foreign Relations. Kissinger, Henry A. USSR. Vladivostok Accord. 1974. *674*
—. Detente. Ford, Gerald R. Vladivostok Accord. 1974. *617*
Brodie, Bernard. Kahn, Herman. Military Strategy. National security. Schelling, Thomas. 1950's-76. *376*
—. Kahn, Herman. Military Strategy. Nuclear Arms. 1950's-82. *358*
Brookings Institution (report). Boston Study Group (report). Budgets. Defense spending. 1979. *321*
Brown, George S. (interview). Acquisition Cost Evaluation project. Air Forces. Inflation. Military Finance. 1970's. *181*
Brown, George S. (views). Joint Chiefs of Staff. Military Strategy. Nuclear Arms. USSR. 1970's. *408*
Brown, Harrison (account). Gromyko, Andrei. Nuclear arms. Scientists. UN. USSR. 1947. *482*

Budgets. Boston Study Group (report). Brookings Institution (report). Defense spending. 1979. *321*
—. Carter, Jimmy (administration). Defense Policy. 1978. *580*
—. Congress. Defense Policy. Military. Research and development. 1976-77. *102*
Bulletin of the Atomic Scientists (periodical). Nuclear science and technology. Rabinowitch, Eugene. 1940's-70's. *751*
Bureaucracies. Air Forces (Air Targets Division). Military intelligence. Morality. Nash, Henry T. (account). Nuclear War. Planning. 1950's-70's. *147*
—. Military Strategy. Nuclear war. 1974. *280*
—. Nuclear Regulatory Commission. 1970's. *746*
Business. Energy. Government. Public Policy. Technology. 1978. *731*
Byrnes, James F. Federal Government. Germany. Memoirs. Nuclear Science and Technology. Szilard, Leo. 1898-1945. *59*
B-1 (aircraft). Airplanes, Military. Lobbyists. Rockwell International Corp. Unemployment. 1975. *149*
—. Congress. Defense budget cuts. Pressure groups. *Trident* submarine. 1970's. *78*
B-29 (aircraft). Japan. Lemay, Curtis. Nuclear War. Surrender. World War II. 1945. *67*
B-47 (aircraft). Airplanes, Military. Peck, Earl G. (account). Strategic Air Command Wing. ca 1960. *153*
B-52 bomber. Airplanes, Military. Technology. Vietnam. 1952-75. *124*
B-58 (aircraft). Airplanes, Military (bombers). 1946-79. *119*

C

California (Livermore, Los Altos). Laboratories. Nuclear Science and Technology. Weapons. 1942-77. *101*
California Nuclear Safeguards Initiative. Nuclear accidents. Radioactive waste. Sabotage. 1976. *735*
California, University of. Military. Nuclear arms. 1945-78. *167*
California, University of, Berkeley. Abelson, Philip H. (reminiscences). Lawrence Radiation Laboratory. Nuclear physics. 1935-40. *1*
—. Alvarez, Luis W. (reminiscences). Lawrence Radiation Laboratory. Nuclear Physics. 1934. *4*
—. Kamen, Martin D. (reminiscences). Lawrence Radiation Laboratory. Nuclear Science and Technology. 1936-39. *30*
Canada. Arms control. China. India. Nuclear Arms. Nuclear power. Regionalism. USA. 1951-78. *387*
—. Attitudes. Nuclear power. USA. 1960's-70's. *717*
—. Diaries. King, William Lyon MacKenzie. Nuclear War. Racism. 1921-48. *49*
—. Foreign Policy. Great Britain. Nuclear Nonproliferation. USSR. 1943-76. *545*
—. Foreign policy. Trusts, industrial. Uranium industry. USA. 1970's. *798*
—. Foreign Relations. Nuclear arms. Pearson, Lester B. (memoirs). USA. 1945. *42*
—. Government secrecy. India. Nuclear Nonproliferation Treaty (1968). Nuclear technology. 1968-70's. *587*
—. NATO. Neutron bomb. Nuclear War. USA. 1977-79. *141*
Cannon, Clarence. House of Representatives. Mexico. Military Service. Spanish-American War. World War I. World War II. 1898-1945. *35*
Capacity, idle. Nuclear power plants. Public utilities. Taxation. 1973-74. *715*
Capitalism. Central Intelligence Agency. Federal government. National security. Nuclear arms. Political power. 1940's-70's. *432*
Carter, Jimmy. Defense Policy. Executive orders. 1973-80. *354*
—. Detente. Disarmament. Human rights. Kissinger, Henry A. 1970-77. *694*
—. Disarmament. Foreign policy. Inaugural Addresses. Morality. 1976. *460*
—. Energy. Nuclear Energy. Public Policy. 1977-78. *698*
—. Foreign policy. 1820's-1970's. *569*
—. Foreign policy. 1977. *422*
—. Foreign Policy. Nuclear nonproliferation. 1968-77. *652*
—. Foreign Policy. Nuclear nonproliferation. Political Leadership. 1977-79. *621*
—. Foreign Policy. Nuclear nonproliferation. Political Leadership. 1979. *480*
—. Strategic Arms Limitation Talks (II). USSR. 1972-77. *666*
Carter, Jimmy (administration). Arms Control and Disarmament Agency. Civil defense. Evacuations. Nuclear War. Public opinion. 1978. *95*
—. Arms race. Cruise missiles. Tomahawk (missile). World War II. 1977-80. *160*
—. Budgets. Defense Policy. 1978. *580*
—. Communist Countries. Foreign Policy. Military Strategy. Reagan, Ronald (administration). 1980. *418*
—. Congress. Energy. Federal Policy. Interest Groups. 1977-78. *756*
—. Congress. Foreign policy. National security. Nixon, Richard M. (administration). Public scrutiny. 1970's. *211*
—. Defense Policy. Diplomacy. Politics. Strategic Arms Limitation Talks (II). USSR. 1978. *487*
—. Energy. Federal Policy. National Energy Plan. 1973-78. *799*
—. Equivalence (definitions). Nuclear arms policy. Strategic Arms Limitation Talks. 1970's. *588*
—. Foreign policy. 1978. *485*
—. Foreign Policy. Nuclear nonproliferation. 1977-81. *655*
—. Foreign policy. Strategic Arms Limitation Talks (II). USSR. 1970. *464*
—. Military Strategy. Nuclear Arms. Presidential Directive No. 59. 1960-80. *139*
—. Nuclear accidents. Pennsylvania. President's Commission on the Accident at Three Mile Island. Three Mile Island power plant. 1979. *761*
—. Strategic Arms Limitation Talks. 1977. *589*
—. Strategic Arms Limitation Talks (II). USSR. 1977. *470*
Carter, Jimmy (views). Foreign policy. USSR. 1977. *499*
Catholic Church. Arms race. Bishops. 1979-81. *818*
Censorship, voluntary. National security. Press. World War II. 1941-80. *515*
Central Intelligence Agency. Capitalism. Federal government. National security. Nuclear arms. Political power. 1940's-70's. *432*
—. Israel. Nuclear Arms. Nuclear Materials and Equipment Corporation. Pennsylvania (Apollo). Uranium. 1960's. *251*

Chicago, University of. Anderson, Herbert L. (reminiscences). Columbia University. Fermi, Enrico. Nuclear Science and Technology. Szilard, Leo. 1933-45. *6*
—. Anderson, Herbert L. (reminiscences). Illinois. Nuclear chain reaction. World War II. 1942. *5*
—. Illinois. Nuclear chain reaction. Wattenberg, Albert (reminiscences). World War II. 1942. *64*
China. Arms control. Canada. India. Nuclear Arms. Nuclear power. Regionalism. USA. 1951-78. *387*
—. Arms control agreements. Detente. Nuclear strategy. USA. USSR. 1970's. *401*
—. Balance of power. Detente. Foreign policy. USSR. 1945-79. *218*
—. Coffey, Joseph I. Military strength. Strategic power. USA. USSR. ca 1965-73. *203*
—. Defense Policy. Strategic Arms Limitation Talks (II). USSR. 1968-79. *534*
—. Nuclear stalemate, function of. USA. USSR. World order. 1970's. *201*
China Syndrome (film). Attitudes. Nuclear Energy. Pennsylvania (Three Mile Island). 1978-79. *764*
Churchill, Winston. Communism. Diaries. Nuclear Arms. Potsdam Conference. Stalin, Joseph. Truman, Harry S. 1945. *37*
—. Diplomacy. Great Britain. Military Strategy. Roosevelt, Franklin D. World War II. 1942-44. *61*
Cities. Energy, alternative. Private utilities. ca 1950-79. *775*
Civil defense. Arms Control and Disarmament Agency. Carter, Jimmy (administration). Evacuations. Nuclear War. Public opinion. 1978. *95*
—. Berlin crisis. Fallout shelters. 1961-62. *129*
—. Defense Policy. Military. Nuclear War. Public Opinion. USSR. 1948-78. *424*
—. Military Strategy. Nuclear War. USSR. 1950's-77. *392*
—. Nuclear War. USSR. 1950-78. *133*
Civil Disobedience. Nuclear energy. 1977-79. *702*
Civil Rights. Martial Law. Nuclear Power Plants. Plutonium. Terrorism. 1970's. *728*
Civilian survivors. Atomic Bomb Casualty Commission. Japan (Hiroshima, Nagasaki). Medical Research. Radiation Effects Research Foundation. 1945-75. *63*
Civil-military relations. Military power. Nuclear Science and Technology. -1973. *362*
Clamshell Alliance. Anti-nuclear power movement. Leftism. Political protest. 1970's. *740*
—. Ideology. New Hampshire (Seabrook). Nuclear power plants. Political Protest. 1978. *741*
Clarke, Duncan L. (review article). Arms Control and Disarmament Agency. 1961-80. *615*
Clausewitz, Karl von (theories). Military Strategy. Nuclear War, limited. War, principles of. 1970's. *342*
Clements, William P., Jr. (interview). Strategic Arms Limitation Talks. USSR. 1970's. *410*
Coal industry. Environment. Nuclear Science and Technology. 1860's-1970's. *772*
Coercion. Deterrence. Diplomacy. International Relations Theory. 1815-1970's. *316*
Coffey, Joseph I. China. Military strength. Strategic power. USA. USSR. ca 1965-73. *203*
Cold War. Afghanistan. Arms race. Foreign Relations. USSR. 1960-80. *355*
—. Arms race. Economic Conditions. International Trade. Truman, Harry S. (administration). 1948-49. *347*
—. Baruch, Bernard. USSR. 1946. *110*
—. Communism. Diplomacy. Middle East. Morgenthau, Hans J. Nuclear war. USSR. 1945-76. *399*
—. Decisionmaking. Disarmament. Eisenhower, Dwight D. Nuclear arms. USSR. 1953-61. *657*
—. Detente. Foreign Relations. Great Powers. USSR. 1945-75. *252*
—. Diplomacy. Foreign policy. Nuclear Arms. USA. USSR. 1942-46. *13*
—. Diplomacy. Nuclear Arms. Roosevelt, Franklin D. 1941-45. *53*
—. Espionage. Nuclear Arms. Rosenberg case. Supreme Court. 1953. *152*
—. Japan (Hiroshima). Nuclear War. World War II. 1945. *31*
—. USSR. 1981. *346*
Cold War (origins). Historiography. 1946-75. *197*
—. Marshall Plan. Nuclear Arms. Second Front. USSR. World War II. 1940's. *429*
Colleges and Universities. Manhattan Project. Nuclear Arms. Oppenheimer, J. Robert. 1904-67. *48*
Colorado. Nuclear Science and Technology (underground explosions). Public Opinion. 1973-74. *188*
Colorado (Boulder). Commerce Department. Maryland (Gaithersburg). National Bureau of Standards. Scientific Experiments and Research. 1901-78. *813*
Columbia University. Anderson, Herbert L. (reminiscences). Chicago, University of. Fermi, Enrico. Nuclear Science and Technology. Szilard, Leo. 1933-45. *6*
Command structure. Defense policy. Nuclear Arms. Reagan, Ronald (administration). USSR. 1980-81. *393*
Commerce Department. Colorado (Boulder). Maryland (Gaithersburg). National Bureau of Standards. Scientific Experiments and Research. 1901-78. *813*
Committees. Allison, Graham *(Essence of Decision)*. Decisionmaking. Foreign policy. Nuclear War. World War II. 1944-45. *54*
Communications, Military. Defense Policy. Nuclear weapons. USSR. 1940-81. *99*
—. Defense Policy. USSR. 1948-82. *174*
Communism. Balance of Power. Foreign Policy. USSR. 1965-78. *336*
—. Churchill, Winston. Diaries. Nuclear Arms. Potsdam Conference. Stalin, Joseph. Truman, Harry S. 1945. *37*
—. Cold War. Diplomacy. Middle East. Morgenthau, Hans J. Nuclear war. USSR. 1945-76. *399*
—. Containment. Military strategy. 1941-70's. *263*
—. *Dr. Strangelove* (film). George, Peter (*Red Alert*). Kubrick, Stanley. Paranoia, patterns of. 1950's. *437*
Communist Countries. Carter, Jimmy (administration). Foreign Policy. Military Strategy. Reagan, Ronald (administration). 1980. *418*
—. Exports. National security. Technology. 1949-78. *242*
Community movements. Anti-nuclear power movement. Political activism. 1970's. *747*
Competition. Arms control. Foreign Policy. Institutions. USSR. 1970-80. *585*
Condor missile system. Congress. Defense Department. 1960's-76. *163*
Conference on Security and Cooperation in Europe. Decisionmaking. Public Opinion. 1975. *526*
—. Europe. Ford, Gerald R. (administration). International Security. Public opinion. USA. 1975. *495*

Conflict avoidance. Defense Policy. Naval presence. Persian Gulf. USA. 1946-75. *237*
Conflict theory. Deterrence. Military Strategy. USA. USSR. 1945-75. *258*
Congress. ABM controversy. Arms control. 1969-75. *685*
—. Budgets. Defense Policy. Military. Research and development. 1976-77. *102*
—. B-1 (aircraft). Defense budget cuts. Pressure groups. *Trident* submarine. 1970's. *78*
—. Carter, Jimmy (administration). Energy. Federal Policy. Interest Groups. 1977-78. *756*
—. Carter, Jimmy (administration). Foreign policy. National security. Nixon, Richard M. (administration). Public scrutiny. 1970's. *211*
—. Condor missile system. Defense Department. 1960's-76. *163*
—. Defense Department. Defense industries. Military. Public Opinion. 1980's. *194*
—. Executive Branch. Foreign Policy. Strategic Arms Limitation Talks. 1972-78. *531*
—. Federation of American Scientists. Nuclear Arms. Presidency. War Powers Resolution (1973). 1975-76. *578*
—. Foreign policy. Panama. Politics. Strategic Arms Limitation Talks. Treaties. 1977-78. *512*
—. Joint Committee on Atomic Energy. Legislative investigations. Nuclear power. 1973-74. *719*
—. Strategic Arms Limitation Talks. 1970's. *600*
Conseil Européen pour la Recherche Nucléaire. Bloch, Felix. Nuclear Science and Technology. 1954-55. *748*
Conservation. Nuclear arms. Nuclear power. Power Resources. 1980. *759*
Conservation of Natural Resources. Energy. 1970's. *789*
Consortium on Peace Research, Education and Development. Peace Research. 1970-81. *458*
Consultation. Decisionmaking. Diplomacy. NATO. 1957-78. *304*
Containment. Communism. Military strategy. 1941-70's. *263*
Contracts. Economic Regulations. General Electric Co. Nuclear power plants. Westinghouse. 1955-78. *711*
Cooper, Joe. Pick, Vernon. Prospecting. Steen, Charles. Uranium rushes. Western States. 1948-57. *144*
Corporations. Politics. Shortages. Stockpiling decisions. Strategic materials. 1946-70's. *103*
Counterforce (concept). Deterrence. Military Strategy. Nuclear Arms. 1970's. *340*
Courts. Industrial safety. Nuclear waste. 1970-80. *806*
Cruise Missiles. Air Warfare. Armaments. 1914-80. *185*
—. Antiwar Sentiment. Great Britain. Neutron Bomb. 1978-81. *441*
—. Arms race. Carter, Jimmy (administration). Tomahawk (missile). World War II. 1977-80. *160*
—. Attitudes. Defense Department. Politics and the Military. Weapons. 1964-76. *140*
—. Balance of Power. Defense Department. Military Capability. Strategic Arms Limitation Talks (II). ca 1950-79. *125*
Cuban Missile Crisis. Balance of power. Kennedy, John F. Military Strategy. USA. USSR. 1962. *236*
—. Bismarck, Otto von. Diplomacy. Escalation theory. Franco-Prussian War. Kennedy, John F. McNamara, Robert S. 1870. 1962. *391*
—. Kennedy, John F. Khrushchev, Nikita. Military strategy. Nuclear war, threat of. 1962. *209*
Currie, Malcolm R. (interview). Defense Policy. 1970's. *417*
Cyclotrons. Denmark. Japan. Nuclear physics. Scientific Experiments and Research. 1935-45. *65*
—. Manhattan Project. New Mexico (Los Alamos). Nuclear Science and Technology. Wilson, Robert R. (reminiscences). 1942-43. *66*
—. Michigan State University. Nuclear Science and Technology. 1958-81. *86*

D

Decisionmaking. Allison, Graham *(Essence of Decision)*. Committees. Foreign policy. Nuclear War. World War II. 1944-45. *54*
—. Alperovitz, Gar (review article). Historiography. Japan (Hiroshima). Nuclear War. World War II. 1945-65. *52*
—. Arms control proposals. 1975. *641*
—. Atomic Energy Commission. Energy. Federal policy. Military. Pressure Groups. 1974. *720*
—. Attitudes. Nuclear power plants. State Government. Voting and Voting Behavior. Western States. 1963-78. *703*
—. Cold War. Disarmament. Eisenhower, Dwight D. Nuclear arms. USSR. 1953-61. *657*
—. Conference on Security and Cooperation in Europe. Public Opinion. 1975. *526*
—. Consultation. Diplomacy. NATO. 1957-78. *304*
—. Defense policy. International Security. Missile defense. Safeguard debate. 1968-72. *175*
—. Divine, Robert A. (review article). Foreign relations. Nuclear arms. USSR. 1954-60. 1978. *195*
—. Foreign policy. 1960-78. *298*
—. Hydrogen bomb. Nuclear Arms policy. York, Herbert (review article). 1949-75. *192*
—. Japan (Hiroshima, attack on). Nuclear War. World War II (Pacific Theater). 1944-45. *50*
—. Japan (Hiroshima, Nagasaki). Military Strategy. Nuclear War. World War II. 1944-45. *23*
Defense. Economic strength. Military systems. 1945-75. *243*
—. ICBM's. Military Capability. Nuclear War. Research and development. USSR. 1970's. *413*
—. Maritime law. Military strategy. Nuclear arms. 1970's. *353*
—. Massachusetts Institute of Technology. 1861-1979. *148*
Defense budget. Arms control objectives. 1975. *481*
—. Detente. USA. USSR. 1959-74. *248*
Defense budget cuts. B-1 (aircraft). Congress. Pressure groups. *Trident* submarine. 1970's. *78*
Defense Department. Atomic Energy Commission. Historical advisory committees. State Department. 1947-74. *754*
—. Attitudes. Cruise missiles. Politics and the Military. Weapons. 1964-76. *140*
—. Balance of Power. Cruise missiles. Military Capability. Strategic Arms Limitation Talks (II). ca 1950-79. *125*
—. Condor missile system. Congress. 1960's-76. *163*
—. Congress. Defense industries. Military. Public Opinion. 1980's. *194*

—. Military Finance. Naval Strategy. Nuclear War. Submarine Warfare. Tidal waves. 1950's-80. *131*
Defense industries. Congress. Defense Department. Military. Public Opinion. 1980's. *194*
Defense Nuclear Agency. Nuclear arms. 1963-76. *180*
Defense Policy. 1945-69. *350*
—. 1945-74. *249*
—. ABM. Strategic Arms Limitation Treaty. USSR. 1982. *610*
—. ABM controversy. Military Strategy. 1965-80. *114*
—. Aerospace systems. Military Strategy. 1977. *128*
—. Air Forces. Military Strategy. North American Air Defense Command. 1970's. *415*
—. Armaments. USSR. 1960's-70's. *396*
—. Armies. Nuclear Arms (tactical). Politics. Rose, John P. (review article). 1950-81. *239*
—. Arms control. 1973. *547*
—. Arms control. Foreign Relations. USSR. 1960-80. *584*
—. Arms control. National security. 1960's-75. *636*
—. Balance of Power. Nuclear Arms. Schlesinger, James R. USSR. 1960's-70's. *380*
—. Balance of Power. Nuclear Arms. USA. USSR. 1974-75. *241*
—. Balance of Power. Sweden. USSR (Kola Peninsula). 1945-74. *322*
—. Budgets. Carter, Jimmy (administration). 1978. *580*
—. Budgets. Congress. Military. Research and development. 1976-77. *102*
—. Carter, Jimmy. Executive orders. 1973-80. *354*
—. Carter, Jimmy (administration). Diplomacy. Politics. Strategic Arms Limitation Talks (II). USSR. 1978. *487*
—. China. Strategic Arms Limitation Talks (II). USSR. 1968-79. *534*
—. Civil defense. Military. Nuclear War. Public Opinion. USSR. 1948-78. *424*
—. Command structure. Nuclear Arms. Reagan, Ronald (administration). USSR. 1980-81. *393*
—. Communications, Military. Nuclear weapons. USSR. 1940-81. *99*
—. Communications, Military. USSR. 1948-82. *174*
—. Conflict avoidance. Naval presence. Persian Gulf. USA. 1946-75. *237*
—. Currie, Malcolm R. (interview). 1970's. *417*
—. Decisionmaking. International Security. Missile defense. Safeguard debate. 1968-72. *175*
—. Detente. Deterrence. USA. USSR. 1975. *379*
—. Deterrence. Europe, Western. NATO. Nuclear arms. Warsaw Pact. 1970's. *326*
—. Deterrence. International stability. Military strategy. 1975. *370*
—. Deterrence. Military strategy. 1950's-72. *202*
—. Deterrence. Mobilization. Planning. 1978. *228*
—. Disarmament. Foreign policy. Nuclear arms. Panel of Consultants on Disarmament (report). USSR. 1953. *486*
—. Documents. Truman, Harry S. (administration). 1946-75. *360*
—. Dulles, John Foster. Eisenhower, Dwight D. Goodpaster, Andrew. Nuclear test ban issue. Strauss, Lewis. USSR. 1958. *240*
—. Economic policy. Strategic response. 1974. *313*
—. Espionage. Groves, Leslie R. Nuclear arms. Secrecy. Truman, Harry S. (administration). USSR. 1945-50. *122*
—. Europe. Flexible response. Neutron bomb. ca 1963-77. *260*
—. Europe. Weapons, strategic. 1973. *447*
—. Europe, Western. Mutual balanced force reductions. USSR. 1970's. *689*
—. Flexible response strategy. Marines. Military strategy. 1953-72. *198*
—. Foreign policy. 1976. *443*
—. Foreign policy. Military Strategy. Strategic Arms Limitation Talks (II). USSR. 1960's-70's. *521*
—. Foreign Relations. Latin America. Nuclear arms. Tlatelolco, Treaty of. 1960's-70's. *637*
—. ICBM's. Johnson, Lyndon B. Kennedy, John F. Kuter, Lawrence S. (account). 1962. *137*
—. ICBM's. Land-mobile system, proposed. Technology. 1950's-74. *80*
—. Manpower. Military. Strategic Arms Limitation Talks. USSR. 1973-74. *384*
—. Massive retaliation. 1950-61. *430*
—. Military Capability. Nuclear Arms. USSR. 1970's. *297*
—. Military Capability. USSR. 1977-80. *374*
—. Military Strategy. Mutual assured destruction. Nuclear War. USA. USSR. 1960-76. *196*
—. Military strategy. Stability, strategic. USSR. 1960-80. *271*
—. Military Strategy. USSR. ca 1960-80. *210*
—. Multiple delivery system. Naval Strategy (Blue Water Strategy). Strategic Arms Limitation Talks. 1974. *683*
—. MX (missile). Nevada. 1979-80. *189*
—. National security. Nuclear War. Peace. 1977. *400*
—. NATO. Nuclear arms. 1967-82. *254*
—. Nixon, Richard M. Nuclear arms. Strategic Arms Limitation Talks (II). 1973. *665*
—. Nuclear arms. Strategic Arms Limitation Talks. Technology. 1975. *231*
—. Nuclear arms. Strategic Arms Limitation Talks. USSR. 1972-79. *582*
—. Nuclear War. Strategic Arms Limitation Talks. USSR. 1969-79. *293*
—. Nuclear War (risk of). USA. USSR. 1945-74. *285*
—. Political Reform. Weapons acquisition. 1945-75. *173*
—. Strategic Arms Limitation Talks. USSR. Weapons. 1972-78. *643*
—. Technology. USSR. Weapons. 1978. *283*
—. USA. USSR. 1974. *381*
Defense research. Military. Research Board for National Security. Science and Government. 1944-50. *132*
Defense spending *See also* Defense budget.
—. Balance of Power. Manpower. Military Capability. 1975-76. *452*
—. Boston Study Group (report). Brookings Institution (report). Budgets. 1979. *321*
—. Disarmament. Weapons. 1978. *206*
—. Economic Planning. National security. Peace research. 1960's-77. *616*
—. Eisenhower, Dwight D. Military strategy. 1952-60. *303*
—. ICBM's. Military Capability. Research and development. USSR. 1970's. *182*
—. Intelligence Service. Military capability. USSR. 1960's-76. *434*
—. Military Strategy. National security. NATO. USSR. 1980. *359*
—. Missile capability. Nuclear arms race. USA. USSR. 1970's. *433*
—. Public Opinion. 1970's. *287*
Defensive war argument. Foreign Policy. Ideology. War. 1941-73. *398*

DeLeon, Peter. Electric Power. Nuclear Energy (review article). Ramsay, William. Rolph, Elizabeth S. 1960-79. *752*
Democracy. Attitudes. Europe. Interest Groups. Japan. Nuclear Science and Technology. USSR. 1970's. *739*
Denmark. Cyclotrons. Japan. Nuclear physics. Scientific Experiments and Research. 1935-45. *65*
Detente. ca 1970-80. *492*
—. Arms control. Austria (Vienna). Mutual Balanced Force Reductions. 1973-80. *522*
—. Arms control. Foreign relations. USA. USSR. 1961-75. *574*
—. Arms control. Mutual Balanced Force Reductions. USA. USSR. 1974. *634*
—. Arms control agreements. China. Nuclear strategy. USA. USSR. 1970's. *401*
—. Arms race. Disarmament. NATO. 1950's-76. *605*
—. Arms race. Nuclear strategy. USA. USSR. 1972-74. *319*
—. Balance of power. China. Foreign policy. USSR. 1945-79. *218*
—. Brezhnev, Leonid. Ford, Gerald R. Vladivostok Accord. 1974. *617*
—. Carter, Jimmy. Disarmament. Human rights. Kissinger, Henry A. 1970-77. *694*
—. Cold War. Foreign Relations. Great Powers. USSR. 1945-75. *252*
—. Defense budget. USA. USSR. 1959-74. *248*
—. Defense policy. Deterrence. USA. USSR. 1975. *379*
—. Diplomacy. Europe. International Security. USSR. 1970-74. *510*
—. Disarmament. NATO. 1970's. *357*
—. Disarmament. Nuclear Arms. USA. USSR. 1970's. *573*
—. Disarmament. Political Science. 1970's. *627*
—. Disarmament. USA. USSR. 1976-77. *693*
—. Economic relations. Nixon, Richard M. (administration). Strategic arms limitation. USSR. 1969-73. *648*
—. Europe. Military Strategy. USSR. 1950's-70's. *444*
—. Europe. NATO. Nuclear War. USA. 1970's. *269*
—. Europe. USA. USSR. 1879-1906. 1945-75. *372*
—. Europe, Western. European Security Conference. Propaganda. USSR. 1960's-73. *463*
—. European Security Conference (measures of confidence). Military. 1968-77. *544*
—. Exports. Nuclear arms. 1978. *630*
—. Foreign Policy. Human rights. USSR. 1973-79. *552*
—. Military strategy. Missile development. Strategic Arms Limitation Talks (II). 1974-75. *548*
—. Strategic Arms Limitation Talks. USA. USSR. 1950's-70's. *679*
—. Strategic Arms Limitation Talks. USSR. 1973-77. *677*
—. USA. USSR. 20c. *624*
—. USSR. 1960's-70's. *676*
Deterrence. Arms control. Foreign Relations. International organizations. 1960-75. *594*
—. Arms race. Balance of Power. Foreign Relations. USSR. 1970's. *312*
—. Atomic Energy Commission. Gillet, Edward B. (interview). Military Strategy. Nuclear Arms (warheads). Research. 1970's. *411*
—. Battles. Morale. Nuclear War. World War II. 1944-81. *130*
—. Coercion. Diplomacy. International Relations Theory. 1815-1970's. *316*
—. Conflict theory. Military Strategy. USA. USSR. 1945-75. *258*
—. Counterforce (concept). Military Strategy. Nuclear Arms. 1970's. *340*
—. Defense policy. Detente. USA. USSR. 1975. *379*
—. Defense Policy. Europe, Western. NATO. Nuclear arms. Warsaw Pact. 1970's. *326*
—. Defense Policy. International stability. Military strategy. 1975. *370*
—. Defense policy. Military strategy. 1950's-72. *202*
—. Defense Policy. Mobilization. Planning. 1978. *228*
—. ICBM's. Modernization. MX (missile). 1957-76. *169*
—. ICBM's. USSR. 1960's-70's. *416*
—. Military Capability. Nuclear arms. USSR. 1945-50. *337*
—. Military policy. Strategy. -1972. *291*
—. Military Strategy. 1950-75. *328*
—. Military Strategy. 1974. *227*
—. Military Strategy. Nuclear Arms. Technological innovations. USSR. 1969-78. *338*
—. Military strategy. Nuclear arms. USSR. 1950-80. *318*
—. Military Strategy. Nuclear War. Schlesinger, James R. 1974. *305*
—. NATO. Neutron bomb. Nuclear Arms. Warsaw Pact. 1954-78. *83*
—. Nuclear Arms. Strategic Arms Limitation Talks. USA. USSR. 1973. *623*
—. Nuclear arms. Strategic Arms Limitation Talks. USSR. 1969-78. *496*
—. Nuclear Arms. USSR. 1952-80. *259*
—. Nuclear stockpiling. USA. USSR. 1960's-70's. *440*
—. Strategic stability. USA. USSR. 1975. *348*
—. USSR. 1963-80. *288*
Deterrence logic. Disarmament. War, risks of. World War II (antecedents). 1933-62. *307*
Deterrence (military). Disarmament. Nuclear Arms. Strategic studies. 1973. *364*
—. Flexible response strategy. Nuclear Arms. 1972-74. *385*
Deterrence theory. Military Strategy. Nuclear Arms. 1966-76. *270*
Developing nations. Arms control. Foreign Policy. Nuclear arms. USA. 1975. *468*
—. Arms control. International organization. Nuclear capability. USA. 1950-75. *501*
—. Balance of power. Foreign policy. USSR. 1961-79. *556*
—. Foreign policy. USSR. 1970-79. *282*
—. Foreign Relations. Great Powers. Nuclear Nonproliferation Treaty. 1970's. *662*
—. Foreign Relations. Nuclear arms. 1960-80. *301*
—. Foreign Relations. Nuclear arms proliferation. 1974-79. *244*
—. Heilbroner, Robert. Nuclear Arms. Peace, prospects for. Violence. 1976. *535*
—. Nuclear arms. 1979-81. *261*
Diaries. Accidents. Nuclear power plants. Pennsylvania (Three Mile Island). 1979. *700*
—. Canada. King, William Lyon MacKenzie. Nuclear War. Racism. 1921-48. *49*
—. Churchill, Winston. Communism. Nuclear Arms. Potsdam Conference. Stalin, Joseph. Truman, Harry S. 1945. *37*
Diplomacy. Alperovitz, Gar. Historiography. Nuclear Arms. World War II. 1943-46. 1965-73. *323*
—. Arms race. Negotiations. Nuclear test ban. Richardson models. USSR. 1962-63. *558*
—. Atomic bomb development. Nuclear Science and Technology. World War II. 1942-45. *46*

—. Balance of Power. Nuclear arms. Politics. 1945-79. *217*
—. Bismarck, Otto von. Cuban missile crisis. Escalation theory. Franco-Prussian War. Kennedy, John F. McNamara, Robert S. 1870. 1962. *391*
—. Brazil. Germany. NATO. Nuclear nonproliferation. 1968-77. *571*
—. Carter, Jimmy (administration). Defense Policy. Politics. Strategic Arms Limitation Talks (II). USSR. 1978. *487*
—. Churchill, Winston. Great Britain. Military Strategy. Roosevelt, Franklin D. World War II. 1942-44. *61*
—. Coercion. Deterrence. International Relations Theory. 1815-1970's. *316*
—. Cold War. Communism. Middle East. Morgenthau, Hans J. Nuclear war. USSR. 1945-76. *399*
—. Cold War. Foreign policy. Nuclear Arms. USA. USSR. 1942-46. *13*
—. Cold War. Nuclear Arms. Roosevelt, Franklin D. 1941-45. *53*
—. Consultation. Decisionmaking. NATO. 1957-78. *304*
—. Detente. Europe. International Security. USSR. 1970-74. *510*
—. Disarmament. Nuclear nonproliferation. UN. 1945-77. *467*
—. European Security Conference. Human rights. 1975. *607*
—. Foreign Policy. 1969-81. *555*
—. Foreign Policy. Nuclear War. Roosevelt, Franklin D. Truman, Harry S. 1945-51. *14*
—. Nuclear arms. Strategic Arms Limitation Talks (II). USSR. 1972-76. *639*
—. Nuclear proliferation. 1945-75. *439*
—. Personal narratives. Rowny, Edward L. Strategic Arms Limitation Talks. 1973-79. *638*
—. Strategic Arms Limitation Talks. USSR. 1968-76. *537*
—. Strategic Arms Limitation Talks (II). Talbott, Strobe (review article). USSR. 1972-79. *629*
Diplomacy (review article). USSR. 1933-80. *532*
Disarmament. 1945-82. *543*
—. 1974-75. *691*
—. 1982. *640*
—. ABM. ICBM's. MIRV. Strategic Arms Limitation Talks. 1969-73. *100*
—. Armaments. Nuclear proliferation. Wars, significance of. 1945-75. *205*
—. Arms control. Foreign Policy. Military strategy. USSR. 1945-70's. *214*
—. Arms control. Foreign Relations. USA. USSR. 1970's. *664*
—. Arms control. Seabed talks. USSR. 1967-70. *633*
—. Arms control. Treaties. USSR. 1977. *570*
—. Arms control. USA. USSR. 1974. *649*
—. Arms Control and Disarmament Agency. Federal Government. General Advisory Committee on Arms Control and Disarmament. 1961-73. *507*
—. Arms control (definitions). 18c-20c. *489*
—. Arms race. Detente. NATO. 1950's-76. *605*
—. Arms race. Political Theory. 1946-79. *317*
—. Arms race. USA. USSR. 1945-70's. *277*
—. Carter, Jimmy. Detente. Human rights. Kissinger, Henry A. 1970-77. *694*
—. Carter, Jimmy. Foreign policy. Inaugural Addresses. Morality. 1976. *460*
—. Cold War. Decisionmaking. Eisenhower, Dwight D. Nuclear arms. USSR. 1953-61. *657*
—. Defense Policy. Foreign policy. Nuclear arms. Panel of Consultants on Disarmament (report). USSR. 1953. *486*
—. Defense spending. Weapons. 1978. *206*
—. Detente. NATO. 1970's. *357*
—. Detente. Nuclear Arms. USA. USSR. 1970's. *573*
—. Detente. Political Science. 1970's. *627*
—. Detente. USA. USSR. 1976-77. *693*
—. Deterrence logic. War, risks of. World War II (antecedents). 1933-62. *307*
—. Deterrence (military). Nuclear Arms. Strategic studies. 1973. *364*
—. Diplomacy. Nuclear nonproliferation. UN. 1945-77. *467*
—. Europe. Mutual Balanced Force Reductions. USSR. 1973. *586*
—. Foreign policy. Nuclear Arms. USA. USSR. 1970's. *300*
—. Foreign Relations. 1945-70's. *690*
—. Foreign Relations. Military. Political power. USSR. 1945-82. *333*
—. Foreign Relations. USA. USSR. 1945-70's. *561*
—. Government secrecy. Nuclear Arms. USA. 1945-70's. *562*
—. Great Britain. Nuclear Test Ban Treaty (1963). USSR. 1954-70's. *675*
—. Military (force reductions). NATO. USSR. 1955-73. *667*
—. National Security. Strategic Arms Limitation Talks. USA. USSR. Verification methods. 1972-76. *593*
—. NATO. Nuclear arms. Theater Nuclear Forces. 1977-79. *278*
—. NATO. USSR. 1932-82. *626*
—. Nuclear arms. Reagan, Ronald (administration). 1981. *246*
—. Nuclear arms. Strategic Arms Limitation Talks. 1974-76. *513*
—. Nuclear Arms. Strategic arms race. USA. USSR. 1960-74. *446*
—. Nuclear Arms. USA. USSR. 1940's-74. *563*
—. Nuclear Arms. USSR. 1943-78. *294*
—. Nuclear War. Public Policy. Scientists. 1939-54. *71*
—. Space. USSR. 1957-81. *659*
Disarmament (term). Arms control (term). 1970-82. *488*
Disasters. Psychology. Survivors. War. 1941-72. *34*
Divine, Robert A. (review article). Decisionmaking. Foreign relations. Nuclear arms. USSR. 1954-60. 1978. *195*
Dr. Strangelove (film). Anti-Communist Movements. Films. Films. Humor. Kubrick, Stanley. Liberalism. Nuclear Arms. 1950-64. *599*
—. Communism. George, Peter (*Red Alert*). Kubrick, Stanley. Paranoia, patterns of. 1950's. *437*
Documents. Armaments. Federal Policy. Nuclear nonproliferation. 1981. *692*
—. Defense Policy. Truman, Harry S. (administration). 1946-75. *360*
—. Military Strategy. Nuclear Arms. USSR. 1954-55. *373*
Domestic Policy. Arms control. USA. USSR. Weapons. 1963-75. *462*
—. Foreign policy. Historiography. Liberalism. Truman, Harry S. 1945-52. *275*
—. Italy. Nuclear power plants. Pennsylvania (Three Mile Island). 1970's. *705*
Drones. Fahrney, Delmar S. (account). Missiles. Navy Bureau of Aeronautics. N2C-2 (aircraft). Radio. Research and development. TG-2 (aircraft). World War II. 1936-45. *107*

Dropshot (plan). Military Strategy. USSR. 1949-57. *87*
Dulles, John Foster. Defense Policy. Eisenhower, Dwight D. Goodpaster, Andrew. Nuclear test ban issue. Strauss, Lewis. USSR. 1958. *240*

E

Economic Conditions. Arms race. Cold War. International Trade. Truman, Harry S. (administration). 1948-49. *347*
—. Electric Power. Nuclear power plants. 1973-78. *796*
—. Energy. NATO. Nuclear War. 1950's-70's. *216*
—. Federal government. Nuclear power plants. Subsidies. 1954-78. *718*
Economic development. Attitudes. Military Strategy. Nuclear arms. Technology. War. 1914-68. *351*
Economic Planning. Defense spending. National security. Peace research. 1960's-77. *616*
Economic policy. Defense Policy. Strategic response. 1974. *313*
Economic Regulations. Contracts. General Electric Co. Nuclear power plants. Westinghouse. 1955-78. *711*
Economic relations. Detente. Nixon, Richard M. (administration). Strategic arms limitation. USSR. 1969-73. *648*
Economic strength. Defense. Military systems. 1945-75. *243*
Economics. Attitudes. Foreign relations. Nuclear arms limitation. USA. USSR. 1972-73. *465*
Education. Nuclear Science and Technology. Value systems. 1974. *109*
—. Peace research. 1969-74. *516*
—. Public Opinion. War. 1944-63. *146*
Eglin Air Force Base. Air Forces. Armament Development and Test Center. Florida. 1960's-70's. *179*
Einstein, Albert. Germany. Nuclear arms. Pacifism. Zionism. 1939-55. *21*
—. Hungarian Americans. Nuclear Arms (program). Szilard, Leo. USA. World War II. 1941-74. *26*
—. Institute for Advanced Study. New Jersey (Princeton). 1933-55. *10*
—. Oppenheimer, J. Robert (reminiscences). Pacifism. Physics. 1919-55. *44*
—. Pacifism. Physics. Science and Society. 1919-55. *38*
Eisenhower, Dwight D. Cold War. Decisionmaking. Disarmament. Nuclear arms. USSR. 1953-61. *657*
—. Defense Policy. Dulles, John Foster. Goodpaster, Andrew. Nuclear test ban issue. Strauss, Lewis. USSR. 1958. *240*
—. Defense spending. Military strategy. 1952-60. *303*
—. Foreign Relations. Nonaligned Nations. Nuclear energy. Nuclear proliferation. 1945-79. *311*
Elections. Initiatives. Nuclear safety. 1976. *734*
Electric Power. Armaments. Nuclear Science and Technology. 1950-75. *221*
—. DeLeon, Peter. Nuclear Energy (review article). Ramsay, William. Rolph, Elizabeth S. 1960-79. *752*
—. Economic Conditions. Nuclear power plants. 1973-78. *796*
Elites. National Security. Political Attitudes. Youth. 1945-75. *431*
Employment. Federal Government. Nuclear energy. Private sector. 1963-73. *767*
Employment (dismissal). Engineering. Faulkner, Peter (account). Nuclear power plants. Nuclear Services Corp. Senate. 1973-77. *722*
Energy *See also* Electric power, Nuclear energy.
—. Atomic Energy Commission. Decisionmaking. Federal policy. Military. Pressure Groups. 1974. *720*
—. Business. Government. Public Policy. Technology. 1978. *731*
—. Carter, Jimmy. Nuclear Energy. Public Policy. 1977-78. *698*
—. Carter, Jimmy (administration). Congress. Federal Policy. Interest Groups. 1977-78. *756*
—. Carter, Jimmy (administration). Federal Policy. National Energy Plan. 1973-78. *799*
—. Conservation of Natural Resources. 1970's. *789*
—. Economic Conditions. NATO. Nuclear War. 1950's-70's. *216*
—. Ethics. National Council of Churches. 1976-78. *732*
—. Federal Policy. 1979. *758*
—. Natural Resources. 1973-76. *787*
—. Nuclear Science and Technology. Treaties. 1940's-70's. *476*
Energy, alternative. Cities. Private utilities. ca 1950-79. *775*
Energy crisis. Industry. Nuclear power. Technology. ca 1940-80. *745*
Energy development. Balance of power. Foreign policy. Pacific Basin. USA. 1973-77. *736*
Engineering. Employment (dismissal). Faulkner, Peter (account). Nuclear power plants. Nuclear Services Corp. Senate. 1973-77. *722*
Engineers. Atomic Energy Commission. Manhattan Project. Nuclear technology. Scientists. 1938-50. *28*
Eniwetok (atoll). Marshall Islands. Nuclear testing. Pacific Dependencies (US). Safety. 1946-80. *127*
Enola Gay (aircraft). Air Warfare. Atomic bomb. Wendover Field. World War II (transportation). 1944-45. *60*
Environment. Arms control. Industrialization. International security. Multinational corporations. 1975. *650*
—. Atomic Energy Commission. Nuclear power. Public participation. Technology assessment. 1946-73. *742*
—. Coal industry. Nuclear Science and Technology. 1860's-1970's. *772*
Environmentalism. Experts. New York (Cayuga Lake). Nuclear power plants. Pressure groups. 1973. *770*
Equivalence (definitions). Carter, Jimmy (administration). Nuclear arms policy. Strategic Arms Limitation Talks. 1970's. *588*
Escalation theory. Bismarck, Otto von. Cuban missile crisis. Diplomacy. Franco-Prussian War. Kennedy, John F. McNamara, Robert S. 1870. 1962. *391*
Espionage. Atomic Energy Commission. Federal Bureau of Investigation. Greenglass, David. Nuclear Arms. Rosenberg case. Trials. 1942-74. *142*
—. Cold War. Nuclear Arms. Rosenberg case. Supreme Court. 1953. *152*
—. Defense Policy. Groves, Leslie R. Nuclear arms. Secrecy. Truman, Harry S. (administration). USSR. 1945-50. *122*
—. Nuclear installations. Sabotage. 1966-75. *723*
Ethics. Energy. National Council of Churches. 1976-78. *732*
—. Nuclear arms. Oppenheimer, J. Robert. Scientists. 1945-66. *76*
—. Progress (concept). Scientists. 1970's. *72*

Europe. Allies. Arms control. Balance of power. USSR. 1979. *598*
—. Arms control. USA. 1958-73. *497*
—. Attitudes. Democracy. Interest Groups. Japan. Nuclear Science and Technology. USSR. 1970's. *739*
—. Balance of Power. Strategic Arms Limitation Talks (II). USSR. 1972-78. *542*
—. Conference on Security and Cooperation in Europe. Ford, Gerald R. (administration). International Security. Public opinion. USA. 1975. *495*
—. Defense Policy. Flexible response. Neutron bomb. ca 1963-77. *260*
—. Defense Policy. Weapons, strategic. 1973. *447*
—. Detente. Diplomacy. International Security. USSR. 1970-74. *510*
—. Detente. Military Strategy. USSR. 1950's-70's. *444*
—. Detente. NATO. Nuclear War. USA. 1970's. *269*
—. Detente. USA. USSR. 1879-1906. 1945-75. *372*
—. Disarmament. Mutual Balanced Force Reductions. USSR. 1973. *586*
—. Foreign policy. Kissinger, Henry A. Nixon, Richard M. (administration). Policymaking, patterns of. USA. 1971-75. *579*
—. Foreign policy. Mutual Balanced Force Reductions. USA. 1970-73. *688*
—. Foreign relations. Military strategy. Nuclear arms. USA. USSR. 1950's-70's. *436*
—. Foreign Relations. Nuclear proliferation. 1954-80. *590*
—. International Law. Nuclear materials. 1977-80. *738*
—. International Security. NATO. Nuclear Arms. 1953-77. *224*
—. Military Strategy. NATO. 1945-81. *352*
—. Military Strategy. NATO. North America. Nuclear Arms (tactical, conventional). 1967-74. *112*
—. Modernization. NATO. Nuclear Arms. 1978-80. *302*
—. Strategic Arms Limitation Talks (II). USSR. 1979. *511*
Europe, Central. Game theory. Kahn, Herman. Military strategy. 1945-67. *382*
Europe, Western. Arms Trade. Middle East. Nuclear equipment sales. USA. 1973-75. *306*
—. Defense Policy. Deterrence. NATO. Nuclear arms. Warsaw Pact. 1970's. *326*
—. Defense Policy. Mutual balanced force reductions. USSR. 1970's. *689*
—. Detente. European Security Conference. Propaganda. USSR. 1960's-73. *463*
—. Foreign policy. Nuclear energy. Nuclear nonproliferation. 1945-79. *794*
—. Japan. Mutual Balanced Force Reductions. Strategic Arms Limitation Talks. USA. USSR. 1974. *695*
—. Military Strategy. Missiles. NATO. Pershing II (missile). 1956-80. *151*
—. Military Strategy. NATO. Nuclear Arms. USSR. 1950-78. *85*
—. Military strategy. NATO. Nuclear arms, tactical. USA. USSR. 1970's. *256*
—. NATO. Nuclear arms. USSR. 1979. *402*
—. Strategic Arms Limitation Talks (II). USSR. 1970's. *500*
European Security Conference. Detente. Europe, Western. Propaganda. USSR. 1960's-73. *463*
—. Diplomacy. Human rights. 1975. *607*
European Security Conference (measures of confidence). Detente. Military. 1968-77. *544*
Evacuations. Arms Control and Disarmament Agency. Carter, Jimmy (administration). Civil defense. Nuclear War. Public opinion. 1978. *95*
Executive Branch. Arms control. Public opinion. Treaties. USSR. 1972-82. *517*
—. Congress. Foreign Policy. Strategic Arms Limitation Talks. 1972-78. *531*
Executive Order 10450. Atomic Energy Commission. Loyalty. Oppenheimer, J. Robert. Strauss, Lewis. 1946-54. *117*
Executive orders. Carter, Jimmy. Defense Policy. 1973-80. *354*
Executive Planning Council. Intergovernmental Relations. National Energy Acts. Public Policy. Radioactive waste. ca 1971-78. *757*
Executive Power. Foreign Policy. Political Participation. Public Opinion. 1960's-79. *687*
—. Foreign policy. Truman, Harry S. 1945-53. *43*
Experts. Environmentalism. New York (Cayuga Lake). Nuclear power plants. Pressure groups. 1973. *770*
Exports. Attitudes. France. Great Britain. Nuclear proliferation. Plutonium, reprocessed. 1946-78. *726*
—. Communist Countries. National security. Technology. 1949-78. *242*
—. Detente. Nuclear arms. 1978. *630*
—. Foreign policy. Fuel. India. Law. Nuclear Non-Proliferation Act (US, 1978). 1978. *644*
—. Foreign Policy. Nuclear Arms (control). Technology. 1976. *435*

F

Fahrney, Delmar S. (account). Drones. Missiles. Navy Bureau of Aeronautics. N2C-2 (aircraft). Radio. Research and development. TG-2 (aircraft). World War II. 1936-45. *107*
Falk, Richard A. (views). Foreign Policy. 1976. *525*
Fallout. Attitudes. Nuclear arms. Ozone layer, destruction. 1945-75. *126*
—. Public Health. Science. 1954-64. *134*
Fallout shelters. Berlin crisis. Civil defense. 1961-62. *129*
Faulkner, Peter (account). Employment (dismissal). Engineering. Nuclear power plants. Nuclear Services Corp. Senate. 1973-77. *722*
Federal Bureau of Investigation. Atomic Energy Commission. Espionage. Greenglass, David. Nuclear Arms. Rosenberg case. Trials. 1942-74. *142*
Federal Government. Arms Control and Disarmament Agency. Disarmament. General Advisory Committee on Arms Control and Disarmament. 1961-73. *507*
—. Byrnes, James F. Germany. Memoirs. Nuclear Science and Technology. Szilard, Leo. 1898-1945. *59*
—. Capitalism. Central Intelligence Agency. National security. Nuclear arms. Political power. 1940's-70's. *432*
—. Economic conditions. Nuclear power plants. Subsidies. 1954-78. *718*
—. Employment. Nuclear energy. Private sector. 1963-73. *767*
—. Military Strategy. Nuclear Arms. Public Opinion. Strategic Arms Limitation Talks. 1945-78. *498*
—. Nuclear arms testing. Public Relations. ca 1948-80. *92*

—. Nuclear power plants. Pennsylvania (Shippingport). Private sector. ca 1940-57. *783*
Federal Policy. Armaments. Documents. Nuclear nonproliferation. 1981. *692*
—. Arms race. Germany, West. Nuclear arms. Power resources. 1968-70's. *572*
—. Atomic Energy Commission. Decisionmaking. Energy. Military. Pressure Groups. 1974. *720*
—. Carter, Jimmy (administration). Congress. Energy. Interest Groups. 1977-78. *756*
—. Carter, Jimmy (administration). Energy. National Energy Plan. 1973-78. *799*
—. Energy. 1979. *758*
—. Indian-White Relations. Navajo Indian Reservation. Southwest. Uranium mining. 1946-79. *82*
—. Nuclear Science and Technology. Safety. 1974. *778*
—. Politics. Weapons acquisition. 1950's-70's. *115*
Federal Regulation *See also* Government Regulation.
—. Accidents. Nuclear power plants. Nuclear Regulatory Commission. Pennsylvania (Three Mile Island). Safety. 1979. *780*
—. Atomic Energy Commission. Politics. Public Policy. Science and Government. 1946-70. *706*
—. Atomic Energy Commission. Public Relations. Safety. 1954-56. *763*
—. Licensing. Nuclear power industry. 1946-79. *803*
Federation of American Scientists. Congress. Nuclear Arms. Presidency. War Powers Resolution (1973). 1975-76. *578*
—. Political activity. Scientists. 1946-74. *653*
Fermi, Enrico. Anderson, Herbert L. (reminiscences). Chicago, University of. Columbia University. Nuclear Science and Technology. Szilard, Leo. 1933-45. *6*
Fighter squadrons, night. Baldwin, Paul H. (account). Nuclear War. Philippines. World War II (aerial operations). 1942-45. *9*
Films. Anti-Communist Movements. *Dr. Strangelove* (film). Films. Humor. Kubrick, Stanley. Liberalism. Nuclear Arms. 1950-64. *599*
—. Anti-Communist Movements. *Dr. Strangelove* (film). Films. Humor. Kubrick, Stanley. Liberalism. Nuclear Arms. 1950-64. *599*
—. *The Beginning or The End* (film). Manhattan Project. Nuclear Arms. Public Opinion. Scientists. 1947. *69*
Flexible response. Defense Policy. Europe. Neutron bomb. ca 1963-77. *260*
Flexible response options. Nuclear Arms. Schlesinger, James R. Strategic arms proposals. 1975. *366*
Flexible response strategy. Defense Policy. Marines. Military strategy. 1953-72. *198*
—. Deterrence (military). Nuclear Arms. 1972-74. *385*
Florida. Air Forces. Armament Development and Test Center. Eglin Air Force Base. 1960's-70's. *179*
Ford, Gerald R. Arms control. Brezhnev, Leonid. Foreign Relations. Kissinger, Henry A. USSR. Vladivostok Accord. 1974. *674*
—. Brezhnev, Leonid. Detente. Vladivostok Accord. 1974. *617*
Ford, Gerald R. (administration). Conference on Security and Cooperation in Europe. Europe. International Security. Public opinion. USA. 1975. *495*
Foreign aid. 1917-78. *390*
Foreign Policy. Afghanistan. 1979-80. *367*
—. Allison, Graham *(Essence of Decision).* Committees. Decisionmaking. Nuclear War. World War II. 1944-45. *54*
—. Angola. Mozambique. Nuclear potential. South Africa. USA. 1970-76. *427*
—. Arms control. Competition. Institutions. USSR. 1970-80. *585*
—. Arms control. Developing nations. Nuclear arms. USA. 1975. *468*
—. Arms control. Disarmament. Military strategy. USSR. 1945-70's. *214*
—. Arms control. Military Strategy. Nuclear arms. Strategic Arms Limitation Talks. 1960's-78. *596*
—. Arms control. Nuclear energy. 1945-80. *292*
—. Arms control. USSR. 1979-82. *622*
—. Arms Control and Disarmament Agency. Reagan, Ronald (administration). 1960-81. *505*
—. Balance of power. China. Detente. USSR. 1945-79. *218*
—. Balance of Power. Communism. USSR. 1965-78. *336*
—. Balance of power. Developing Nations. USSR. 1961-79. *556*
—. Balance of power. Energy development. Pacific Basin. USA. 1973-77. *736*
—. Canada. Great Britain. Nuclear Nonproliferation. USSR. 1943-76. *545*
—. Canada. Trusts, industrial. Uranium industry. USA. 1970's. *798*
—. Carter, Jimmy. 1820's-1970's. *569*
—. Carter, Jimmy. 1977. *422*
—. Carter, Jimmy. Disarmament. Inaugural Addresses. Morality. 1976. *460*
—. Carter, Jimmy. Nuclear nonproliferation. 1968-77. *652*
—. Carter, Jimmy. Nuclear nonproliferation. Political Leadership. 1977-79. *621*
—. Carter, Jimmy. Nuclear nonproliferation. Political Leadership. 1979. *480*
—. Carter, Jimmy (administration). 1978. *485*
—. Carter, Jimmy (administration). Communist Countries. Military Strategy. Reagan, Ronald (administration). 1980. *418*
—. Carter, Jimmy (administration). Congress. National security. Nixon, Richard M. (administration). Public scrutiny. 1970's. *211*
—. Carter, Jimmy (administration). Nuclear nonproliferation. 1977-81. *655*
—. Carter, Jimmy (administration). Strategic Arms Limitation Talks (II). USSR. 1970. *464*
—. Carter, Jimmy (views). USSR. 1977. *499*
—. Cold War. Diplomacy. Nuclear Arms. USA. USSR. 1942-46. *13*
—. Congress. Executive Branch. Strategic Arms Limitation Talks. 1972-78. *531*
—. Congress. Panama. Politics. Strategic Arms Limitation Talks. Treaties. 1977-78. *512*
—. Decisionmaking. 1960-78. *298*
—. Defense Policy. 1976. *443*
—. Defense Policy. Disarmament. Nuclear arms. Panel of Consultants on Disarmament (report). USSR. 1953. *486*
—. Defense Policy. Military Strategy. Strategic Arms Limitation Talks (II). USSR. 1960's-70's. *521*
—. Defensive war argument. Ideology. War. 1941-73. *398*
—. Detente. Human rights. USSR. 1973-79. *552*
—. Developing nations. USSR. 1970-79. *282*
—. Diplomacy. 1969-81. *555*
—. Diplomacy. Nuclear War. Roosevelt, Franklin D. Truman, Harry S. 1945-51. *14*

—. Disarmament. Nuclear Arms. USA. USSR. 1970's. *300*
—. Domestic policy. Historiography. Liberalism. Truman, Harry S. 1945-52. *275*
—. Europe. Kissinger, Henry A. Nixon, Richard M. (administration). Policymaking, patterns of. USA. 1971-75. *579*
—. Europe. Mutual Balanced Force Reductions. USA. 1970-73. *688*
—. Europe, Western. Nuclear energy. Nuclear nonproliferation. 1945-79. *794*
—. Executive Power. Political Participation. Public Opinion. 1960's-79. *687*
—. Executive Power. Truman, Harry S. 1945-53. *43*
—. Exports. Fuel. India. Law. Nuclear Non-Proliferation Act (US, 1978). 1978. *644*
—. Exports. Nuclear Arms (control). Technology. 1976. *435*
—. Falk, Richard A. (views). 1976. *525*
—. Gowing, Margaret. Great Britain. Nuclear Science and Technology (review article). Nuclear War. Sherwin, Martin J. 1938-52. *33*
—. Graduated response (policy). Military strategy. Vietnam War. 1950-75. *230*
—. Great Powers. National Security. Nuclear proliferation. 1946-79. *344*
—. Hydrogen bomb testing. Nuclear Arms. USSR. York, Herbert (hypothesis). 1953-55. 1970's. *84*
—. International Nuclear Fuel Cycle Evaluation. Nuclear energy. 1970's. *753*
—. International Trade. Nuclear power industry. Uranium cartel (possible). USA. 1970's. *815*
—. Israel (defense policy). Middle East. Nuclear arms, danger of. USA. 1973-75. *405*
—. Japan. 1970's. *503*
—. Kissinger, Henry A. Newhouse, John (review article). Nuclear arms. Strategic Arms Limitation Talks. 1973. *461*
—. Kissinger, Henry A. Nuclear Arms. 1957-78. *420*
—. Kissinger, Henry A. (review article). Nuclear Arms. 1957-75. *314*
—. Middle East. Strategic Arms Limitation Talks. USSR. 1977. *673*
—. Military Capability. Nitze, Paul. Senate. Strategic Arms Limitation Talks (II). 1945-79. *668*
—. Military Capability. USSR. 1952-82. *329*
—. Military Strategy. 1945-75. *403*
—. Military Strategy. Nuclear Arms. 1941-78. *208*
—. Military Strategy. Nuclear War. 1980. *272*
—. Military Strategy. Nuclear war, targets of. 1974. *375*
—. Models. Presidency. 1947-65. *235*
—. National Security. Nuclear Arms. 1946-50. *162*
—. National Security. Nuclear Arms. Strategic Arms Limitation Talks. USSR. 1950's-70's. *565*
—. National Security. Strategic Arms Limitation Talks (II). USSR. 1974-78. *592*
—. Nuclear arms. 1948-81. *245*
—. Nuclear Arms. 1962-74. *320*
—. Nuclear Arms. 1970's. *456*
—. Nuclear Arms. October War. USSR. 1973. *213*
—. Nuclear Arms. Strategic Arms Limitation Talks. 1969-72. *479*
—. Nuclear Arms. USA. USSR. 1969-74. *438*
—. Nuclear Energy. Nuclear nonproliferation. 1950's-70's. *620*
—. Nuclear nonproliferation. 1953-80. *619*
—. Nuclear proliferation. 1970's. *345*
—. Nuclear proliferation. Technology. 1946-80. *684*
—. Nuclear Proliferation. USA. 1954-74. *454*
—. Nuclear reactors. Technology. 1953-74. *290*
—. Nuclear War. Retaliation, massive. 1954-73. *281*
—. Nuclear-weapon free zones. 1970's. *553*
—. Peace, concepts of. 1940's-70's. *601*
—. South Africa. Uranium. ca 1961-79. *428*
—. Strategic balance. 1974. *266*
—. USA. USSR. 1945-75. *378*
—. USSR. 1960's-75. *421*
Foreign Relations. Afghanistan. Arms race. Cold War. USSR. 1960-80. *355*
—. Alperovitz, Gar. Nuclear War. World War II (review article). 1945. *27*
—. Armies. NATO. 1950-80. *365*
—. Arms control. Brezhnev, Leonid. Ford, Gerald R. Kissinger, Henry A. USSR. Vladivostok Accord. 1974. *674*
—. Arms control. Defense policy. USSR. 1960-80. *584*
—. Arms control. Detente. USA. USSR. 1961-75. *574*
—. Arms control. Deterrence. International organizations. 1960-75. *594*
—. Arms control. Disarmament. USA. USSR. 1970's. *664*
—. Arms control. Great Powers. 1960's-72. *508*
—. Arms race. Balance of Power. Deterrence. USSR. 1970's. *312*
—. Arms race. Great Powers. USSR. 1945-78. *660*
—. Arms race. Technologies. USA. USSR. 1973. *343*
—. Attitudes. Economics. Nuclear arms limitation. USA. USSR. 1972-73. *465*
—. Austria. Nuclear Test Ban Treaty. USSR. 1955. 1963. *581*
—. Balance of Power. USSR. 1960-69. *363*
—. Baruch Plan. Nuclear Energy. Nuclear proliferation. UN. 1945-77. *712*
—. Canada. Nuclear arms. Pearson, Lester B. (memoirs). USA. 1945. *42*
—. Cold War. Detente. Great Powers. USSR. 1945-75. *252*
—. Decisionmaking. Divine, Robert A. (review article). Nuclear arms. USSR. 1954-60. 1978. *195*
—. Defense Policy. Latin America. Nuclear arms. Tlatelolco, Treaty of. 1960's-70's. *637*
—. Developing Nations. Great Powers. Nuclear Nonproliferation Treaty. 1970's. *662*
—. Developing nations. Nuclear arms. 1960-80. *301*
—. Developing nations. Nuclear arms proliferation. 1974-79. *244*
—. Disarmament. 1945-70's. *690*
—. Disarmament. Military. Political power. USSR. 1945-82. *333*
—. Disarmament. USA. USSR. 1945-70's. *561*
—. Eisenhower, Dwight D. Nonaligned Nations. Nuclear energy. Nuclear proliferation. 1945-79. *311*
—. Europe. Military strategy. Nuclear arms. USA. USSR. 1950's-70's. *436*
—. Europe. Nuclear proliferation. 1954-80. *590*
—. Germany, West. 1976-81. *276*
—. Great Britain. Nuclear Arms. Roosevelt, Franklin D. USA. World War II. 1940-45. *16*
—. India. Nuclear nonproliferation. 1977-79. *361*
—. International Trade. Nuclear Energy. Nuclear Nonproliferation Act (US, 1978). Technology. 1978-79. *614*
—. Intervention. Military. Politics. 1946-75. *212*

—. Kissinger, Henry A. Nixon, Richard M. Secret agreements. Strategic Arms Limitation Talks. USA. USSR. 1972-74. *539*
—. Kissinger, Henry A. *(White House Years).* 1968-72. *559*
—. Mandelbaum, Michael. Nuclear Arms (review article). 1945-80. *32*
—. Military Strategy. Naval power. ca 1921-76. *136*
—. Military Strategy. Nuclear Arms. Strategic Arms Limitation Talks. USSR. 1972-77. *656*
—. Military strategy. Strategic Arms Limitation Talks (II). 1972. *583*
—. National security. Nixon Doctrine. Strategic Arms Limitation Talks. 1970-71. *632*
—. NATO. Nuclear Arms. USSR. 1970's-81. *455*
—. Nuclear arms. Strategic Arms Limitation Talks. USA. USSR. 1972-75. *602*
—. Nuclear arms limitation. USA. USSR. 1959-75. *527*
—. Nuclear Nonproliferation Treaty (1968). 1968-75. *484*
—. Nuclear War. 1974. *474*
—. Pakistan. 1970's. *459*
—. USSR. 1933-78. *628*
—. USSR. 1958-73. *591*
—. USSR. 1975-76. *200*
Founding Fathers. Nuclear holocaust, threat of. Political Theory. Posterity, commitment to. 1776-1976. *234*
France. Attitudes. Exports. Great Britain. Nuclear proliferation. Plutonium, reprocessed. 1946-78. *726*
—. Balance of Power. Great Britain. Newspapers. USSR. 1948-73. *262*
—. Great Britain. Nuclear arms testing. Secrecy. World War II. 1939-45. *25*
Franck, James. Atomic bomb. Nuclear arms race. Political Attitudes. Truman, Harry S. (administration). World War II. 1944-45. *62*
—. Atomic bomb. Policymaking. Scientists. Stimson, Henry L. World War II. 1945. *58*
Franco-Prussian War. Bismarck, Otto von. Cuban missile crisis. Diplomacy. Escalation theory. Kennedy, John F. McNamara, Robert S. 1870. 1962. *391*
Freedom of the press. Nuclear arms. Press. *Progressive* (periodical). *United States* v. *The Progressive* (US, 1979). 1979. *800*
Frisch, Otto R. (account). Nuclear Physics. Nuclear War. World War II. 1945. *24*
Fuel. Exports. Foreign policy. India. Law. Nuclear Non-Proliferation Act (US, 1978). 1978. *644*
Fusion, Thermonuclear. Lasers. Nuclear Science and Technology. Scientific Experiments and Research. USSR. USSR. 1971-75. *701*

G

Game theory. Europe, Central. Kahn, Herman. Military strategy. 1945-67. *382*
Garn, Jake. National Security. Political Speeches. Senate. Strategic Arms Limitation Talks (II). 1979. *533*
General Advisory Committee on Arms Control and Disarmament. Arms Control and Disarmament Agency. Disarmament. Federal Government. 1961-73. *507*
General Electric Co. Contracts. Economic Regulations. Nuclear power plants. Westinghouse. 1955-78. *711*
Generations, political. Attitudes. Nuclear war. 1974. *295*
George, Peter (*Red Alert*). Communism. *Dr. Strangelove* (film). Kubrick, Stanley. Paranoia, patterns of. 1950's. *437*
George Washington (vessel). Navies (Special Projects Office). Nuclear Arms. Polaris (missile). Submarine Warfare. Whitmore, William F. (account). 1955-60. *186*
Georgia Nuclear Laboratory. Nuclear Science and Technology. Radioactive waste. 1950's-79. *791*
Germany. Brazil. Diplomacy. NATO. Nuclear nonproliferation. 1968-77. *571*
—. Byrnes, James F. Federal Government. Memoirs. Nuclear Science and Technology. Szilard, Leo. 1898-1945. *59*
—. Einstein, Albert. Nuclear arms. Pacifism. Zionism. 1939-55. *21*
Germany, West. Arms race. Federal Policy. Nuclear arms. Power resources. 1968-70's. *572*
—. Brazil. Nuclear technology, transfer of. USA. 1945-76. *257*
—. Foreign Relations. 1976-81. *276*
Gillet, Edward B. (interview). Atomic Energy Commission. Deterrence. Military Strategy. Nuclear Arms (warheads). Research. 1970's. *411*
Goodpaster, Andrew. Defense Policy. Dulles, John Foster. Eisenhower, Dwight D. Nuclear test ban issue. Strauss, Lewis. USSR. 1958. *240*
Government. Business. Energy. Public Policy. Technology. 1978. *731*
—. Insurance. Liability. Public Opinion. Risk management. Taxation. 1948-78. *730*
Government Regulation *See also* Federal Regulation.
—. Nuclear waste. Safety. South. Transportation. ca 1960-79. *781*
Government secrecy. Canada. India. Nuclear Nonproliferation Treaty (1968). Nuclear technology. 1968-70's. *587*
—. Disarmament. Nuclear Arms. USA. 1945-70's. *562*
Gowing, Margaret. Foreign Policy. Great Britain. Nuclear Science and Technology (review article). Nuclear War. Sherwin, Martin J. 1938-52. *33*
Graduated response (policy). Foreign Policy. Military strategy. Vietnam War. 1950-75. *230*
Grand Canyon. Arizona. Mines. Uranium. 1540-1969. *156*
Great Britain. Antiwar Sentiment. Cruise missiles. Neutron Bomb. 1978-81. *441*
—. Attitudes. Exports. France. Nuclear proliferation. Plutonium, reprocessed. 1946-78. *726*
—. Balance of Power. France. Newspapers. USSR. 1948-73. *262*
—. Canada. Foreign Policy. Nuclear Nonproliferation. USSR. 1943-76. *545*
—. Churchill, Winston. Diplomacy. Military Strategy. Roosevelt, Franklin D. World War II. 1942-44. *61*
—. Disarmament. Nuclear Test Ban Treaty (1963). USSR. 1954-70's. *675*
—. Foreign Policy. Gowing, Margaret. Nuclear Science and Technology (review article). Nuclear War. Sherwin, Martin J. 1938-52. *33*
—. Foreign Relations. Nuclear Arms. Roosevelt, Franklin D. USA. World War II. 1940-45. *16*
—. France. Nuclear arms testing. Secrecy. World War II. 1939-45. *25*
Great Powers. Arms control. Foreign Relations. 1960's-72. *508*
—. Arms race. Foreign Relations. USSR. 1945-78. *660*

—. Cold War. Detente. Foreign Relations. USSR. 1945-75. *252*
—. Developing Nations. Foreign Relations. Nuclear Nonproliferation Treaty. 1970's. *662*
—. Foreign Policy. National Security. Nuclear proliferation. 1946-79. *344*
Grechko, Andrei. Arms race. Beam energy. ICBM's. USSR. 1960's-70's. *414*
Greenglass, David. Atomic Energy Commission. Espionage. Federal Bureau of Investigation. Nuclear Arms. Rosenberg case. Trials. 1942-74. *142*
—. Atomic Energy Commission. Nuclear Science and Technology. Rosenberg case. Secrecy. Trials. 1945-51. *75*
Gromyko, Andrei. Brown, Harrison (account). Nuclear arms. Scientists. UN. USSR. 1947. *482*
Groves, Leslie R. Defense Policy. Espionage. Nuclear arms. Secrecy. Truman, Harry S. (administration). USSR. 1945-50. *122*

H

Heilbroner, Robert. Developing Nations. Nuclear Arms. Peace, prospects for. Violence. 1976. *535*
Hersey, John (*Hiroshima*). Literature. Morality. Nuclear War. 1945-74. *70*
Hijacking. Atomic Energy Commission. Nuclear Arms. 1970-74. *737*
Hiroshima. *See* Japan (Hiroshima).
Historical advisory committees. Atomic Energy Commission. Defense Department. State Department. 1947-74. *754*
Historiography. Alperovitz, Gar. Diplomacy. Nuclear Arms. World War II. 1943-46. 1965-73. *323*
—. Alperovitz, Gar (review article). Decisionmaking. Japan (Hiroshima). Nuclear War. World War II. 1945-65. *52*
—. Cold War (origins). 1946-75. *197*
—. Domestic policy. Foreign policy. Liberalism. Truman, Harry S. 1945-52. *275*
History Teaching. Japan. Nuclear War. Simulation and Games. World War II. 1945. 1978. *19*
House of Representatives. Cannon, Clarence. Mexico. Military Service. Spanish-American War. World War I. World War II. 1898-1945. *35*
Housing. Nuclear accidents. Pennsylvania. Prices. Three Mile Island power plant. 1977-79. *771*
Human rights. Carter, Jimmy. Detente. Disarmament. Kissinger, Henry A. 1970-77. *694*
—. Detente. Foreign Policy. USSR. 1973-79. *552*
—. Diplomacy. European Security Conference. 1975. *607*
Humor. Anti-Communist Movements. *Dr. Strangelove* (film). Films. Films. Kubrick, Stanley. Liberalism. Nuclear Arms. 1950-64. *599*
Hungarian Americans. Einstein, Albert. Nuclear Arms (program). Szilard, Leo. USA. World War II. 1941-74. *26*
Hydrogen bomb. Attitudes. Scientists. 1950-55. *105*
—. Bikini Atoll. Japanese. *Lucky Dragon* (vessel). Pacific Dependencies (US). 1954-75. *164*
—. Decisionmaking. Nuclear Arms policy. York, Herbert (review article). 1949-75. *192*
—. Joint Chiefs of Staff. Military Strategy. Nuclear Arms. Truman, Harry S. USSR. 1945-50. *161*
Hydrogen bomb testing. Foreign policy. Nuclear Arms. USSR. York, Herbert (hypothesis). 1953-55. 1970's. *84*

I

ICBM's. ABM. Disarmament. MIRV. Strategic Arms Limitation Talks. 1969-73. *100*
—. American Aeronautics and Astronautics Associations. USSR. 1970's. *406*
—. Arms race. Beam energy. Grechko, Andrei. USSR. 1960's-70's. *414*
—. Balance of Power. USSR. 1972-75. *453*
—. Defense. Military Capability. Nuclear War. Research and development. USSR. 1970's. *413*
—. Defense Policy. Johnson, Lyndon B. Kennedy, John F. Kuter, Lawrence S. (account). 1962. *137*
—. Defense Policy. Land-mobile system, proposed. Technology. 1950's-74. *80*
—. Defense spending. Military Capability. Research and development. USSR. 1970's. *182*
—. Deterrence. Modernization. MX (missile). 1957-76. *169*
—. Deterrence. USSR. 1960's-70's. *416*
—. Military strategy. Nuclear arms. Schlesinger, James R. Strategic Arms Limitation Talks. 1974. *327*
—. Military strategy. Nuclear arms limitation. Technology. Vladivostok Accord. 1970's. *611*
—. MX (missile). National security. 1978. *113*
—. Nuclear Science and Technology. Terrorism. 1960's-70's. *704*
ICBM's (survivability). Nuclear War. USSR. 1960's-70's. *332*
Ideology. Clamshell Alliance. New Hampshire (Seabrook). Nuclear power plants. Political Protest. 1978. *741*
—. Defensive war argument. Foreign Policy. War. 1941-73. *398*
Illinois. Anderson, Herbert L. (reminiscences). Chicago, University of. Nuclear chain reaction. World War II. 1942. *5*
—. Chicago, University of. Nuclear chain reaction. Wattenberg, Albert (reminiscences). World War II. 1942. *64*
Inaugural Addresses. Carter, Jimmy. Disarmament. Foreign policy. Morality. 1976. *460*
India. Arms control. Canada. China. Nuclear Arms. Nuclear power. Regionalism. USA. 1951-78. *387*
—. Canada. Government secrecy. Nuclear Nonproliferation Treaty (1968). Nuclear technology. 1968-70's. *587*
—. Exports. Foreign policy. Fuel. Law. Nuclear Non-Proliferation Act (US, 1978). 1978. *644*
—. Foreign Relations. Nuclear nonproliferation. 1977-79. *361*
—. Nuclear Power. Trade. Treaties. Uranium. 1963-80. *774*
Indian Ocean and Area. Mediterranean Sea and Area. NATO. Naval strategy. USA. 1970's. *397*
Indian-White Relations. Federal Policy. Navajo Indian Reservation. Southwest. Uranium mining. 1946-79. *82*
Industrial safety. Courts. Nuclear waste. 1970-80. *806*
Industrialization. Arms control. Environment. International security. Multinational corporations. 1975. *650*
Industry. Energy crisis. Nuclear power. Technology. ca 1940-80. *745*

Inflation. Acquisition Cost Evaluation project. Air Forces. Brown, George S. (interview). Military Finance. 1970's. *181*
Information. Nuclear energy. Oak Ridge Associated Universities, Inc. Public Opinion. 1970-74. *785*
Initiatives. Elections. Nuclear safety. 1976. *734*
Institute for Advanced Study. Einstein, Albert. New Jersey (Princeton). 1933-55. *10*
Institute of Defense Analysis (Jason group). Attitudes. Scientific Experiments and Research. Vietnam War. 1966. *223*
Institutions. Arms control. Competition. Foreign Policy. USSR. 1970-80. *585*
Insurance. Government. Liability. Public Opinion. Risk management. Taxation. 1948-78. *730*
Intellectuals. Kissinger, Henry A. Political Attitudes. 1957-74. *341*
Intelligence Service. Defense spending. Military capability. USSR. 1960's-76. *434*
—. Nuclear arms. Strategic Arms Limitation Talks. USSR. 1970-79. *642*
Intercontinental Ballistic Missiles. *See* ICBM's.
Interest Groups. Attitudes. Democracy. Europe. Japan. Nuclear Science and Technology. USSR. 1970's. *739*
—. Carter, Jimmy (administration). Congress. Energy. Federal Policy. 1977-78. *756*
—. Nuclear energy. Oil Industry and Trade. Politics. 1979. *822*
—. Nuclear Regulatory Commission. Radioactive waste. Uranium Mill Tailings Radiation Control Act (US, 1978). 1978. *801*
Intergovernmental Relations. Annexation. New Mexico (Albuquerque). Nuclear Science and Technology. Weapons. 1940-78. *157*
—. Executive Planning Council. National Energy Acts. Public Policy. Radioactive waste. ca 1971-78. *757*
International Law. Europe. Nuclear materials. 1977-80. *738*
—. Strategic Arms Limitation Talks. Strategic vicinity (concept). USSR. 1972-74. *523*
International Nuclear Fuel Cycle Evaluation. Foreign Policy. Nuclear energy. 1970's. *753*
—. Nuclear Arms. 1978-79. *678*
International organization. Arms control. Developing nations. Nuclear capability. USA. 1950-75. *501*
International organizations. Arms control. Deterrence. Foreign Relations. 1960-75. *594*
International politics. Nuclear Arms (review article). 1945-75. *389*
International Relations (discipline). Action-reaction theory. Arms race. Models. USSR. 1970's. *331*
International relations theory. Arms race. Models. USSR. 1948-70. *289*
—. Coercion. Deterrence. Diplomacy. 1815-1970's. *316*
International security. Arms control. Environment. Industrialization. Multinational corporations. 1975. *650*
—. Arms control. World government. 1975. *524*
—. Arms control (myths of). Strategic balance. USA. USSR. 1975. *286*
—. Arms control (qualitative). Strategic Arms Limitation Talks. 1974-75. *514*
—. Arms race, qualitative. Military innovation. Nuclear test ban. Strategic Arms Limitation Treaty (1972). 1972-75. *220*
—. Arms races, rationales for. 1974. *273*
—. Conference on Security and Cooperation in Europe. Europe. Ford, Gerald R. (administration). Public opinion. USA. 1975. *495*
—. Decisionmaking. Defense policy. Missile defense. Safeguard debate. 1968-72. *175*
—. Detente. Diplomacy. Europe. USSR. 1970-74. *510*
—. Europe. NATO. Nuclear Arms. 1953-77. *224*
—. Mutual Balanced Force Reductions. 1970's. *686*
—. USSR. 1980. *681*
International stability. Defense Policy. Deterrence. Military strategy. 1975. *370*
International Trade. Arms race. Cold War. Economic Conditions. Truman, Harry S. (administration). 1948-49. *347*
—. Foreign policy. Nuclear power industry. Uranium cartel (possible). USA. 1970's. *815*
—. Foreign Relations. Nuclear Energy. Nuclear Nonproliferation Act (US, 1978). Technology. 1978-79. *614*
Intervention. Foreign Relations. Military. Politics. 1946-75. *212*
Israel. Central Intelligence Agency. Nuclear Arms. Nuclear Materials and Equipment Corporation. Pennsylvania (Apollo). Uranium. 1960's. *251*
Israel (defense policy). Foreign policy. Middle East. Nuclear arms, danger of. USA. 1973-75. *405*
Italy. Domestic Policy. Nuclear power plants. Pennsylvania (Three Mile Island). 1970's. *705*

J

Japan. Atomic bombs. Public opinion. USA. World War II. 1945-49. *68*
—. Attitudes. Democracy. Europe. Interest Groups. Nuclear Science and Technology. USSR. 1970's. *739*
—. Bombing. World War II. 1937-45. *18*
—. B-29 (aircraft). Lemay, Curtis. Nuclear War. Surrender. World War II. 1945. *67*
—. Cyclotrons. Denmark. Nuclear physics. Scientific Experiments and Research. 1935-45. *65*
—. Europe, Western. Mutual Balanced Force Reductions. Strategic Arms Limitation Talks. USA. USSR. 1974. *695*
—. Foreign policy. 1970's. *503*
—. History Teaching. Nuclear War. Simulation and Games. World War II. 1945. 1978. *19*
—. Leftists. Nuclear War. USSR. World War II. 1945. *17*
—. Military Strategy. Nuclear energy. Smyth Report, 1945. USA. World War II. 1940-45. *56*
—. Military strategy. Nuclear War. USA. World War II. 1945. *29*
—. Nuclear War. Politics. Surrender offer. Truman, Harry S. (administration). World War II. 1945. *12*
Japan (Hiroshima). Alperovitz, Gar (review article). Decisionmaking. Historiography. Nuclear War. World War II. 1945-65. *52*
—. Cold War. Nuclear War. World War II. 1945. *31*
Japan (Hiroshima, attack on). Decisionmaking. Nuclear War. World War II (Pacific Theater). 1944-45. *50*
Japan (Hiroshima, Nagasaki). Atomic Bomb Casualty Commission. Civilian survivors. Medical Research. Radiation Effects Research Foundation. 1945-75. *63*
—. Decisionmaking. Military Strategy. Nuclear War. World War II. 1944-45. *23*

—. Military Strategy. Nuclear War. World War II. 1945. *20*
—. Nuclear War. USA. USSR. World War II. 1942-47. *15*
—. Nuclear War. USA. World War II (strategy). 1945. *11*
—. Nuclear War. World War II. 1945. *40*
Japan (Kyoto). Nuclear War. World War II. 1945. *45*
Japanese. Bikini Atoll. Hydrogen bomb. *Lucky Dragon* (vessel). Pacific Dependencies (US). 1954-75. *164*
Johnson, Lyndon B. Defense Policy. ICBM's. Kennedy, John F. Kuter, Lawrence S. (account). 1962. *137*
Joint Chiefs of Staff. Brown, George S. (views). Military Strategy. Nuclear Arms. USSR. 1970's. *408*
—. Hydrogen bomb. Military Strategy. Nuclear Arms. Truman, Harry S. USSR. 1945-50. *161*
—. Military Strategy. Nuclear Arms. Single Integrated Operations Plan. Strategic Arms Limitation Talks. Target planning. 1972-74. *383*
Joint Committee on Atomic Energy. Atomic Energy Commission. Politics. 1939-79. *820*
—. Congress. Legislative investigations. Nuclear power. 1973-74. *719*
Journalism. Army Corps of Engineers. Nuclear energy. Oak Ridge *Journal*. Tennessee. 1943-49. *2*

K

Kahn, Herman. Brodie, Bernard. Military Strategy. National security. Schelling, Thomas. 1950's-76. *376*
—. Brodie, Bernard. Military Strategy. Nuclear Arms. 1950's-82. *358*
—. Europe, Central. Game theory. Military strategy. 1945-67. *382*
Kamen, Martin D. (reminiscences). California, University of, Berkeley. Lawrence Radiation Laboratory. Nuclear Science and Technology. 1936-39. *30*
Kansas. Michigan. Radioactive waste. 1970's. *697*
Kemeny Report. Accidents. Nuclear power plants. Pennsylvania (Three Mile Island). 1950-79. *824*
Kennedy, John F. Balance of power. Cuban Missile Crisis. Military Strategy. USA. USSR. 1962. *236*
—. Bismarck, Otto von. Cuban missile crisis. Diplomacy. Escalation theory. Franco-Prussian War. McNamara, Robert S. 1870. 1962. *391*
—. Cuban Missile Crisis. Khrushchev, Nikita. Military strategy. Nuclear war, threat of. 1962. *209*
—. Defense Policy. ICBM's. Johnson, Lyndon B. Kuter, Lawrence S. (account). 1962. *137*
Khrushchev, Nikita. Cuban Missile Crisis. Kennedy, John F. Military strategy. Nuclear war, threat of. 1962. *209*
King, William Lyon MacKenzie. Canada. Diaries. Nuclear War. Racism. 1921-48. *49*
Kissinger, Henry A. Arms control. Brezhnev, Leonid. Ford, Gerald R. Foreign Relations. USSR. Vladivostok Accord. 1974. *674*
—. Carter, Jimmy. Detente. Disarmament. Human rights. 1970-77. *694*
—. Europe. Foreign policy. Nixon, Richard M. (administration). Policymaking, patterns of. USA. 1971-75. *579*
—. Foreign Policy. Newhouse, John (review article). Nuclear arms. Strategic Arms Limitation Talks. 1973. *461*
—. Foreign policy. Nuclear Arms. 1957-78. *420*
—. Foreign Relations. Nixon, Richard M. Secret agreements. Strategic Arms Limitation Talks. USA. USSR. 1972-74. *539*
—. Intellectuals. Political Attitudes. 1957-74. *341*
Kissinger, Henry A. (review article). Foreign Policy. Nuclear Arms. 1957-75. *314*
Kissinger, Henry A. *(White House Years).* Foreign Relations. 1968-72. *559*
Kubrick, Stanley. Anti-Communist Movements. *Dr. Strangelove* (film). Films. Films. Humor. Liberalism. Nuclear Arms. 1950-64. *599*
—. Communism. *Dr. Strangelove* (film). George, Peter (*Red Alert*). Paranoia, patterns of. 1950's. *437*
Kuter, Lawrence S. (account). Defense Policy. ICBM's. Johnson, Lyndon B. Kennedy, John F. 1962. *137*

L

Laboratories. California (Livermore, Los Altos). Nuclear Science and Technology. Weapons. 1942-77. *101*
Land-mobile system, proposed. Defense Policy. ICBM's. Technology. 1950's-74. *80*
Laser fusion. Nuclear Arms. 1960-79. *171*
Lasers. Fusion, Thermonuclear. Nuclear Science and Technology. Scientific Experiments and Research. USSR. USSR. 1971-75. *701*
Latin America. Defense Policy. Foreign Relations. Nuclear arms. Tlatelolco, Treaty of. 1960's-70's. *637*
—. Nuclear arms. Tlatelolco, Treaty of (protocol). 1973-77. *520*
Law. Atomic Energy Commission. Security classification. 1945-75. *729*
—. Exports. Foreign policy. Fuel. India. Nuclear Non-Proliferation Act (US, 1978). 1978. *644*
Lawrence Livermore laboratory. Nuclear arms. York, Herbert. ca 1949-53. *191*
Lawrence Radiation Laboratory. Abelson, Philip H. (reminiscences). California, University of, Berkeley. Nuclear physics. 1935-40. *1*
—. Alvarez, Luis W. (reminiscences). California, University of, Berkeley. Nuclear Physics. 1934. *4*
—. California, University of, Berkeley. Kamen, Martin D. (reminiscences). Nuclear Science and Technology. 1936-39. *30*
Leftism. Anti-nuclear power movement. Clamshell Alliance. Political protest. 1970's. *740*
Leftists. Japan. Nuclear War. USSR. World War II. 1945. *17*
Legislation. Atomic Energy Commission. Safety. State Government. 1954-70. *808*
Legislative investigations. Congress. Joint Committee on Atomic Energy. Nuclear power. 1973-74. *719*
Lemay, Curtis. Air Forces. Military Capability. Nuclear Arms. Strategic Air Command. 1945-48. *88*
—. B-29 (aircraft). Japan. Nuclear War. Surrender. World War II. 1945. *67*
Letters. New Mexico (Los Alamos). Oppenheimer, J. Robert. Scientists. 1932-34. 1943-45. *55*
—. Nuclear Science and Technology. Oppenheimer, J. Robert (reminiscences). Physics. Smith, Alice Kimball. Weiner, Charles. ca 1900-45. 1980. *47*
Liability. Government. Insurance. Public Opinion. Risk management. Taxation. 1948-78. *730*

Liberalism. Anti-Communist Movements. *Dr. Strangelove* (film). Films. Films. Humor. Kubrick, Stanley. Nuclear Arms. 1950-64. *599*
—. Domestic policy. Foreign policy. Historiography. Truman, Harry S. 1945-52. *275*
Licensing. Federal Regulation. Nuclear power industry. 1946-79. *803*
—. Nuclear accidents. Reform. Three Mile Island power plant. 1977-79. *802*
Literature. Hersey, John (*Hiroshima*). Morality. Nuclear War. 1945-74. *70*
Little Theatre Group. Manhattan Project. New Mexico (Los Alamos). Nuclear Arms. Theater Production and Direction. World War II. 1943-46. *3*
Lobbyists. Airplanes, Military. B-1 (aircraft). Rockwell International Corp. Unemployment. 1975. *149*
Loyalty. Atomic Energy Commission. Executive Order 10450. Oppenheimer, J. Robert. Strauss, Lewis. 1946-54. *117*
Lucky Dragon (vessel). Bikini Atoll. Hydrogen bomb. Japanese. Pacific Dependencies (US). 1954-75. *164*

M

Magnets, floating. Atomic structure, theory of. Mayer, Alfred Marshall. Physics. Scientific Experiments and Research. 1878-1907. *57*
Management. Arms race. Military. Scientific Experiments and Research. Technology. 1945-75. *190*
Mandelbaum, Michael. Foreign Relations. Nuclear Arms (review article). 1945-80. *32*
Mandelbaum, Michael (review article). Nuclear Arms. 1946-79. *94*
Manhattan Project. Atomic Energy Commission. Engineers. Nuclear technology. Scientists. 1938-50. *28*
—. Bainbridge, Kenneth T. (reminiscences). New Mexico (Los Alamos). Nuclear Arms. World War II. 1945. *8*
—. *The Beginning or The End* (film). Films. Nuclear Arms. Public Opinion. Scientists. 1947. *69*
—. Colleges and Universities. Nuclear Arms. Oppenheimer, J. Robert. 1904-67. *48*
—. Cyclotrons. New Mexico (Los Alamos). Nuclear Science and Technology. Wilson, Robert R. (reminiscences). 1942-43. *66*
—. Little Theatre Group. New Mexico (Los Alamos). Nuclear Arms. Theater Production and Direction. World War II. 1943-46. *3*
Manley, John H. (reminiscences). New Mexico (Los Alamos). Nuclear Science and Technology. World War II. 1942-45. *36*
Manpower. Balance of Power. Defense spending. Military Capability. 1975-76. *452*
—. Defense Policy. Military. Strategic Arms Limitation Talks. USSR. 1973-74. *384*
Manufactures. American Atomics Corp. Arizona (Tucson). Public Health. Radiation. Tritium. Working Conditions. 1979. *709*
—. Nuclear Arms. South. ca 1942-79. *159*
Marines. Defense Policy. Flexible response strategy. Military strategy. 1953-72. *198*
Maritime law. Defense. Military strategy. Nuclear arms. 1970's. *353*
Marshall Islands. Atomic Energy Commission. Nuclear Science and Technology. Radiation exposure. 1942-72. *81*
—. Eniwetok (atoll). Nuclear testing. Pacific Dependencies (US). Safety. 1946-80. *127*
Marshall Islands (Bikini) *See also* Bikini Atoll, Eniwetok (Atoll).
—. Nuclear arms testing. Pacific Dependencies (US). Refugees. 1946-80. *184*
Marshall Plan. Cold War (origins). Nuclear Arms. Second Front. USSR. World War II. 1940's. *429*
Martial Law. Civil Rights. Nuclear Power Plants. Plutonium. Terrorism. 1970's. *728*
Maryland (Gaithersburg). Colorado (Boulder). Commerce Department. National Bureau of Standards. Scientific Experiments and Research. 1901-78. *813*
Massachusetts Institute of Technology. Defense. 1861-1979. *148*
Massive retaliation. Defense Policy. 1950-61. *430*
Mayer, Alfred Marshall. Atomic structure, theory of. Magnets, floating. Physics. Scientific Experiments and Research. 1878-1907. *57*
McDaniel, Boyce (reminiscences). New Mexico (Los Alamos). Nuclear Arms. Plutonium. World War II. 1940-45. *39*
McNamara, Robert S. Bismarck, Otto von. Cuban missile crisis. Diplomacy. Escalation theory. Franco-Prussian War. Kennedy, John F. 1870. 1962. *391*
Mechanism (philosophy). Arms control. Military strategy. Social Conditions. 1950's-60's. *661*
Medical Research. Atomic Bomb Casualty Commission. Civilian survivors. Japan (Hiroshima, Nagasaki). Radiation Effects Research Foundation. 1945-75. *63*
Mediterranean Sea and Area. Indian Ocean and Area. NATO. Naval strategy. USA. 1970's. *397*
Memoirs. Byrnes, James F. Federal Government. Germany. Nuclear Science and Technology. Szilard, Leo. 1898-1945. *59*
Mental health. Mothers. Nuclear accidents. Pennsylvania. Three Mile Island power plant. 1979. *708*
Metaphysics. Balance of Power. Nuclear Arms. USSR. 1970's. *222*
Mexico. Cannon, Clarence. House of Representatives. Military Service. Spanish-American War. World War I. World War II. 1898-1945. *35*
Meyer, John C. (interview). Balance of Power. Military Strategy. USSR. 1970's. *412*
Michigan. Kansas. Radioactive waste. 1970's. *697*
Michigan State University. Cyclotrons. Nuclear Science and Technology. 1958-81. *86*
Middle classes. Nuclear Arms. Peace Movements. 1980-82. *727*
Middle East. Arms Trade. Europe, Western. Nuclear equipment sales. USA. 1973-75. *306*
—. Cold War. Communism. Diplomacy. Morgenthau, Hans J. Nuclear war. USSR. 1945-76. *399*
—. Foreign policy. Israel (defense policy). Nuclear arms, danger of. USA. 1973-75. *405*
—. Foreign Policy. Strategic Arms Limitation Talks. USSR. 1977. *673*
Military. Arms race. Management. Scientific Experiments and Research. Technology. 1945-75. *190*
—. Atomic Energy Commission. Decisionmaking. Energy. Federal policy. Pressure Groups. 1974. *720*
—. Balance of Power. 1953-75. *450*
—. Balance of Power. NATO. Warsaw Pact. 1962-75. *449*
—. Budgets. Congress. Defense Policy. Research and development. 1976-77. *102*

—. California, University of. Nuclear arms. 1945-78. *167*
—. Civil defense. Defense Policy. Nuclear War. Public Opinion. USSR. 1948-78. *424*
—. Congress. Defense Department. Defense industries. Public Opinion. 1980's. *194*
—. Defense Policy. Manpower. Strategic Arms Limitation Talks. USSR. 1973-74. *384*
—. Defense research. Research Board for National Security. Science and Government. 1944-50. *132*
—. Detente. European Security Conference (measures of confidence). 1968-77. *544*
—. Disarmament. Foreign Relations. Political power. USSR. 1945-82. *333*
—. Foreign Relations. Intervention. Politics. 1946-75. *212*
—. Missiles. USSR. 1955-80. *170*
—. MX (missile). Politics. 1960's-81. *104*
—. National security. Research and development. Weapons. 1978. *123*
—. Strategic Arms Limitation Talks (II). USSR. 1973. *566*
Military Capability. Air Forces. Lemay, Curtis. Nuclear Arms. Strategic Air Command. 1945-48. *88*
—. Air Forces. Weapons. 1970's. *183*
—. Arms race. Nuclear Arms. USSR. 1960-80. *404*
—. Balance of Power. Cruise missiles. Defense Department. Strategic Arms Limitation Talks (II). ca 1950-79. *125*
—. Balance of Power. Defense spending. Manpower. 1975-76. *452*
—. Balance of Power. USSR. 1980. *238*
—. Defense. ICBM's. Nuclear War. Research and development. USSR. 1970's. *413*
—. Defense Policy. Nuclear Arms. USSR. 1970's. *297*
—. Defense Policy. USSR. 1977-80. *374*
—. Defense spending. ICBM's. Research and development. USSR. 1970's. *182*
—. Defense spending. Intelligence Service. USSR. 1960's-76. *434*
—. Deterrence. Nuclear arms. USSR. 1945-50. *337*
—. Foreign Policy. Nitze, Paul. Senate. Strategic Arms Limitation Talks (II). 1945-79. *668*
—. Foreign Policy. USSR. 1952-82. *329*
—. NATO. Nuclear arms. Technology. USA. USSR. 1970's. *98*
—. NATO. Nuclear Arms. Warsaw Pact. 1970. *226*
—. Nuclear arms. USA. USSR. 1946-72. *89*
—. Strategic Arms Limitation Talks (II). Treaties. USSR. 1950's-70's. *613*
—. USSR. 1973-74. *451*
—. USSR. 1974-75. *448*
—. USSR. 1979. *442*
Military Finance. Acquisition Cost Evaluation project. Air Forces. Brown, George S. (interview). Inflation. 1970's. *181*
—. Defense Department. Naval Strategy. Nuclear War. Submarine Warfare. Tidal waves. 1950's-80. *131*
Military (force reductions). Disarmament. NATO. USSR. 1955-73. *667*
Military Ground Forces. Balance of Power. Nuclear Arms. USSR. 1969-77. *193*
Military innovation. Arms race, qualitative. International security. Nuclear test ban. Strategic Arms Limitation Treaty (1972). 1972-75. *220*
Military intelligence. Air Forces (Air Targets Division). Bureaucracies. Morality. Nash, Henry T. (account). Nuclear War. Planning. 1950's-70's. *147*
—. Arms control. USSR. 1960-82. *518*
—. Nuclear submarines. 1979. *138*
—. Strategic Arms Limitation Talks (II). Telemetry (encryption). Verification. 1977-79. *663*
Military officers, role of. Arms Control and Disarmament Agency. 1974. *506*
Military policy. Deterrence. Strategy. -1972. *291*
Military potential. Arms race. USA. USSR. 1940-74. *445*
Military power. Civil-military relations. Nuclear Science and Technology. -1973. *362*
Military Service. Cannon, Clarence. House of Representatives. Mexico. Spanish-American War. World War I. World War II. 1898-1945. *35*
Military Strategy. ABM controversy. Defense Policy. 1965-80. *114*
—. Aerospace systems. Defense Policy. 1977. *128*
—. Air Force Association (symposium). Balance of Power. National Security. 1964-76. *407*
—. Air Force Association (symposium). Strategic Arms Limitation Talks (II). Weapons. 1976. *670*
—. Air Forces. Defense Policy. North American Air Defense Command. 1970's. *415*
—. Armaments. 1945-78. *335*
—. Armies. 1970's. *284*
—. Arms control. Disarmament. Foreign Policy. USSR. 1945-70's. *214*
—. Arms control. Foreign Policy. Nuclear arms. Strategic Arms Limitation Talks. 1960's-78. *596*
—. Arms control. Mechanism (philosophy). Social Conditions. 1950's-60's. *661*
—. Atomic Energy Commission. Deterrence. Gillet, Edward B. (interview). Nuclear Arms (warheads). Research. 1970's. *411*
—. Attitudes. Economic development. Nuclear arms. Technology. War. 1914-68. *351*
—. Balance of power. Cuban Missile Crisis. Kennedy, John F. USA. USSR. 1962. *236*
—. Balance of Power. Meyer, John C. (interview). USSR. 1970's. *412*
—. Brodie, Bernard. Kahn, Herman. National security. Schelling, Thomas. 1950's-76. *376*
—. Brodie, Bernard. Kahn, Herman. Nuclear Arms. 1950's-82. *358*
—. Brown, George S. (views). Joint Chiefs of Staff. Nuclear Arms. USSR. 1970's. *408*
—. Bureaucracies. Nuclear war. 1974. *280*
—. Carter, Jimmy (administration). Communist Countries. Foreign Policy. Reagan, Ronald (administration). 1980. *418*
—. Carter, Jimmy (administration). Nuclear Arms. Presidential Directive No. 59. 1960-80. *139*
—. Churchill, Winston. Diplomacy. Great Britain. Roosevelt, Franklin D. World War II. 1942-44. *61*
—. Civil defense. Nuclear War. USSR. 1950's-77. *392*
—. Clausewitz, Karl von (theories). Nuclear War, limited. War, principles of. 1970's. *342*
—. Communism. Containment. 1941-70's. *263*
—. Conflict theory. Deterrence. USA. USSR. 1945-75. *258*
—. Counterforce (concept). Deterrence. Nuclear Arms. 1970's. *340*
—. Cuban Missile Crisis. Kennedy, John F. Khrushchev, Nikita. Nuclear war, threat of. 1962. *209*
—. Decisionmaking. Japan (Hiroshima, Nagasaki). Nuclear War. World War II. 1944-45. *23*
—. Defense. Maritime law. Nuclear arms. 1970's. *353*
—. Defense policy. Deterrence. 1950's-72. *202*
—. Defense Policy. Deterrence. International stability. 1975. *370*

—. Defense Policy. Flexible response strategy. Marines. 1953-72. *198*
—. Defense Policy. Foreign policy. Strategic Arms Limitation Talks (II). USSR. 1960's-70's. *521*
—. Defense policy. Mutual assured destruction. Nuclear War. USA. USSR. 1960-76. *196*
—. Defense Policy. Stability, strategic. USSR. 1960-80. *271*
—. Defense Policy. USSR. ca 1960-80. *210*
—. Defense spending. Eisenhower, Dwight D. 1952-60. *303*
—. Defense spending. National security. NATO. USSR. 1980. *359*
—. Detente. Europe. USSR. 1950's-70's. *444*
—. Detente. Missile development. Strategic Arms Limitation Talks (II). 1974-75. *548*
—. Deterrence. 1950-75. *328*
—. Deterrence. 1974. *227*
—. Deterrence. Nuclear Arms. Technological innovations. USSR. 1969-78. *338*
—. Deterrence. Nuclear arms. USSR. 1950-80. *318*
—. Deterrence. Nuclear War. Schlesinger, James R. 1974. *305*
—. Deterrence theory. Nuclear Arms. 1966-76. *270*
—. Documents. Nuclear Arms. USSR. 1954-55. *373*
—. Dropshot (plan). USSR. 1949-57. *87*
—. Europe. Foreign relations. Nuclear arms. USA. USSR. 1950's-70's. *436*
—. Europe. NATO. 1945-81. *352*
—. Europe. NATO. North America. Nuclear Arms (tactical, conventional). 1967-74. *112*
—. Europe, Central. Game theory. Kahn, Herman. 1945-67. *382*
—. Europe, Western. Missiles. NATO. Pershing II (missile). 1956-80. *151*
—. Europe, Western. NATO. Nuclear Arms. USSR. 1950-78. *85*
—. Europe, Western. NATO. Nuclear arms, tactical. USA. USSR. 1970's. *256*
—. Federal Government. Nuclear Arms. Public Opinion. Strategic Arms Limitation Talks. 1945-78. *498*
—. Foreign Policy. 1945-75. *403*
—. Foreign Policy. Graduated response (policy). Vietnam War. 1950-75. *230*
—. Foreign policy. Nuclear Arms. 1941-78. *208*
—. Foreign Policy. Nuclear War. 1980. *272*
—. Foreign policy. Nuclear war, targets of. 1974. *375*
—. Foreign Relations. Naval power. ca 1921-76. *136*
—. Foreign Relations. Nuclear Arms. Strategic Arms Limitation Talks. USSR. 1972-77. *656*
—. Foreign relations. Strategic Arms Limitation Talks (II). 1972. *583*
—. Hydrogen bomb. Joint Chiefs of Staff. Nuclear Arms. Truman, Harry S. USSR. 1945-50. *161*
—. ICBM's. Nuclear arms. Schlesinger, James R. Strategic Arms Limitation Talks. 1974. *327*
—. ICBM's. Nuclear arms limitation. Technology. Vladivostok Accord. 1970's. *611*
—. Japan. Nuclear energy. Smyth Report, 1945. USA. World War II. 1940-45. *56*
—. Japan. Nuclear War. USA. World War II. 1945. *29*
—. Japan (Hiroshima, Nagasaki). Nuclear War. World War II. 1945. *20*
—. Joint Chiefs of Staff. Nuclear Arms. Single Integrated Operations Plan. Strategic Arms Limitation Talks. Target planning. 1972-74. *383*
—. Missiles. 1945-80. *97*
—. Missiles. Nuclear Arms. 1945-76. *166*
—. Missiles (vulnerability). USA. USSR. 1975. *394*
—. Mutual Balanced Force Reductions. NATO. Warsaw Pact (negotiations). 1973-74. *606*
—. National Characteristics. Nuclear Arms. 20c. *268*
—. National security. Nuclear Arms. USA. USSR. 1954-74. *309*
—. NATO. 1960-69. *371*
—. NATO. Nuclear arms. 1949-82. *388*
—. NATO. Nuclear Arms (tactical). 1974. *368*
—. NATO. Nuclear Arms (tactical). 1974. *369*
—. NATO. Nuclear arms, tactical. USA. Weapons, conventional. 1970's. *225*
—. NATO. USA (role of). Warsaw Pact. 1960's-75. *247*
—. Nuclear arms. 1960's-73. *233*
—. Nuclear Arms. 1970-80. *199*
—. Nuclear Arms. 1974-81. *250*
—. Nuclear Arms. 1978-81. *207*
—. Nuclear arms. Schlesinger, James R. USA. USSR. 1970's. *229*
—. Nuclear Arms. SLBM. Strategic arms control. 1950's-70's. *79*
—. Nuclear Arms. Strategic Arms Limitation Talks. 1970's. *551*
—. Nuclear arms. Strategic Arms Limitation Talks. USSR. 1969-78. *567*
—. Nuclear Arms. Strategic Arms Limitation Talks. USSR. 1970-81. *204*
—. Nuclear arms. USSR. 1945-78. *219*
—. Nuclear Arms. USSR. 1950's-79. *296*
—. Nuclear Arms. USSR. 1964-81. *377*
—. Nuclear energy. Publishers and Publishing. Smyth Report, 1945. War Department. 1946-73. *172*
—. Nuclear War. 1945-80. *255*
—. Nuclear War. 1950-80. *334*
—. Nuclear War. Targeting. 1980. *74*
—. Nuclear War. USSR. 1977. *356*
—. Planning. USSR. 1945-79. *339*
—. Sea-launched ballistic missiles. Weapons. 1969-82. *187*
—. Technology. Weapons. 1970's. *409*
—. USSR. 1950's-80. *308*
Military Strategy (studies). Vietnam War (ramifications). 1945-73. *232*
Military strength. China. Coffey, Joseph I. Strategic power. USA. USSR. ca 1965-73. *203*
Military systems. Defense. Economic strength. 1945-75. *243*
Miller, Walter *(A Canticle for Leibowitz).* Nuclear Energy. Religion. Science fiction. 1959-81. *795*
Mines. Arizona. Grand Canyon. Uranium. 1540-1969. *156*
MIRV. ABM. Disarmament. ICBM's. Strategic Arms Limitation Talks. 1969-73. *100*
—. Strategic Arms Limitation Talks (II). USA. USSR. 1970's. *645*
Missile capability. Defense spending. Nuclear arms race. USA. USSR. 1970's. *433*
Missile defense. Decisionmaking. Defense policy. International Security. Safeguard debate. 1968-72. *175*
Missile development. Detente. Military strategy. Strategic Arms Limitation Talks (II). 1974-75. *548*
Missiles. Air Forces. Space Vehicles. 1954-74. *118*
—. Arms race. Navies. Nuclear submarines. Trident (missile). 1960-80. *158*
—. Drones. Fahrney, Delmar S. (account). Navy Bureau of Aeronautics. N2C-2 (aircraft). Radio. Research and development. TG-2 (aircraft). World War II. 1936-45. *107*

—. Europe, Western. Military Strategy. NATO. Pershing II (missile). 1956-80. *151*
—. Military. USSR. 1955-80. *170*
—. Military Strategy. 1945-80. *97*
—. Military Strategy. Nuclear Arms. 1945-76. *166*
—. Naval Strategy. 1970-81. *90*
—. Naval Strategy. Strategic Arms Limitation Talks (II). 1979. *541*
—. Navies. 1936-52. *106*
—. Nuclear Arms. 1954-79. *96*
—. Nuclear Arms. 1955-80. *77*
Missiles (vulnerability). Military Strategy. USA. USSR. 1975. *394*
Mobilization. Defense Policy. Deterrence. Planning. 1978. *228*
Models. Action-reaction theory. Arms race. International Relations (discipline). USSR. 1970's. *331*
—. Arms race. International relations theory. USSR. 1948-70. *289*
—. Foreign policy. Presidency. 1947-65. *235*
Modernization. Deterrence. ICBM's. MX (missile). 1957-76. *169*
—. Europe. NATO. Nuclear Arms. 1978-80. *302*
Monopolies. Nuclear Science and Technology. 1970's. *707*
Morale. Battles. Deterrence. Nuclear War. World War II. 1944-81. *130*
Morality. Air Forces (Air Targets Division). Bureaucracies. Military intelligence. Nash, Henry T. (account). Nuclear War. Planning. 1950's-70's. *147*
—. Carter, Jimmy. Disarmament. Foreign policy. Inaugural Addresses. 1976. *460*
—. Hersey, John (*Hiroshima*). Literature. Nuclear War. 1945-74. *70*
Morgenthau, Hans J. Cold War. Communism. Diplomacy. Middle East. Nuclear war. USSR. 1945-76. *399*
Mothers. Mental health. Nuclear accidents. Pennsylvania. Three Mile Island power plant. 1979. *708*
Mozambique. Angola. Foreign policy. Nuclear potential. South Africa. USA. 1970-76. *427*
Multinational corporations. Arms control. Environment. Industrialization. International security. 1975. *650*
Multiple delivery system. Defense Policy. Naval Strategy (Blue Water Strategy). Strategic Arms Limitation Talks. 1974. *683*
Mutual assured destruction. Defense policy. Military Strategy. Nuclear War. USA. USSR. 1960-76. *196*
—. National security. Nonintervention. Strategic Arms Limitation Talks (II). 1980. *625*
Mutual Balanced Force Reductions. Arms control. Austria (Vienna). Detente. 1973-80. *522*
—. Arms control. Detente. USA. USSR. 1974. *634*
—. Arms control. NATO. Warsaw Pact. 1973-74. *635*
—. Defense Policy. Europe, Western. USSR. 1970's. *689*
—. Disarmament. Europe. USSR. 1973. *586*
—. Europe. Foreign policy. USA. 1970-73. *688*
—. Europe, Western. Japan. Strategic Arms Limitation Talks. USA. USSR. 1974. *695*
—. International Security. 1970's. *686*
—. Military Strategy. NATO. Warsaw Pact (negotiations). 1973-74. *606*
MX (missile). Defense Policy. Nevada. 1979-80. *189*
—. Deterrence. ICBM's. Modernization. 1957-76. *169*
—. ICBM's. National security. 1978. *113*
—. Military. Politics. 1960's-81. *104*

N

Nash, Henry T. (account). Air Forces (Air Targets Division). Bureaucracies. Military intelligence. Morality. Nuclear War. Planning. 1950's-70's. *147*
National Bureau of Standards. Colorado (Boulder). Commerce Department. Maryland (Gaithersburg). Scientific Experiments and Research. 1901-78. *813*
National Characteristics. Military Strategy. Nuclear Arms. 20c. *268*
National Council of Churches. Energy. Ethics. 1976-78. *732*
National Energy Acts. Executive Planning Council. Intergovernmental Relations. Public Policy. Radioactive waste. ca 1971-78. *757*
National Energy Plan. Carter, Jimmy (administration). Energy. Federal Policy. 1973-78. *799*
National Security. Air Force Association (symposium). Balance of Power. Military Strategy. 1964-76. *407*
—. Arms control. Defense Policy. 1960's-75. *636*
—. Bibliographies. 1960's-70's. *274*
—. Brodie, Bernard. Kahn, Herman. Military Strategy. Schelling, Thomas. 1950's-76. *376*
—. Capitalism. Central Intelligence Agency. Federal government. Nuclear arms. Political power. 1940's-70's. *432*
—. Carter, Jimmy (administration). Congress. Foreign policy. Nixon, Richard M. (administration). Public scrutiny. 1970's. *211*
—. Censorship, voluntary. Press. World War II. 1941-80. *515*
—. Communist Countries. Exports. Technology. 1949-78. *242*
—. Defense policy. Nuclear War. Peace. 1977. *400*
—. Defense spending. Economic Planning. Peace research. 1960's-77. *616*
—. Defense spending. Military Strategy. NATO. USSR. 1980. *359*
—. Disarmament. Strategic Arms Limitation Talks. USA. USSR. Verification methods. 1972-76. *593*
—. Elites. Political Attitudes. Youth. 1945-75. *431*
—. Foreign Policy. Great Powers. Nuclear proliferation. 1946-79. *344*
—. Foreign policy. Nuclear Arms. 1946-50. *162*
—. Foreign Policy. Nuclear Arms. Strategic Arms Limitation Talks. USSR. 1950's-70's. *565*
—. Foreign Policy. Strategic Arms Limitation Talks (II). USSR. 1974-78. *592*
—. Foreign Relations. Nixon Doctrine. Strategic Arms Limitation Talks. 1970-71. *632*
—. Garn, Jake. Political Speeches. Senate. Strategic Arms Limitation Talks (II). 1979. *533*
—. ICBM's. MX (missile). 1978. *113*
—. Military. Research and development. Weapons. 1978. *123*
—. Military Strategy. Nuclear Arms. USA. USSR. 1954-74. *309*
—. Mutual Assured Destruction. Nonintervention. Strategic Arms Limitation Talks (II). 1980. *625*
—. Nuclear arms. 1977. *325*
—. Nuclear Arms. Strategic Arms Limitation Talks. USSR. 1972-78. *696*

—. Politics. Strategic Arms Limitation Talks (II). Weapons. 1977-78. *612*
—. Strategic weapons limitation (review article). 1958-74. *575*
National Security (review article). Arms control. Stockholm International Peace Research Institute. 1979. *604*
NATO. 1960-80. *253*
—. Armies. Foreign Relations. 1950-80. *365*
—. Arms control. Mutual Balanced Force Reductions. Warsaw Pact. 1973-74. *635*
—. Arms race. Detente. Disarmament. 1950's-76. *605*
—. Balance of Power. Military. Warsaw Pact. 1962-75. *449*
—. Brazil. Diplomacy. Germany. Nuclear nonproliferation. 1968-77. *571*
—. Canada. Neutron bomb. Nuclear War. USA. 1977-79. *141*
—. Consultation. Decisionmaking. Diplomacy. 1957-78. *304*
—. Defense Policy. Deterrence. Europe, Western. Nuclear arms. Warsaw Pact. 1970's. *326*
—. Defense Policy. Nuclear arms. 1967-82. *254*
—. Defense spending. Military Strategy. National security. USSR. 1980. *359*
—. Detente. Disarmament. 1970's. *357*
—. Detente. Europe. Nuclear War. USA. 1970's. *269*
—. Deterrence. Neutron bomb. Nuclear Arms. Warsaw Pact. 1954-78. *83*
—. Disarmament. Military (force reductions). USSR. 1955-73. *667*
—. Disarmament. Nuclear arms. Theater Nuclear Forces. 1977-79. *278*
—. Disarmament. USSR. 1932-82. *626*
—. Economic Conditions. Energy. Nuclear War. 1950's-70's. *216*
—. Europe. International Security. Nuclear Arms. 1953-77. *224*
—. Europe. Military Strategy. 1945-81. *352*
—. Europe. Military Strategy. North America. Nuclear Arms (tactical, conventional). 1967-74. *112*
—. Europe. Modernization. Nuclear Arms. 1978-80. *302*
—. Europe, Western. Military Strategy. Missiles. Pershing II (missile). 1956-80. *151*
—. Europe, Western. Military Strategy. Nuclear Arms. USSR. 1950-78. *85*
—. Europe, Western. Military strategy. Nuclear arms, tactical. USA. USSR. 1970's. *256*
—. Europe, Western. Nuclear arms. USSR. 1979. *402*
—. Foreign Relations. Nuclear Arms. USSR. 1970's-81. *455*
—. Indian Ocean and Area. Mediterranean Sea and Area. Naval strategy. USA. 1970's. *397*
—. Military Capability. Nuclear arms. Technology. USA. USSR. 1970's. *98*
—. Military Capability. Nuclear Arms. Warsaw Pact. 1970. *226*
—. Military strategy. 1960-69. *371*
—. Military Strategy. Mutual Balanced Force Reductions. Warsaw Pact (negotiations). 1973-74. *606*
—. Military Strategy. Nuclear arms. 1949-82. *388*
—. Military Strategy. Nuclear Arms (tactical). 1974. *368*
—. Military Strategy. Nuclear Arms (tactical). 1974. *369*
—. Military strategy. Nuclear arms, tactical. USA. Weapons, conventional. 1970's. *225*
—. Military Strategy. USA (role of). Warsaw Pact. 1960's-75. *247*
—. Naval Strategy. 1980. *73*
—. Nuclear Arms. 1950-82. *386*
—. Nuclear Arms. Theater nuclear forces. 1950's. *395*
Natural Resources. Energy. 1973-76. *787*
Navajo Indian Reservation. Federal Policy. Indian-White Relations. Southwest. Uranium mining. 1946-79. *82*
Naval Air Forces. Nuclear Arms. P2V (aircraft). 1949-50. *120*
Naval construction. USA. USSR. Western nations. 1965-76. *143*
Naval power. Foreign Relations. Military Strategy. ca 1921-76. *136*
Naval presence. Conflict avoidance. Defense Policy. Persian Gulf. USA. 1946-75. *237*
Naval Strategy. Aircraft carriers. 1945-74. *177*
—. Defense Department. Military Finance. Nuclear War. Submarine Warfare. Tidal waves. 1950's-80. *131*
—. Indian Ocean and Area. Mediterranean Sea and Area. NATO. USA. 1970's. *397*
—. Missiles. 1970-81. *90*
—. Missiles. Strategic Arms Limitation Talks (II). 1979. *541*
—. NATO. 1980. *73*
—. Nuclear arms. USSR. 1950's-80. *330*
—. Submarines. USA. USSR. 1970's. *279*
Naval Strategy (Blue Water Strategy). Defense Policy. Multiple delivery system. Strategic Arms Limitation Talks. 1974. *683*
Naval Vessels. Weapons. 1930-76. *178*
Navies. Antarctic (McMurdo Sound). Nuclear reactors. Radioactive waste. 1960-71. *814*
—. Arms race. Missiles. Nuclear submarines. Trident (missile). 1960-80. *158*
—. Bikini Atoll. Nuclear testing. Operation Crossroads. 1946. *116*
—. Missiles. 1936-52. *106*
—. *Norton Sound* (vessel). Research. Technology. Weapons. 1943-79. *155*
—. Nuclear Arms. Submarines. 1975. *111*
—. Nuclear weapons. USA. USSR. 1945-74. *264*
Navies (Special Projects Office). *George Washington* (vessel). Nuclear Arms. Polaris (missile). Submarine Warfare. Whitmore, William F. (account). 1955-60. *186*
Navy Bureau of Aeronautics. Drones. Fahrney, Delmar S. (account). Missiles. N2C-2 (aircraft). Radio. Research and development. TG-2 (aircraft). World War II. 1936-45. *107*
Negotiations. Arms race. Diplomacy. Nuclear test ban. Richardson models. USSR. 1962-63. *558*
Neutron Bomb. Antiwar Sentiment. Cruise missiles. Great Britain. 1978-81. *441*
—. Canada. NATO. Nuclear War. USA. 1977-79. *141*
—. Defense Policy. Europe. Flexible response. ca 1963-77. *260*
—. Deterrence. NATO. Nuclear Arms. Warsaw Pact. 1954-78. *83*
Nevada. Defense Policy. MX (missile). 1979-80. *189*
New Hampshire (Seabrook). Clamshell Alliance. Ideology. Nuclear power plants. Political Protest. 1978. *741*
New Jersey (Princeton). Einstein, Albert. Institute for Advanced Study. 1933-55. *10*
New Mexico (Albuquerque). Annexation. Intergovernmental Relations. Nuclear Science and Technology. Weapons. 1940-78. *157*
New Mexico (Los Alamos). Atomic bomb. World War II. 1943-46. *41*
—. Bainbridge, Kenneth T. (reminiscences). Manhattan Project. Nuclear Arms. World War II. 1945. *8*

—. Cyclotrons. Manhattan Project. Nuclear Science and Technology. Wilson, Robert R. (reminiscences). 1942-43. *66*
—. Letters. Oppenheimer, J. Robert. Scientists. 1932-34. 1943-45. *55*
—. Little Theatre Group. Manhattan Project. Nuclear Arms. Theater Production and Direction. World War II. 1943-46. *3*
—. Manley, John H. (reminiscences). Nuclear Science and Technology. World War II. 1942-45. *36*
—. McDaniel, Boyce (reminiscences). Nuclear Arms. Plutonium. World War II. 1940-45. *39*
—. Nuclear Arms (first testing). 1945. *22*
New York (Cayuga Lake). Environmentalism. Experts. Nuclear power plants. Pressure groups. 1973. *770*
New York (Love Canal). Waste, toxic. 1894-1980. *819*
New York (West Valley). Atomic Energy Commission. Radioactive waste. 1954-78. *786*
Newhouse, John (review article). Foreign Policy. Kissinger, Henry A. Nuclear arms. Strategic Arms Limitation Talks. 1973. *461*
Newspapers. Balance of Power. France. Great Britain. USSR. 1948-73. *262*
Nitze, Paul. Foreign Policy. Military Capability. Senate. Strategic Arms Limitation Talks (II). 1945-79. *668*
Nixon Doctrine. Foreign Relations. National security. Strategic Arms Limitation Talks. 1970-71. *632*
Nixon, Richard M. Defense Policy. Nuclear arms. Strategic Arms Limitation Talks (II). 1973. *665*
—. Foreign Relations. Kissinger, Henry A. Secret agreements. Strategic Arms Limitation Talks. USA. USSR. 1972-74. *539*
Nixon, Richard M. (administration). Carter, Jimmy (administration). Congress. Foreign policy. National security. Public scrutiny. 1970's. *211*
—. Detente. Economic relations. Strategic arms limitation. USSR. 1969-73. *648*
—. Europe. Foreign policy. Kissinger, Henry A. Policymaking, patterns of. USA. 1971-75. *579*
Nonaligned Nations. Eisenhower, Dwight D. Foreign Relations. Nuclear energy. Nuclear proliferation. 1945-79. *311*
Nonintervention. Mutual Assured Destruction. National security. Strategic Arms Limitation Talks (II). 1980. *625*
North America. Europe. Military Strategy. NATO. Nuclear Arms (tactical, conventional). 1967-74. *112*
North American Air Defense Command. Air Forces. Defense Policy. Military Strategy. 1970's. *415*
Norton Sound (vessel). Navies. Research. Technology. Weapons. 1943-79. *155*
Nuclear accidents. California Nuclear Safeguards Initiative. Radioactive waste. Sabotage. 1976. *735*
—. Carter, Jimmy (administration). Pennsylvania. President's Commission on the Accident at Three Mile Island. Three Mile Island power plant. 1979. *761*
—. Housing. Pennsylvania. Prices. Three Mile Island power plant. 1977-79. *771*
—. Licensing. Reform. Three Mile Island power plant. 1977-79. *802*
—. Mental health. Mothers. Pennsylvania. Three Mile Island power plant. 1979. *708*
—. Nuclear Regulatory Commission. President's Commission on the Accident at Three Mile Island. 1979-81. *744*
—. Pennsylvania. Political Protest. Three Mile Island power plant. 1979-80. *810*
—. Pennsylvania. Politics. Research. Social Problems. Three Mile Island power plant. 1969-80. *823*
—. Pennsylvania. Public Relations. Reporters and Reporting. Three Mile Island power plant. 1976-79. *724*
—. Pennsylvania. Three Mile Island power plant. 1977-81. *776*
—. Reporters and Reporting. Television. Three Mile Island power plant. 1979. *773*
—. Rhetoric. Three Mile Island power plant. 1979. *721*
Nuclear arms *See also* Nuclear Weapons.
—. 1950's-75. *165*
—. Air Forces. Lemay, Curtis. Military Capability. Strategic Air Command. 1945-48. *88*
—. Alperovitz, Gar. Diplomacy. Historiography. World War II. 1943-46. 1965-73. *323*
—. Anti-Communist Movements. *Dr. Strangelove* (film). Films. Films. Humor. Kubrick, Stanley. Liberalism. 1950-64. *599*
—. Anti-nuclear power movement. Peace Movements. 1960-80. *769*
—. Arms control. Canada. China. India. Nuclear power. Regionalism. USA. 1951-78. *387*
—. Arms control. Developing nations. Foreign Policy. USA. 1975. *468*
—. Arms control. Foreign Policy. Military Strategy. Strategic Arms Limitation Talks. 1960's-78. *596*
—. Arms race. Federal Policy. Germany, West. Power resources. 1968-70's. *572*
—. Arms race. Military Capability. USSR. 1960-80. *404*
—. Arms race. Research. 1946-74. *267*
—. Arms race. Technology. Values, human. 1973. *528*
—. Atomic Energy Commission. Espionage. Federal Bureau of Investigation. Greenglass, David. Rosenberg case. Trials. 1942-74. *142*
—. Atomic Energy Commission. Hijacking. 1970-74. *737*
—. Atomic Scientists of Chicago. Nuclear energy. Simpson, John A. (views). 1945-80. *792*
—. Attitudes. Economic development. Military Strategy. Technology. War. 1914-68. *351*
—. Attitudes. Fallout. Ozone layer, destruction. 1945-75. *126*
—. Bainbridge, Kenneth T. (reminiscences). Manhattan Project. New Mexico (Los Alamos). World War II. 1945. *8*
—. Balance of Power. Defense Policy. Schlesinger, James R. USSR. 1960's-70's. *380*
—. Balance of Power. Defense Policy. USA. USSR. 1974-75. *241*
—. Balance of Power. Diplomacy. Politics. 1945-79. *217*
—. Balance of Power. Metaphysics. USSR. 1970's. *222*
—. Balance of Power. Military Ground Forces. USSR. 1969-77. *193*
—. *The Beginning or The End* (film). Films. Manhattan Project. Public Opinion. Scientists. 1947. *69*
—. Brodie, Bernard. Kahn, Herman. Military Strategy. 1950's-82. *358*
—. Brown, George S. (views). Joint Chiefs of Staff. Military Strategy. USSR. 1970's. *408*
—. Brown, Harrison (account). Gromyko, Andrei. Scientists. UN. USSR. 1947. *482*
—. California, University of. Military. 1945-78. *167*

—. Canada. Foreign Relations. Pearson, Lester B. (memoirs). USA. 1945. *42*
—. Capitalism. Central Intelligence Agency. Federal government. National security. Political power. 1940's-70's. *432*
—. Carter, Jimmy (administration). Military Strategy. Presidential Directive No. 59. 1960-80. *139*
—. Central Intelligence Agency. Israel. Nuclear Materials and Equipment Corporation. Pennsylvania (Apollo). Uranium. 1960's. *251*
—. Churchill, Winston. Communism. Diaries. Potsdam Conference. Stalin, Joseph. Truman, Harry S. 1945. *37*
—. Cold War. Decisionmaking. Disarmament. Eisenhower, Dwight D. USSR. 1953-61. *657*
—. Cold War. Diplomacy. Foreign policy. USA. USSR. 1942-46. *13*
—. Cold War. Diplomacy. Roosevelt, Franklin D. 1941-45. *53*
—. Cold War. Espionage. Rosenberg case. Supreme Court. 1953. *152*
—. Cold War (origins). Marshall Plan. Second Front. USSR. World War II. 1940's. *429*
—. Colleges and Universities. Manhattan Project. Oppenheimer, J. Robert. 1904-67. *48*
—. Command structure. Defense policy. Reagan, Ronald (administration). USSR. 1980-81. *393*
—. Congress. Federation of American Scientists. Presidency. War Powers Resolution (1973). 1975-76. *578*
—. Conservation. Nuclear power. Power Resources. 1980. *759*
—. Counterforce (concept). Deterrence. Military Strategy. 1970's. *340*
—. Decisionmaking. Divine, Robert A. (review article). Foreign relations. USSR. 1954-60. 1978. *195*
—. Defense. Maritime law. Military strategy. 1970's. *353*
—. Defense Nuclear Agency. 1963-76. *180*
—. Defense Policy. Deterrence. Europe, Western. NATO. Warsaw Pact. 1970's. *326*
—. Defense Policy. Disarmament. Foreign policy. Panel of Consultants on Disarmament (report). USSR. 1953. *486*
—. Defense Policy. Espionage. Groves, Leslie R. Secrecy. Truman, Harry S. (administration). USSR. 1945-50. *122*
—. Defense Policy. Foreign Relations. Latin America. Tlatelolco, Treaty of. 1960's-70's. *637*
—. Defense Policy. Military Capability. USSR. 1970's. *297*
—. Defense Policy. NATO. 1967-82. *254*
—. Defense Policy. Nixon, Richard M. Strategic Arms Limitation Talks (II). 1973. *665*
—. Defense Policy. Strategic Arms Limitation Talks. Technology. 1975. *231*
—. Defense Policy. Strategic Arms Limitation Talks. USSR. 1972-79. *582*
—. Detente. Disarmament. USA. USSR. 1970's. *573*
—. Detente. Exports. 1978. *630*
—. Deterrence. Military Capability. USSR. 1945-50. *337*
—. Deterrence. Military Strategy. Technological innovations. USSR. 1969-78. *338*
—. Deterrence. Military strategy. USSR. 1950-80. *318*
—. Deterrence. NATO. Neutron bomb. Warsaw Pact. 1954-78. *83*
—. Deterrence. Strategic Arms Limitation Talks. USA. USSR. 1973. *623*
—. Deterrence. Strategic Arms Limitation Talks. USSR. 1969-78. *496*
—. Deterrence. USSR. 1952-80. *259*
—. Deterrence (military). Disarmament. Strategic studies. 1973. *364*
—. Deterrence (military). Flexible response strategy. 1972-74. *385*
—. Deterrence theory. Military Strategy. 1966-76. *270*
—. Developing nations. 1979-81. *261*
—. Developing nations. Foreign Relations. 1960-80. *301*
—. Developing Nations. Heilbroner, Robert. Peace, prospects for. Violence. 1976. *535*
—. Diplomacy. Strategic Arms Limitation Talks (II). USSR. 1972-76. *639*
—. Disarmament. Foreign policy. USA. USSR. 1970's. *300*
—. Disarmament. Government secrecy. USA. 1945-70's. *562*
—. Disarmament. NATO. Theater Nuclear Forces. 1977-79. *278*
—. Disarmament. Reagan, Ronald (administration). 1981. *246*
—. Disarmament. Strategic Arms Limitation Talks. 1974-76. *513*
—. Disarmament. Strategic arms race. USA. USSR. 1960-74. *446*
—. Disarmament. USA. USSR. 1940's-74. *563*
—. Disarmament. USSR. 1943-78. *294*
—. Documents. Military Strategy. USSR. 1954-55. *373*
—. Einstein, Albert. Germany. Pacifism. Zionism. 1939-55. *21*
—. Ethics. Oppenheimer, J. Robert. Scientists. 1945-66. *76*
—. Europe. Foreign relations. Military strategy. USA. USSR. 1950's-70's. *436*
—. Europe. International Security. NATO. 1953-77. *224*
—. Europe. Modernization. NATO. 1978-80. *302*
—. Europe, Western. Military Strategy. NATO. USSR. 1950-78. *85*
—. Europe, Western. NATO. USSR. 1979. *402*
—. Federal Government. Military Strategy. Public Opinion. Strategic Arms Limitation Talks. 1945-78. *498*
—. Flexible response options. Schlesinger, James R. Strategic arms proposals. 1975. *366*
—. Foreign Policy. 1948-81. *245*
—. Foreign Policy. 1962-74. *320*
—. Foreign Policy. 1970's. *456*
—. Foreign policy. Hydrogen bomb testing. USSR. York, Herbert (hypothesis). 1953-55. 1970's. *84*
—. Foreign policy. Kissinger, Henry A. 1957-78. *420*
—. Foreign Policy. Kissinger, Henry A. Newhouse, John (review article). Strategic Arms Limitation Talks. 1973. *461*
—. Foreign Policy. Kissinger, Henry A. (review article). 1957-75. *314*
—. Foreign policy. Military Strategy. 1941-78. *208*
—. Foreign policy. National Security. 1946-50. *162*
—. Foreign Policy. National Security. Strategic Arms Limitation Talks. USSR. 1950's-70's. *565*
—. Foreign Policy. October War. USSR. 1973. *213*
—. Foreign Policy. Strategic Arms Limitation Talks. 1969-72. *479*
—. Foreign Policy. USA. USSR. 1969-74. *438*
—. Foreign Relations. Great Britain. Roosevelt, Franklin D. USA. World War II. 1940-45. *16*

—. Foreign Relations. Military Strategy. Strategic Arms Limitation Talks. USSR. 1972-77. *656*
—. Foreign Relations. NATO. USSR. 1970's-81. *455*
—. Foreign Relations. Strategic Arms Limitation Talks. USA. USSR. 1972-75. *602*
—. Freedom of the press. Press. *Progressive* (periodical). *United States* v. *The Progressive* (US, 1979). 1979. *800*
—. *George Washington* (vessel). Navies (Special Projects Office). Polaris (missile). Submarine Warfare. Whitmore, William F. (account). 1955-60. *186*
—. Hydrogen bomb. Joint Chiefs of Staff. Military Strategy. Truman, Harry S. USSR. 1945-50. *161*
—. ICBM's. Military strategy. Schlesinger, James R. Strategic Arms Limitation Talks. 1974. *327*
—. Intelligence Service. Strategic Arms Limitation Talks. USSR. 1970-79. *642*
—. International Nuclear Fuel Cycle Evaluation. 1978-79. *678*
—. Joint Chiefs of Staff. Military Strategy. Single Integrated Operations Plan. Strategic Arms Limitation Talks. Target planning. 1972-74. *383*
—. Laser fusion. 1960-79. *171*
—. Latin America. Tlatelolco, Treaty of (protocol). 1973-77. *520*
—. Lawrence Livermore laboratory. York, Herbert. ca 1949-53. *191*
—. Little Theatre Group. Manhattan Project. New Mexico (Los Alamos). Theater Production and Direction. World War II. 1943-46. *3*
—. Mandelbaum, Michael (review article). 1946-79. *94*
—. Manufactures. South. ca 1942-79. *159*
—. McDaniel, Boyce (reminiscences). New Mexico (Los Alamos). Plutonium. World War II. 1940-45. *39*
—. Middle classes. Peace Movements. 1980-82. *727*
—. Military Capability. NATO. Technology. USA. USSR. 1970's. *98*
—. Military Capability. NATO. Warsaw Pact. 1970. *226*
—. Military Capability. USA. USSR. 1946-72. *89*
—. Military Strategy. 1960's-73. *233*
—. Military Strategy. 1970-80. *199*
—. Military Strategy. 1974-81. *250*
—. Military Strategy. 1978-81. *207*
—. Military Strategy. Missiles. 1945-76. *166*
—. Military Strategy. National Characteristics. 20c. *268*
—. Military Strategy. National security. USA. USSR. 1954-74. *309*
—. Military Strategy. NATO. 1949-82. *388*
—. Military strategy. Schlesinger, James R. USA. USSR. 1970's. *229*
—. Military Strategy. SLBM. Strategic arms control. 1950's-70's. *79*
—. Military Strategy. Strategic Arms Limitation Talks. 1970's. *551*
—. Military Strategy. Strategic Arms Limitation Talks. USSR. 1969-78. *567*
—. Military strategy. Strategic Arms Limitation Talks. USSR. 1970-81. *204*
—. Military Strategy. USSR. 1945-78. *219*
—. Military Strategy. USSR. 1950's-79. *296*
—. Military Strategy. USSR. 1964-81. *377*
—. Missiles. 1954-79. *96*
—. Missiles. 1955-80. *77*
—. National Security. 1977. *325*
—. National security. Strategic Arms Limitation Talks. USSR. 1972-78. *696*
—. NATO. 1950-82. *386*
—. NATO. Theater nuclear forces. 1950's. *395*
—. Naval Air Forces. P2V (aircraft). 1949-50. *120*
—. Naval strategy. USSR. 1950's-80. *330*
—. Navies. Submarines. 1975. *111*
—. Nuclear power plants. Pollution. 1970's. *760*
—. Peace. USSR. 1948-80. *423*
—. Population vulnerability. Strategic Arms Limitation Talks. 1974. *502*
—. Schlesinger, James R. Strategic adaptability. 1970's. *419*
—. Social Psychology. 1945-79. *324*
—. Space. Treaties. USSR. 1962-80. *536*
—. Strategic Arms Limitation Talks. 1974. *682*
—. Strategic Arms Limitation Talks. 1978. *577*
—. Strategic Arms Limitation Talks. USA. USSR. 1972. *549*
—. Strategic Arms Limitation Talks. War. 1805-1972. *597*
—. Terrorism. ca 1972-74. *749*
—. UN. 1961-78. *519*
—. USA. USSR. 1965-73. *215*
Nuclear Arms (control). Exports. Foreign Policy. Technology. 1976. *435*
Nuclear arms, danger of. Foreign policy. Israel (defense policy). Middle East. USA. 1973-75. *405*
Nuclear Arms (first testing). New Mexico (Los Alamos). 1945. *22*
Nuclear arms limitation. Attitudes. Economics. Foreign relations. USA. USSR. 1972-73. *465*
—. Foreign relations. USA. USSR. 1959-75. *527*
—. ICBM's. Military strategy. Technology. Vladivostok Accord. 1970's. *611*
Nuclear arms, offensive. Strategic Arms Limitation Talks (II). USA. USSR. 1970. *490*
Nuclear arms policy. Carter, Jimmy (administration). Equivalence (definitions). Strategic Arms Limitation Talks. 1970's. *588*
—. Decisionmaking. Hydrogen bomb. York, Herbert (review article). 1949-75. *192*
Nuclear Arms (program). Einstein, Albert. Hungarian Americans. Szilard, Leo. USA. World War II. 1941-74. *26*
Nuclear arms proliferation. Developing nations. Foreign Relations. 1974-79. *244*
Nuclear arms race. Atomic bomb. Franck, James. Political Attitudes. Truman, Harry S. (administration). World War II. 1944-45. *62*
—. Defense spending. Missile capability. USA. USSR. 1970's. *433*
Nuclear Arms (review article). Foreign Relations. Mandelbaum, Michael. 1945-80. *32*
—. International politics. 1945-75. *389*
Nuclear Arms (tactical). 1980. *93*
—. Armies. Defense Policy. Politics. Rose, John P. (review article). 1950-81. *239*
—. Europe, Western. Military strategy. NATO. USA. USSR. 1970's. *256*
—. Military Strategy. NATO. 1974. *368*
—. Military Strategy. NATO. 1974. *369*
—. Military strategy. NATO. USA. Weapons, conventional. 1970's. *225*
Nuclear Arms (tactical, conventional). Europe. Military Strategy. NATO. North America. 1967-74. *112*
Nuclear arms testing. Federal Government. Public Relations. ca 1948-80. *92*
—. France. Great Britain. Secrecy. World War II. 1939-45. *25*
—. Marshall Islands (Bikini). Pacific Dependencies (US). Refugees. 1946-80. *184*

Nuclear Arms (warheads). Atomic Energy Commission. Deterrence. Gillet, Edward B. (interview). Military Strategy. Research. 1970's. *411*
Nuclear capability. Arms control. Developing nations. International organization. USA. 1950-75. *501*
Nuclear chain reaction. Anderson, Herbert L. (reminiscences). Chicago, University of. Illinois. World War II. 1942. *5*
—. Chicago, University of. Illinois. Wattenberg, Albert (reminiscences). World War II. 1942. *64*
Nuclear Energy. 1945-77. *765*
—. 1950's-79. *768*
—. 1973-80. *710*
—. Arms control. Foreign Policy. 1945-80. *292*
—. Army Corps of Engineers. Journalism. Oak Ridge *Journal*. Tennessee. 1943-49. *2*
—. Atomic Scientists of Chicago. Nuclear arms. Simpson, John A. (views). 1945-80. *792*
—. Attitudes. *China Syndrome* (film). Pennsylvania (Three Mile Island). 1978-79. *764*
—. Baruch Plan. Foreign Relations. Nuclear proliferation. UN. 1945-77. *712*
—. Bates, Albert (review article). 1970-79. *804*
—. Bohr, Niels. World War II. 1943. *51*
—. Carter, Jimmy. Energy. Public Policy. 1977-78. *698*
—. Civil Disobedience. 1977-79. *702*
—. Eisenhower, Dwight D. Foreign Relations. Nonaligned Nations. Nuclear proliferation. 1945-79. *311*
—. Employment. Federal Government. Private sector. 1963-73. *767*
—. Europe, Western. Foreign policy. Nuclear nonproliferation. 1945-79. *794*
—. Foreign Policy. International Nuclear Fuel Cycle Evaluation. 1970's. *753*
—. Foreign Policy. Nuclear nonproliferation. 1950's-70's. *620*
—. Foreign Relations. International Trade. Nuclear Nonproliferation Act (US, 1978). Technology. 1978-79. *614*
—. Information. Oak Ridge Associated Universities, Inc. Public Opinion. 1970-74. *785*
—. Interest groups. Oil Industry and Trade. Politics. 1979. *822*
—. Japan. Military Strategy. Smyth Report, 1945. USA. World War II. 1940-45. *56*
—. Military Strategy. Publishers and Publishing. Smyth Report, 1945. War Department. 1946-73. *172*
—. Miller, Walter *(A Canticle for Leibowitz)*. Religion. Science fiction. 1959-81. *795*
—. *Plutonium: An Element of Risk* (documentary). Television. 1970's. *743*
—. Pressure Groups. 1950's-81. *766*
—. Public Opinion. Spain. 1950-82. *725*
—. Public Policy. Strategic studies. 1490-1977. *265*
—. South. States' rights. 1954-62. *809*
—. Technology. 1946-77. *805*
Nuclear Energy (review article). DeLeon, Peter. Electric Power. Ramsay, William. Rolph, Elizabeth S. 1960-79. *752*
Nuclear equipment sales. Arms Trade. Europe, Western. Middle East. USA. 1973-75. *306*
Nuclear holocaust, threat of. Founding Fathers. Political Theory. Posterity, commitment to. 1776-1976. *234*
Nuclear installations. Espionage. Sabotage. 1966-75. *723*
Nuclear materials. Europe. International Law. 1977-80. *738*
Nuclear Materials and Equipment Corporation. Central Intelligence Agency. Israel. Nuclear Arms. Pennsylvania (Apollo). Uranium. 1960's. *251*
Nuclear materials theft, possible. 1975. *816*
Nuclear nonproliferation. Armaments. Documents. Federal Policy. 1981. *692*
—. Brazil. Diplomacy. Germany. NATO. 1968-77. *571*
—. Canada. Foreign Policy. Great Britain. USSR. 1943-76. *545*
—. Carter, Jimmy. Foreign Policy. 1968-77. *652*
—. Carter, Jimmy. Foreign Policy. Political Leadership. 1977-79. *621*
—. Carter, Jimmy. Foreign Policy. Political Leadership. 1979. *480*
—. Carter, Jimmy (administration). Foreign Policy. 1977-81. *655*
—. Diplomacy. Disarmament. UN. 1945-77. *467*
—. Europe, Western. Foreign policy. Nuclear energy. 1945-79. *794*
—. Foreign Policy. 1953-80. *619*
—. Foreign Policy. Nuclear Energy. 1950's-70's. *620*
—. Foreign Relations. India. 1977-79. *361*
Nuclear Non-Proliferation Act (US, 1978). Exports. Foreign policy. Fuel. India. Law. 1978. *644*
—. Foreign Relations. International Trade. Nuclear Energy. Technology. 1978-79. *614*
Nuclear Nonproliferation Treaty. Developing Nations. Foreign Relations. Great Powers. 1970's. *662*
Nuclear Nonproliferation Treaty (1968). Canada. Government secrecy. India. Nuclear technology. 1968-70's. *587*
—. Foreign Relations. 1968-75. *484*
Nuclear physics. Abelson, Philip H. (reminiscences). California, University of, Berkeley. Lawrence Radiation Laboratory. 1935-40. *1*
—. Alvarez, Luis W. (reminiscences). California, University of, Berkeley. Lawrence Radiation Laboratory. 1934. *4*
—. Cyclotrons. Denmark. Japan. Scientific Experiments and Research. 1935-45. *65*
—. Frisch, Otto R. (account). Nuclear War. World War II. 1945. *24*
Nuclear potential. Angola. Foreign policy. Mozambique. South Africa. USA. 1970-76. *427*
Nuclear power. 20c. *811*
—. Arms control. Canada. China. India. Nuclear Arms. Regionalism. USA. 1951-78. *387*
—. Atomic Energy Commission. Environment. Public participation. Technology assessment. 1946-73. *742*
—. Attitudes. Canada. USA. 1960's-70's. *717*
—. Congress. Joint Committee on Atomic Energy. Legislative investigations. 1973-74. *719*
—. Conservation. Nuclear arms. Power Resources. 1980. *759*
—. Energy crisis. Industry. Technology. ca 1940-80. *745*
—. India. Trade. Treaties. Uranium. 1963-80. *774*
—. Submarines. USSR. 1958-80. *154*
Nuclear power industry. Federal Regulation. Licensing. 1946-79. *803*
—. Foreign policy. International Trade. Uranium cartel (possible). USA. 1970's. *815*
—. Power Resources. Technology. 1954-73. *716*
Nuclear power plants. 1974. *713*
—. Accidents. Diaries. Pennsylvania (Three Mile Island). 1979. *700*

—. Accidents. Federal Regulation. Nuclear Regulatory Commission. Pennsylvania (Three Mile Island). Safety. 1979. *780*
—. Accidents. Kemeny Report. Pennsylvania (Three Mile Island). 1950-79. *824*
—. Allied General Nuclear Services. Barnwell Nuclear Fuel Plant. South Carolina. ca 1963-79. *784*
—. Atomic Energy Commission. Nuclear Regulatory Commission. Pennsylvania (Three Mile Island). 1979. *750*
—. Attitudes. Decisionmaking. State Government. Voting and Voting Behavior. Western States. 1963-78. *703*
—. Capacity, idle. Public utilities. Taxation. 1973-74. *715*
—. Civil Rights. Martial Law. Plutonium. Terrorism. 1970's. *728*
—. Clamshell Alliance. Ideology. New Hampshire (Seabrook). Political Protest. 1978. *741*
—. Contracts. Economic Regulations. General Electric Co. Westinghouse. 1955-78. *711*
—. Domestic Policy. Italy. Pennsylvania (Three Mile Island). 1970's. *705*
—. Economic Conditions. Electric Power. 1973-78. *796*
—. Economic conditions. Federal government. Subsidies. 1954-78. *718*
—. Employment (dismissal). Engineering. Faulkner, Peter (account). Nuclear Services Corp. Senate. 1973-77. *722*
—. Environmentalism. Experts. New York (Cayuga Lake). Pressure groups. 1973. *770*
—. Federal Government. Pennsylvania (Shippingport). Private sector. ca 1940-57. *783*
—. Nuclear arms. Pollution. 1970's. *760*
—. Ohio. Referendum. Western States. 1976. *812*
—. Plutonium. Private sector. 1974. *797*
—. Radiation. Shipyards. 1954-79. *108*
—. Rasmussen, Norman (interview). Safety. WASH-1400 (report). 1970-79. *779*
—. Terrorism. 1974. *788*
Nuclear proliferation. Armaments. Disarmament. Wars, significance of. 1945-75. *205*
—. Attitudes. Exports. France. Great Britain. Plutonium, reprocessed. 1946-78. *726*
—. Baruch Plan. Foreign Relations. Nuclear Energy. UN. 1945-77. *712*
—. Diplomacy. 1945-75. *439*
—. Eisenhower, Dwight D. Foreign Relations. Nonaligned Nations. Nuclear energy. 1945-79. *311*
—. Europe. Foreign Relations. 1954-80. *590*
—. Foreign Policy. 1970's. *345*
—. Foreign Policy. Great Powers. National Security. 1946-79. *344*
—. Foreign Policy. Technology. 1946-80. *684*
—. Foreign Policy. USA. 1954-74. *454*
Nuclear reactors. 1947-78. *817*
—. Antarctic (McMurdo Sound). Navies. Radioactive waste. 1960-71. *814*
—. Foreign policy. Technology. 1953-74. *290*
Nuclear reactors, problems of. 1975. *793*
Nuclear Regulatory Commission. Accidents. Federal Regulation. Nuclear power plants. Pennsylvania (Three Mile Island). Safety. 1979. *780*
—. Atomic Energy Commission. Nuclear power plants. Pennsylvania (Three Mile Island). 1979. *750*
—. Bureaucracies. 1970's. *746*
—. Interest groups. Radioactive waste. Uranium Mill Tailings Radiation Control Act (US, 1978). 1978. *801*
—. Nuclear accidents. President's Commission on the Accident at Three Mile Island. 1979-81. *744*
—. Power Resources. Rasmussen report. 1970's. *807*
Nuclear safety. Elections. Initiatives. 1976. *734*
Nuclear Science and Technology. Anderson, Herbert L. (reminiscences). Chicago, University of. Columbia University. Fermi, Enrico. Szilard, Leo. 1933-45. *6*
—. Annexation. Intergovernmental Relations. New Mexico (Albuquerque). Weapons. 1940-78. *157*
—. Armaments. Electric Power. 1950-75. *221*
—. Atomic bomb development. Diplomacy. World War II. 1942-45. *46*
—. Atomic Energy Commission. Greenglass, David. Rosenberg case. Secrecy. Trials. 1945-51. *75*
—. Atomic Energy Commission. Marshall Islands. Radiation exposure. 1942-72. *81*
—. Attitudes. Democracy. Europe. Interest Groups. Japan. USSR. 1970's. *739*
—. Bloch, Felix. Conseil Européen pour la Recherche Nucléaire. 1954-55. *748*
—. *Bulletin of the Atomic Scientists* (periodical). Rabinowitch, Eugene. 1940's-70's. *751*
—. Byrnes, James F. Federal Government. Germany. Memoirs. Szilard, Leo. 1898-1945. *59*
—. California (Livermore, Los Altos). Laboratories. Weapons. 1942-77. *101*
—. California, University of, Berkeley. Kamen, Martin D. (reminiscences). Lawrence Radiation Laboratory. 1936-39. *30*
—. Civil-military relations. Military power. -1973. *362*
—. Coal industry. Environment. 1860's-1970's. *772*
—. Cyclotrons. Manhattan Project. New Mexico (Los Alamos). Wilson, Robert R. (reminiscences). 1942-43. *66*
—. Cyclotrons. Michigan State University. 1958-81. *86*
—. Education. Value systems. 1974. *109*
—. Energy. Treaties. 1940's-70's. *476*
—. Federal policy. Safety. 1974. *778*
—. Fusion, Thermonuclear. Lasers. Scientific Experiments and Research. USSR. USSR. 1971-75. *701*
—. Georgia Nuclear Laboratory. Radioactive waste. 1950's-79. *791*
—. ICBM's. Terrorism. 1960's-70's. *704*
—. Letters. Oppenheimer, J. Robert (reminiscences). Physics. Smith, Alice Kimball. Weiner, Charles. ca 1900-45. 1980. *47*
—. Manley, John H. (reminiscences). New Mexico (Los Alamos). World War II. 1942-45. *36*
—. Monopolies. 1970's. *707*
—. Pugwash Conferences. Rabinowitch, Eugene. Scientists. 1957-73. *529*
Nuclear Science and Technology (review article). Foreign Policy. Gowing, Margaret. Great Britain. Nuclear War. Sherwin, Martin J. 1938-52. *33*
Nuclear Science and Technology (underground explosions). Colorado. Public Opinion. 1973-74. *188*
Nuclear Services Corp. Employment (dismissal). Engineering. Faulkner, Peter (account). Nuclear power plants. Senate. 1973-77. *722*
Nuclear stalemate, function of. China. USA. USSR. World order. 1970's. *201*
Nuclear stockpiling. Deterrence. USA. USSR. 1960's-70's. *440*
Nuclear strategy. Arms control agreements. China. Detente. USA. USSR. 1970's. *401*

—. Arms race. Detente. USA. USSR. 1972-74. *319*
Nuclear submarines. Arms race. Missiles. Navies. Trident (missile). 1960-80. *158*
—. Military Intelligence. 1979. *138*
—. Shipyards (capacity). 1955-75. *176*
Nuclear technology. Atomic Energy Commission. Engineers. Manhattan Project. Scientists. 1938-50. *28*
—. Canada. Government secrecy. India. Nuclear Nonproliferation Treaty (1968). 1968-70's. *587*
Nuclear technology, transfer of. Brazil. Germany, West. USA. 1945-76. *257*
Nuclear test ban. Arms race. Diplomacy. Negotiations. Richardson models. USSR. 1962-63. *558*
—. Arms race, qualitative. International security. Military innovation. Strategic Arms Limitation Treaty (1972). 1972-75. *220*
Nuclear test ban issue. Defense Policy. Dulles, John Foster. Eisenhower, Dwight D. Goodpaster, Andrew. Strauss, Lewis. USSR. 1958. *240*
Nuclear Test Ban Treaty. Austria. Foreign Relations. USSR. 1955. 1963. *581*
Nuclear Test Ban Treaty (1963). Disarmament. Great Britain. USSR. 1954-70's. *675*
Nuclear testing. Bikini Atoll. Navies. Operation Crossroads. 1946. *116*
—. Eniwetok (atoll). Marshall Islands. Pacific Dependencies (US). Safety. 1946-80. *127*
—. Treaties. USSR. 1963-73. *631*
Nuclear Vessels. *Ohio* (vessel). Submarines. 1981. *91*
Nuclear War. Air Forces (Air Targets Division). Bureaucracies. Military intelligence. Morality. Nash, Henry T. (account). Planning. 1950's-70's. *147*
—. Allison, Graham *(Essence of Decision).* Committees. Decisionmaking. Foreign policy. World War II. 1944-45. *54*
—. Alperovitz, Gar. Foreign Relations. World War II (review article). 1945. *27*
—. Alperovitz, Gar (review article). Decisionmaking. Historiography. Japan (Hiroshima). World War II. 1945-65. *52*
—. Arms Control and Disarmament Agency. Carter, Jimmy (administration). Civil defense. Evacuations. Public opinion. 1978. *95*
—. Attitudes. Generations, political. 1974. *295*
—. Baldwin, Paul H. (account). Fighter squadrons, night. Philippines. World War II (aerial operations). 1942-45. *9*
—. Battles. Deterrence. Morale. World War II. 1944-81. *130*
—. Bureaucracies. Military Strategy. 1974. *280*
—. B-29 (aircraft). Japan. Lemay, Curtis. Surrender. World War II. 1945. *67*
—. Canada. Diaries. King, William Lyon MacKenzie. Racism. 1921-48. *49*
—. Canada. NATO. Neutron bomb. USA. 1977-79. *141*
—. Civil defense. Defense Policy. Military. Public Opinion. USSR. 1948-78. *424*
—. Civil defense. Military Strategy. USSR. 1950's-77. *392*
—. Civil defense. USSR. 1950-78. *133*
—. Cold War. Communism. Diplomacy. Middle East. Morgenthau, Hans J. USSR. 1945-76. *399*
—. Cold War. Japan (Hiroshima). World War II. 1945. *31*
—. Decisionmaking. Japan (Hiroshima, attack on). World War II (Pacific Theater). 1944-45. *50*
—. Decisionmaking. Japan (Hiroshima, Nagasaki). Military Strategy. World War II. 1944-45. *23*
—. Defense. ICBM's. Military Capability. Research and development. USSR. 1970's. *413*
—. Defense Department. Military Finance. Naval Strategy. Submarine Warfare. Tidal waves. 1950's-80. *131*
—. Defense policy. Military Strategy. Mutual assured destruction. USA. USSR. 1960-76. *196*
—. Defense policy. National security. Peace. 1977. *400*
—. Defense policy. Strategic Arms Limitation Talks. USSR. 1969-79. *293*
—. Detente. Europe. NATO. USA. 1970's. *269*
—. Deterrence. Military Strategy. Schlesinger, James R. 1974. *305*
—. Diplomacy. Foreign Policy. Roosevelt, Franklin D. Truman, Harry S. 1945-51. *14*
—. Disarmament. Public Policy. Scientists. 1939-54. *71*
—. Economic Conditions. Energy. NATO. 1950's-70's. *216*
—. Foreign Policy. Gowing, Margaret. Great Britain. Nuclear Science and Technology (review article). Sherwin, Martin J. 1938-52. *33*
—. Foreign Policy. Military Strategy. 1980. *272*
—. Foreign Policy. Retaliation, massive. 1954-73. *281*
—. Foreign relations. 1974. *474*
—. Frisch, Otto R. (account). Nuclear Physics. World War II. 1945. *24*
—. Hersey, John (*Hiroshima*). Literature. Morality. 1945-74. *70*
—. History Teaching. Japan. Simulation and Games. World War II. 1945. 1978. *19*
—. ICBM's (survivability). USSR. 1960's-70's. *332*
—. Japan. Leftists. USSR. World War II. 1945. *17*
—. Japan. Military strategy. USA. World War II. 1945. *29*
—. Japan. Politics. Surrender offer. Truman, Harry S. (administration). World War II. 1945. *12*
—. Japan (Hiroshima, Nagasaki). Military Strategy. World War II. 1945. *20*
—. Japan (Hiroshima, Nagasaki). USA. USSR. World War II. 1942-47. *15*
—. Japan (Hiroshima, Nagasaki). USA. World War II (strategy). 1945. *11*
—. Japan (Hiroshima, Nagasaki). World War II. 1945. *40*
—. Japan (Kyoto). World War II. 1945. *45*
—. Military Strategy. 1945-80. *255*
—. Military Strategy. 1950-80. *334*
—. Military Strategy. Targeting. 1980. *74*
—. Military Strategy. USSR. 1977. *356*
—. Public Opinion. 1945-71. *150*
Nuclear war, likelihood of. Strategic Arms Limitation Talks. 1974. *457*
Nuclear War, limited. Clausewitz, Karl von (theories). Military Strategy. War, principles of. 1970's. *342*
Nuclear war, prevention of. 1945-74. *564*
Nuclear War (risk of). Defense Policy. USA. USSR. 1945-74. *285*
Nuclear war, targets of. Foreign policy. Military Strategy. 1974. *375*
Nuclear War (threat of). Arms Control and Disarmament Agency. 1960's-70's. *504*
—. Cuban Missile Crisis. Kennedy, John F. Khrushchev, Nikita. Military strategy. 1962. *209*
Nuclear waste *See also* Radioactive waste.
—. Accountability. Safety. Washington (Hanford). 1970's. *699*
—. Courts. Industrial safety. 1970-80. *806*

—. Government Regulation. Safety. South. Transportation. ca 1960-79. *781*
Nuclear waste disposal. South. ca 1945-79. *782*
Nuclear weapons *See also* Nuclear arms.
—. Communications, Military. Defense Policy. USSR. 1940-81. *99*
—. Navies. USA. USSR. 1945-74. *264*
Nuclear-weapon free zones. Foreign policy. 1970's. *553*
N2C-2 (aircraft). Drones. Fahrney, Delmar S. (account). Missiles. Navy Bureau of Aeronautics. Radio. Research and development. TG-2 (aircraft). World War II. 1936-45. *107*

O

Oak Ridge Associated Universities, Inc. Information. Nuclear energy. Public Opinion. 1970-74. *785*
Oak Ridge *Journal.* Army Corps of Engineers. Journalism. Nuclear energy. Tennessee. 1943-49. *2*
October War. Foreign Policy. Nuclear Arms. USSR. 1973. *213*
Ohio. Nuclear power plants. Referendum. Western States. 1976. *812*
Ohio (vessel). Nuclear Vessels. Submarines. 1981. *91*
Oil Industry and Trade. Interest groups. Nuclear energy. Politics. 1979. *822*
Operation Crossroads. Bikini Atoll. Navies. Nuclear testing. 1946. *116*
Oppenheimer, J. Robert. Atomic Energy Commission. Executive Order 10450. Loyalty. Strauss, Lewis. 1946-54. *117*
—. Colleges and Universities. Manhattan Project. Nuclear Arms. 1904-67. *48*
—. Ethics. Nuclear arms. Scientists. 1945-66. *76*
—. Letters. New Mexico (Los Alamos). Scientists. 1932-34. 1943-45. *55*
Oppenheimer, J. Robert (reminiscences). Einstein, Albert. Pacifism. Physics. 1919-55. *44*
—. Letters. Nuclear Science and Technology. Physics. Smith, Alice Kimball. Weiner, Charles. ca 1900-45. 1980. *47*
Ozone layer, destruction. Attitudes. Fallout. Nuclear arms. 1945-75. *126*

P

Pacific Basin. Balance of power. Energy development. Foreign policy. USA. 1973-77. *736*
Pacific Dependencies (US). Bikini Atoll. Hydrogen bomb. Japanese. *Lucky Dragon* (vessel). 1954-75. *164*
—. Eniwetok (atoll). Marshall Islands. Nuclear testing. Safety. 1946-80. *127*
—. Marshall Islands (Bikini). Nuclear arms testing. Refugees. 1946-80. *184*
Pacifism. Einstein, Albert. Germany. Nuclear arms. Zionism. 1939-55. *21*
—. Einstein, Albert. Oppenheimer, J. Robert (reminiscences). Physics. 1919-55. *44*
—. Einstein, Albert. Physics. Science and Society. 1919-55. *38*
Pakistan. Foreign Relations. 1970's. *459*
Panama. Congress. Foreign policy. Politics. Strategic Arms Limitation Talks. Treaties. 1977-78. *512*
Panel of Consultants on Disarmament (report). Defense Policy. Disarmament. Foreign policy. Nuclear arms. USSR. 1953. *486*
Paranoia, patterns of. Communism. *Dr. Strangelove* (film). George, Peter (*Red Alert*). Kubrick, Stanley. 1950's. *437*
Patterns. Scientific discoveries. Secrecy. 1970-80. *755*
Peace. Defense policy. National security. Nuclear War. 1977. *400*
—. Nuclear Arms. USSR. 1948-80. *423*
Peace, concepts of. Foreign policy. 1940's-70's. *601*
Peace movement. 1973. *557*
Peace Movements. Anti-nuclear power movement. Nuclear Arms. 1960-80. *769*
—. Middle classes. Nuclear Arms. 1980-82. *727*
Peace, prospects for. Developing Nations. Heilbroner, Robert. Nuclear Arms. Violence. 1976. *535*
Peace research. Boulding, Kenneth (review article). Sociopolitical theory. 1974-75. 1977. *473*
—. Consortium on Peace Research, Education and Development. 1970-81. *458*
—. Defense spending. Economic Planning. National security. 1960's-77. *616*
—. Education. 1969-74. *516*
Pearson, Lester B. (memoirs). Canada. Foreign Relations. Nuclear arms. USA. 1945. *42*
Peck, Earl G. (account). Airplanes, Military. B-47 (aircraft). Strategic Air Command Wing. ca 1960. *153*
Pennsylvania. Carter, Jimmy (administration). Nuclear accidents. President's Commission on the Accident at Three Mile Island. Three Mile Island power plant. 1979. *761*
—. Housing. Nuclear accidents. Prices. Three Mile Island power plant. 1977-79. *771*
—. Mental health. Mothers. Nuclear accidents. Three Mile Island power plant. 1979. *708*
—. Nuclear accidents. Political Protest. Three Mile Island power plant. 1979-80. *810*
—. Nuclear accidents. Politics. Research. Social Problems. Three Mile Island power plant. 1969-80. *823*
—. Nuclear Accidents. Public Relations. Reporters and Reporting. Three Mile Island power plant. 1976-79. *724*
—. Nuclear accidents. Three Mile Island power plant. 1977-81. *776*
Pennsylvania (Apollo). Central Intelligence Agency. Israel. Nuclear Arms. Nuclear Materials and Equipment Corporation. Uranium. 1960's. *251*
Pennsylvania (Shippingport). Federal Government. Nuclear power plants. Private sector. ca 1940-57. *783*
Pennsylvania (Three Mile Island). Accidents. Diaries. Nuclear power plants. 1979. *700*
—. Accidents. Federal Regulation. Nuclear power plants. Nuclear Regulatory Commission. Safety. 1979. *780*
—. Accidents. Kemeny Report. Nuclear power plants. 1950-79. *824*
—. Atomic Energy Commission. Nuclear power plants. Nuclear Regulatory Commission. 1979. *750*
—. Attitudes. *China Syndrome* (film). Nuclear Energy. 1978-79. *764*
—. Domestic Policy. Italy. Nuclear power plants. 1970's. *705*
Pershing II (missile). Europe, Western. Military Strategy. Missiles. NATO. 1956-80. *151*
Persian Gulf. Conflict avoidance. Defense Policy. Naval presence. USA. 1946-75. *237*
Personal narratives. Diplomacy. Rowny, Edward L. Strategic Arms Limitation Talks. 1973-79. *638*
Philippines. Baldwin, Paul H. (account). Fighter squadrons, night. Nuclear War. World War II (aerial operations). 1942-45. *9*

Physics. Atomic structure, theory of. Magnets, floating. Mayer, Alfred Marshall. Scientific Experiments and Research. 1878-1907. *57*
—. Einstein, Albert. Oppenheimer, J. Robert (reminiscences). Pacifism. 1919-55. *44*
—. Einstein, Albert. Pacifism. Science and Society. 1919-55. *38*
—. Letters. Nuclear Science and Technology. Oppenheimer, J. Robert (reminiscences). Smith, Alice Kimball. Weiner, Charles. ca 1900-45. 1980. *47*
Pick, Vernon. Cooper, Joe. Prospecting. Steen, Charles. Uranium rushes. Western States. 1948-57. *144*
Planning. Air Forces (Air Targets Division). Bureaucracies. Military intelligence. Morality. Nash, Henry T. (account). Nuclear War. 1950's-70's. *147*
—. Defense Policy. Deterrence. Mobilization. 1978. *228*
—. Military Strategy. USSR. 1945-79. *339*
Plutonium. Civil Rights. Martial Law. Nuclear Power Plants. Terrorism. 1970's. *728*
—. McDaniel, Boyce (reminiscences). New Mexico (Los Alamos). Nuclear Arms. World War II. 1940-45. *39*
—. Nuclear power plants. Private sector. 1974. *797*
Plutonium: An Element of Risk (documentary). Nuclear energy. Television. 1970's. *743*
Plutonium, reprocessed. Attitudes. Exports. France. Great Britain. Nuclear proliferation. 1946-78. *726*
Polaris (missile). *George Washington* (vessel). Navies (Special Projects Office). Nuclear Arms. Submarine Warfare. Whitmore, William F. (account). 1955-60. *186*
Policymaking. Atomic bomb. Franck, James. Scientists. Stimson, Henry L. World War II. 1945. *58*
Policymaking, patterns of. Europe. Foreign policy. Kissinger, Henry A. Nixon, Richard M. (administration). USA. 1971-75. *579*
Political activism. Anti-nuclear power movement. Community movements. 1970's. *747*
Political activity. Federation of American Scientists. Scientists. 1946-74. *653*
Political Attitudes. Atomic bomb. Franck, James. Nuclear arms race. Truman, Harry S. (administration). World War II. 1944-45. *62*
—. Elites. National Security. Youth. 1945-75. *431*
—. Intellectuals. Kissinger, Henry A. 1957-74. *341*
Political change. Arms control. USA. USSR. 1945-73. *603*
Political Leadership. Carter, Jimmy. Foreign Policy. Nuclear nonproliferation. 1977-79. *621*
—. Carter, Jimmy. Foreign Policy. Nuclear nonproliferation. 1979. *480*
Political Participation. Executive Power. Foreign Policy. Public Opinion. 1960's-79. *687*
Political power. Capitalism. Central Intelligence Agency. Federal government. National security. Nuclear arms. 1940's-70's. *432*
—. Disarmament. Foreign Relations. Military. USSR. 1945-82. *333*
Political protest. Anti-nuclear power movement. Clamshell Alliance. Leftism. 1970's. *740*
—. Clamshell Alliance. Ideology. New Hampshire (Seabrook). Nuclear power plants. 1978. *741*
—. Nuclear accidents. Pennsylvania. Three Mile Island power plant. 1979-80. *810*
Political Reform. Defense Policy. Weapons acquisition. 1945-75. *173*
Political Science. Detente. Disarmament. 1970's. *627*
Political Speeches. Garn, Jake. National Security. Senate. Strategic Arms Limitation Talks (II). 1979. *533*
Political Surveys. Presidents. Strategic Arms Limitation Talks (II). USSR. 1979. *608*
Political Theory. Arms race. Disarmament. 1946-79. *317*
—. Founding Fathers. Nuclear holocaust, threat of. Posterity, commitment to. 1776-1976. *234*
Politics. Armies. Defense Policy. Nuclear Arms (tactical). Rose, John P. (review article). 1950-81. *239*
—. Atomic Energy Commission. Federal regulation. Public Policy. Science and Government. 1946-70. *706*
—. Atomic Energy Commission. Joint Committee on Atomic Energy. 1939-79. *820*
—. Balance of Power. Diplomacy. Nuclear arms. 1945-79. *217*
—. Carter, Jimmy (administration). Defense Policy. Diplomacy. Strategic Arms Limitation Talks (II). USSR. 1978. *487*
—. Congress. Foreign policy. Panama. Strategic Arms Limitation Talks. Treaties. 1977-78. *512*
—. Corporations. Shortages. Stockpiling decisions. Strategic materials. 1946-70's. *103*
—. Federal Policy. Weapons acquisition. 1950's-70's. *115*
—. Foreign Relations. Intervention. Military. 1946-75. *212*
—. Interest groups. Nuclear energy. Oil Industry and Trade. 1979. *822*
—. Japan. Nuclear War. Surrender offer. Truman, Harry S. (administration). World War II. 1945. *12*
—. Military. MX (missile). 1960's-81. *104*
—. National Security. Strategic Arms Limitation Talks (II). Weapons. 1977-78. *612*
—. Nuclear accidents. Pennsylvania. Research. Social Problems. Three Mile Island power plant. 1969-80. *823*
Politics and the Military. Attitudes. Cruise missiles. Defense Department. Weapons. 1964-76. *140*
Pollution. Nuclear arms. Nuclear power plants. 1970's. *760*
Popular Culture. Public Opinion. Radioactivity. Radium. Scientific discoveries. 1898-1935. *7*
Population vulnerability. Nuclear Arms. Strategic Arms Limitation Talks. 1974. *502*
Posterity, commitment to. Founding Fathers. Nuclear holocaust, threat of. Political Theory. 1776-1976. *234*
Potsdam Conference. Churchill, Winston. Communism. Diaries. Nuclear Arms. Stalin, Joseph. Truman, Harry S. 1945. *37*
Power resources. Arms race. Federal Policy. Germany, West. Nuclear arms. 1968-70's. *572*
—. Conservation. Nuclear arms. Nuclear power. 1980. *759*
—. Nuclear power industry. Technology. 1954-73. *716*
—. Nuclear Regulatory Commission. Rasmussen report. 1970's. *807*
Power Resources (alternatives). Technology. 1900's-1985. *777*
Preparedness. Arms control. 1970's. *546*
Presidency. Congress. Federation of American Scientists. Nuclear Arms. War Powers Resolution (1973). 1975-76. *578*
—. Foreign policy. Models. 1947-65. *235*
Presidential Directive No. 59. Carter, Jimmy (administration). Military Strategy. Nuclear Arms. 1960-80. *139*

Presidents. Political Surveys. Strategic Arms Limitation Talks (II). USSR. 1979. *608*
President's Commission on the Accident at Three Mile Island. Carter, Jimmy (administration). Nuclear accidents. Pennsylvania. Three Mile Island power plant. 1979. *761*
—. Nuclear accidents. Nuclear Regulatory Commission. 1979-81. *744*
Press. Censorship, voluntary. National security. World War II. 1941-80. *515*
—. Freedom of the press. Nuclear arms. *Progressive* (periodical). *United States* v. *The Progressive* (US, 1979). 1979. *800*
Pressure Groups. Atomic Energy Commission. Decisionmaking. Energy. Federal policy. Military. 1974. *720*
—. B-1 (aircraft). Congress. Defense budget cuts. *Trident* submarine. 1970's. *78*
—. Environmentalism. Experts. New York (Cayuga Lake). Nuclear power plants. 1973. *770*
—. Nuclear energy. 1950's-81. *766*
Prices. Housing. Nuclear accidents. Pennsylvania. Three Mile Island power plant. 1977-79. *771*
Private sector. Employment. Federal Government. Nuclear energy. 1963-73. *767*
—. Federal Government. Nuclear power plants. Pennsylvania (Shippingport). ca 1940-57. *783*
—. Nuclear power plants. Plutonium. 1974. *797*
Private utilities. Cities. Energy, alternative. ca 1950-79. *775*
Progress (concept). Ethics. Scientists. 1970's. *72*
Progressive (periodical). Freedom of the press. Nuclear arms. Press. *United States* v. *The Progressive* (US, 1979). 1979. *800*
Propaganda. Detente. Europe, Western. European Security Conference. USSR. 1960's-73. *463*
Prospecting. Cooper, Joe. Pick, Vernon. Steen, Charles. Uranium rushes. Western States. 1948-57. *144*
Psychology. Disasters. Survivors. War. 1941-72. *34*
Public Health. American Atomics Corp. Arizona (Tucson). Manufactures. Radiation. Tritium. Working Conditions. 1979. *709*
—. Fallout. Science. 1954-64. *134*
—. Radiation. Safety. 1980. *714*
Public opinion. Arms control. Executive Branch. Treaties. USSR. 1972-82. *517*
—. Arms Control and Disarmament Agency. Carter, Jimmy (administration). Civil defense. Evacuations. Nuclear War. 1978. *95*
—. Atomic bombs. Japan. USA. World War II. 1945-49. *68*
—. *The Beginning or The End* (film). Films. Manhattan Project. Nuclear Arms. Scientists. 1947. *69*
—. Civil defense. Defense Policy. Military. Nuclear War. USSR. 1948-78. *424*
—. Colorado. Nuclear Science and Technology (underground explosions). 1973-74. *188*
—. Conference on Security and Cooperation in Europe. Decisionmaking. 1975. *526*
—. Conference on Security and Cooperation in Europe. Europe. Ford, Gerald R. (administration). International Security. USA. 1975. *495*
—. Congress. Defense Department. Defense industries. Military. 1980's. *194*
—. Defense spending. 1970's. *287*
—. Education. War. 1944-63. *146*
—. Executive Power. Foreign Policy. Political Participation. 1960's-79. *687*
—. Federal Government. Military Strategy. Nuclear Arms. Strategic Arms Limitation Talks. 1945-78. *498*
—. Government. Insurance. Liability. Risk management. Taxation. 1948-78. *730*
—. Information. Nuclear energy. Oak Ridge Associated Universities, Inc. 1970-74. *785*
—. Nuclear energy. Spain. 1950-82. *725*
—. Nuclear war. 1945-71. *150*
—. Popular Culture. Radioactivity. Radium. Scientific discoveries. 1898-1935. *7*
—. Public policy. Radioactive waste. Theft. 1970's. *821*
Public participation. Atomic Energy Commission. Environment. Nuclear power. Technology assessment. 1946-73. *742*
Public Policy. Atomic Energy Commission. Federal regulation. Politics. Science and Government. 1946-70. *706*
—. Business. Energy. Government. Technology. 1978. *731*
—. Carter, Jimmy. Energy. Nuclear Energy. 1977-78. *698*
—. Disarmament. Nuclear War. Scientists. 1939-54. *71*
—. Executive Planning Council. Intergovernmental Relations. National Energy Acts. Radioactive waste. ca 1971-78. *757*
—. Nuclear Energy. Strategic studies. 1490-1977. *265*
—. Public opinion. Radioactive waste. Theft. 1970's. *821*
Public protest. Scientists (mobilization of). 1970. *466*
Public Relations. Atomic Energy Commission. Federal Regulation. Safety. 1954-56. *763*
—. Federal Government. Nuclear arms testing. ca 1948-80. *92*
—. Nuclear Accidents. Pennsylvania. Reporters and Reporting. Three Mile Island power plant. 1976-79. *724*
Public scrutiny. Carter, Jimmy (administration). Congress. Foreign policy. National security. Nixon, Richard M. (administration). 1970's. *211*
Public utilities. Capacity, idle. Nuclear power plants. Taxation. 1973-74. *715*
Publishers and Publishing. Military Strategy. Nuclear energy. Smyth Report, 1945. War Department. 1946-73. *172*
Pugwash Conferences. Nuclear Science and Technology. Rabinowitch, Eugene. Scientists. 1957-73. *529*
P2V (aircraft). Naval Air Forces. Nuclear Arms. 1949-50. *120*

R

Rabinowitch, Eugene. *Bulletin of the Atomic Scientists* (periodical). Nuclear science and technology. 1940's-70's. *751*
—. Nuclear Science and Technology. Pugwash Conferences. Scientists. 1957-73. *529*
Racism. Canada. Diaries. King, William Lyon MacKenzie. Nuclear War. 1921-48. *49*
Radiation. American Atomics Corp. Arizona (Tucson). Manufactures. Public Health. Tritium. Working Conditions. 1979. *709*
—. Nuclear power plants. Shipyards. 1954-79. *108*
—. Public Health. Safety. 1980. *714*
Radiation Effects Research Foundation. Atomic Bomb Casualty Commission. Civilian survivors. Japan (Hiroshima, Nagasaki). Medical Research. 1945-75. *63*
Radiation exposure. Atomic Energy Commission. Marshall Islands. Nuclear Science and Technology. 1942-72. *81*

Radio. Drones. Fahrney, Delmar S. (account). Missiles. Navy Bureau of Aeronautics. N2C-2 (aircraft). Research and development. TG-2 (aircraft). World War II. 1936-45. *107*
Radioactive waste. Antarctic (McMurdo Sound). Navies. Nuclear reactors. 1960-71. *814*
—. Atomic Energy Commission. New York (West Valley). 1954-78. *786*
—. California Nuclear Safeguards Initiative. Nuclear accidents. Sabotage. 1976. *735*
—. Executive Planning Council. Intergovernmental Relations. National Energy Acts. Public Policy. ca 1971-78. *757*
—. Georgia Nuclear Laboratory. Nuclear Science and Technology. 1950's-79. *791*
—. Interest groups. Nuclear Regulatory Commission. Uranium Mill Tailings Radiation Control Act (US, 1978). 1978. *801*
—. Kansas. Michigan. 1970's. *697*
—. Public opinion. Public policy. Theft. 1970's. *821*
Radioactivity. Popular Culture. Public Opinion. Radium. Scientific discoveries. 1898-1935. *7*
Radium. Popular Culture. Public Opinion. Radioactivity. Scientific discoveries. 1898-1935. *7*
Ramsay, William. DeLeon, Peter. Electric Power. Nuclear Energy (review article). Rolph, Elizabeth S. 1960-79. *752*
Rasmussen, Norman (interview). Nuclear Power Plants. Safety. WASH-1400 (report). 1970-79. *779*
Rasmussen report. Nuclear Regulatory Commission. Power Resources. 1970's. *807*
Reagan, Ronald (administration). Arms Control and Disarmament Agency. Foreign policy. 1960-81. *505*
—. Carter, Jimmy (administration). Communist Countries. Foreign Policy. Military Strategy. 1980. *418*
—. Command structure. Defense policy. Nuclear Arms. USSR. 1980-81. *393*
—. Disarmament. Nuclear arms. 1981. *246*
—. Strategic Arms Limitation Talks. USSR. 1981. *493*
Referendum. Nuclear power plants. Ohio. Western States. 1976. *812*
Reform. Licensing. Nuclear accidents. Three Mile Island power plant. 1977-79. *802*
Refugees. Marshall Islands (Bikini). Nuclear arms testing. Pacific Dependencies (US). 1946-80. *184*
Regionalism. Arms control. Canada. China. India. Nuclear Arms. Nuclear power. USA. 1951-78. *387*
Religion. Miller, Walter *(A Canticle for Leibowitz).* Nuclear Energy. Science fiction. 1959-81. *795*
Reporters and Reporting. Nuclear Accidents. Pennsylvania. Public Relations. Three Mile Island power plant. 1976-79. *724*
—. Nuclear accidents. Television. Three Mile Island power plant. 1979. *773*
Research. Arms race. Nuclear Arms. 1946-74. *267*
—. Atomic Energy Commission. Deterrence. Gillet, Edward B. (interview). Military Strategy. Nuclear Arms (warheads). 1970's. *411*
—. Navies. *Norton Sound* (vessel). Technology. Weapons. 1943-79. *155*
—. Nuclear accidents. Pennsylvania. Politics. Social Problems. Three Mile Island power plant. 1969-80. *823*
—. Science and Government. 1945. *299*
Research and development. Budgets. Congress. Defense Policy. Military. 1976-77. *102*
—. Defense. ICBM's. Military Capability. Nuclear War. USSR. 1970's. *413*
—. Defense spending. ICBM's. Military Capability. USSR. 1970's. *182*
—. Drones. Fahrney, Delmar S. (account). Missiles. Navy Bureau of Aeronautics. N2C-2 (aircraft). Radio. TG-2 (aircraft). World War II. 1936-45. *107*
—. Military. National security. Weapons. 1978. *123*
Research Board for National Security. Defense research. Military. Science and Government. 1944-50. *132*
Retaliation, massive. Foreign Policy. Nuclear War. 1954-73. *281*
Rhetoric. Nuclear Accidents. Three Mile Island power plant. 1979. *721*
Richardson models. Arms race. Diplomacy. Negotiations. Nuclear test ban. USSR. 1962-63. *558*
Risk management. Government. Insurance. Liability. Public Opinion. Taxation. 1948-78. *730*
Rockwell International Corp. Airplanes, Military. B-1 (aircraft). Lobbyists. Unemployment. 1975. *149*
Rolph, Elizabeth S. DeLeon, Peter. Electric Power. Nuclear Energy (review article). Ramsay, William. 1960-79. *752*
Roosevelt, Franklin D. Churchill, Winston. Diplomacy. Great Britain. Military Strategy. World War II. 1942-44. *61*
—. Cold War. Diplomacy. Nuclear Arms. 1941-45. *53*
—. Diplomacy. Foreign Policy. Nuclear War. Truman, Harry S. 1945-51. *14*
—. Foreign Relations. Great Britain. Nuclear Arms. USA. World War II. 1940-45. *16*
Rose, John P. (review article). Armies. Defense Policy. Nuclear Arms (tactical). Politics. 1950-81. *239*
Rosenberg case. Atomic Energy Commission. Espionage. Federal Bureau of Investigation. Greenglass, David. Nuclear Arms. Trials. 1942-74. *142*
—. Atomic Energy Commission. Greenglass, David. Nuclear Science and Technology. Secrecy. Trials. 1945-51. *75*
—. Cold War. Espionage. Nuclear Arms. Supreme Court. 1953. *152*
Rowny, Edward L. Diplomacy. Personal narratives. Strategic Arms Limitation Talks. 1973-79. *638*

S

Sabotage. California Nuclear Safeguards Initiative. Nuclear accidents. Radioactive waste. 1976. *735*
—. Espionage. Nuclear installations. 1966-75. *723*
Safeguard debate. Decisionmaking. Defense policy. International Security. Missile defense. 1968-72. *175*
Safety. Accidents. Federal Regulation. Nuclear power plants. Nuclear Regulatory Commission. Pennsylvania (Three Mile Island). 1979. *780*
—. Accountability. Nuclear waste. Washington (Hanford). 1970's. *699*
—. Atomic Energy Commission. Federal Regulation. Public Relations. 1954-56. *763*
—. Atomic Energy Commission. Legislation. State Government. 1954-70. *808*
—. Eniwetok (atoll). Marshall Islands. Nuclear testing. Pacific Dependencies (US). 1946-80. *127*

—. Federal policy. Nuclear Science and Technology. 1974. *778*
—. Government Regulation. Nuclear waste. South. Transportation. ca 1960-79. *781*
—. Nuclear Power Plants. Rasmussen, Norman (interview). WASH-1400 (report). 1970-79. *779*
—. Public Health. Radiation. 1980. *714*
Schelling, Thomas. Brodie, Bernard. Kahn, Herman. Military Strategy. National security. 1950's-76. *376*
Schlesinger, James R. Balance of Power. Defense Policy. Nuclear Arms. USSR. 1960's-70's. *380*
—. Deterrence. Military Strategy. Nuclear War. 1974. *305*
—. Flexible response options. Nuclear Arms. Strategic arms proposals. 1975. *366*
—. ICBM's. Military strategy. Nuclear arms. Strategic Arms Limitation Talks. 1974. *327*
—. Military strategy. Nuclear arms. USA. USSR. 1970's. *229*
—. Nuclear arms. Strategic adaptability. 1970's. *419*
Science. Fallout. Public Health. 1954-64. *134*
—. Technical cooperation. USSR. 1972. *790*
Science and Government. Atomic Energy Commission. Federal regulation. Politics. Public Policy. 1946-70. *706*
—. Defense research. Military. Research Board for National Security. 1944-50. *132*
—. Research. 1945. *299*
Science and society. Attitudes. 19c-20c. *762*
—. Einstein, Albert. Pacifism. Physics. 1919-55. *38*
Science fiction. 20c. *121*
—. Miller, Walter *(A Canticle for Leibowitz).* Nuclear Energy. Religion. 1959-81. *795*
Scientific discoveries. Patterns. Secrecy. 1970-80. *755*
—. Popular Culture. Public Opinion. Radioactivity. Radium. 1898-1935. *7*
Scientific Experiments and Research. Arms race. Management. Military. Technology. 1945-75. *190*
—. Atomic structure, theory of. Magnets, floating. Mayer, Alfred Marshall. Physics. 1878-1907. *57*
—. Attitudes. Institute of Defense Analysis (Jason group). Vietnam War. 1966. *223*
—. Colorado (Boulder). Commerce Department. Maryland (Gaithersburg). National Bureau of Standards. 1901-78. *813*
—. Cyclotrons. Denmark. Japan. Nuclear physics. 1935-45. *65*
—. Fusion, Thermonuclear. Lasers. Nuclear Science and Technology. USSR. USSR. 1971-75. *701*
Scientists. Atomic bomb. Franck, James. Policymaking. Stimson, Henry L. World War II. 1945. *58*
—. Atomic Energy Commission. Engineers. Manhattan Project. Nuclear technology. 1938-50. *28*
—. Attitudes. Hydrogen bomb. 1950-55. *105*
—. *The Beginning or The End* (film). Films. Manhattan Project. Nuclear Arms. Public Opinion. 1947. *69*
—. Brown, Harrison (account). Gromyko, Andrei. Nuclear arms. UN. USSR. 1947. *482*
—. Disarmament. Nuclear War. Public Policy. 1939-54. *71*
—. Ethics. Nuclear arms. Oppenheimer, J. Robert. 1945-66. *76*
—. Ethics. Progress (concept). 1970's. *72*
—. Federation of American Scientists. Political activity. 1946-74. *653*
—. Letters. New Mexico (Los Alamos). Oppenheimer, J. Robert. 1932-34. 1943-45. *55*
—. Nuclear Science and Technology. Pugwash Conferences. Rabinowitch, Eugene. 1957-73. *529*
Scientists (mobilization of). Public protest. 1970. *466*
Seabed talks. Arms control. Disarmament. USSR. 1967-70. *633*
Sea-launched ballistic missiles. Military Strategy. Weapons. 1969-82. *187*
Second Front. Cold War (origins). Marshall Plan. Nuclear Arms. USSR. World War II. 1940's. *429*
Secrecy. Atomic Energy Commission. Greenglass, David. Nuclear Science and Technology. Rosenberg case. Trials. 1945-51. *75*
—. Defense Policy. Espionage. Groves, Leslie R. Nuclear arms. Truman, Harry S. (administration). USSR. 1945-50. *122*
—. France. Great Britain. Nuclear arms testing. World War II. 1939-45. *25*
—. Patterns. Scientific discoveries. 1970-80. *755*
Secret agreements. Foreign Relations. Kissinger, Henry A. Nixon, Richard M. Strategic Arms Limitation Talks. USA. USSR. 1972-74. *539*
Security classification. Atomic Energy Commission. 1944-75. *733*
—. Atomic Energy Commission. Law. 1945-75. *729*
Senate. Employment (dismissal). Engineering. Faulkner, Peter (account). Nuclear power plants. Nuclear Services Corp. 1973-77. *722*
—. Foreign Policy. Military Capability. Nitze, Paul. Strategic Arms Limitation Talks (II). 1945-79. *668*
—. Garn, Jake. National Security. Political Speeches. Strategic Arms Limitation Talks (II). 1979. *533*
—. Strategic Arms Limitation Talks (II). 1970's. *472*
Sherwin, Martin J. Foreign Policy. Gowing, Margaret. Great Britain. Nuclear Science and Technology (review article). Nuclear War. 1938-52. *33*
Shipyards. Nuclear power plants. Radiation. 1954-79. *108*
Shipyards (capacity). Nuclear Submarines. 1955-75. *176*
Shortages. Corporations. Politics. Stockpiling decisions. Strategic materials. 1946-70's. *103*
Simpson, John A. (views). Atomic Scientists of Chicago. Nuclear arms. Nuclear energy. 1945-80. *792*
Simulation and Games. History Teaching. Japan. Nuclear War. World War II. 1945. 1978. *19*
Single Integrated Operations Plan. Joint Chiefs of Staff. Military Strategy. Nuclear Arms. Strategic Arms Limitation Talks. Target planning. 1972-74. *383*
SLBM. Military Strategy. Nuclear Arms. Strategic arms control. 1950's-70's. *79*
Smith, Alice Kimball. Letters. Nuclear Science and Technology. Oppenheimer, J. Robert (reminiscences). Physics. Weiner, Charles. ca 1900-45. 1980. *47*
Smyth Report, 1945. Japan. Military Strategy. Nuclear energy. USA. World War II. 1940-45. *56*
—. Military Strategy. Nuclear energy. Publishers and Publishing. War Department. 1946-73. *172*
Social Conditions. Arms control. Mechanism (philosophy). Military strategy. 1950's-60's. *661*

Social Problems. Nuclear accidents. Pennsylvania. Politics. Research. Three Mile Island power plant. 1969-80. *823*
Social Psychology. Nuclear arms. 1945-79. *324*
Sociopolitical theory. Boulding, Kenneth (review article). Peace research. 1974-75. 1977. *473*
South. Government Regulation. Nuclear waste. Safety. Transportation. ca 1960-79. *781*
—. Manufactures. Nuclear Arms. ca 1942-79. *159*
—. Nuclear energy. States' rights. 1954-62. *809*
—. Nuclear waste disposal. ca 1945-79. *782*
South Africa. Angola. Foreign policy. Mozambique. Nuclear potential. USA. 1970-76. *427*
—. Foreign policy. Uranium. ca 1961-79. *428*
South Carolina. Allied General Nuclear Services. Barnwell Nuclear Fuel Plant. Nuclear power plants. ca 1963-79. *784*
Southwest. Federal Policy. Indian-White Relations. Navajo Indian Reservation. Uranium mining. 1946-79. *82*
Space. Antisatellite weapons. Armaments. USSR. 1960-80. *145*
—. Disarmament. USSR. 1957-81. *659*
—. Nuclear arms. Treaties. USSR. 1962-80. *536*
Space Vehicles. Air Forces. Missiles. 1954-74. *118*
Spain. Nuclear energy. Public Opinion. 1950-82. *725*
Spanish-American War. Cannon, Clarence. House of Representatives. Mexico. Military Service. World War I. World War II. 1898-1945. *35*
Stability, strategic. Defense Policy. Military strategy. USSR. 1960-80. *271*
Stalin, Joseph. Churchill, Winston. Communism. Diaries. Nuclear Arms. Potsdam Conference. Truman, Harry S. 1945. *37*
State Department. Atomic Energy Commission. Defense Department. Historical advisory committees. 1947-74. *754*
State Government. Atomic Energy Commission. Legislation. Safety. 1954-70. *808*
—. Attitudes. Decisionmaking. Nuclear power plants. Voting and Voting Behavior. Western States. 1963-78. *703*
States' rights. Nuclear energy. South. 1954-62. *809*
Steen, Charles. Cooper, Joe. Pick, Vernon. Prospecting. Uranium rushes. Western States. 1948-57. *144*
Stimson, Henry L. Atomic bomb. Franck, James. Policymaking. Scientists. World War II. 1945. *58*
Stockholm International Peace Research Institute. Arms control. National Security (review article). 1979. *604*
Stockpiling decisions. Corporations. Politics. Shortages. Strategic materials. 1946-70's. *103*
Strategic adaptability. Nuclear arms. Schlesinger, James R. 1970's. *419*
Strategic Air Command. Air Forces. Lemay, Curtis. Military Capability. Nuclear Arms. 1945-48. *88*
Strategic Air Command Wing. Airplanes, Military. B-47 (aircraft). Peck, Earl G. (account). ca 1960. *153*
Strategic arms control. Military Strategy. Nuclear Arms. SLBM. 1950's-70's. *79*
Strategic arms limitation. Detente. Economic relations. Nixon, Richard M. (administration). USSR. 1969-73. *648*
Strategic Arms Limitation Talks. 1960's-72. *538*
—. ABM. Disarmament. ICBM's. MIRV. 1969-73. *100*
—. Aeronautics, Military. Bombers, long-range. USSR. 1950-77. *651*
—. Arms control. Foreign Policy. Military Strategy. Nuclear arms. 1960's-78. *596*
—. Arms control (qualitative). International Security. 1974-75. *514*
—. Balance of power. USA. USSR. 1972. *554*
—. Carter, Jimmy (administration). 1977. *589*
—. Carter, Jimmy (administration). Equivalence (definitions). Nuclear arms policy. 1970's. *588*
—. Clements, William P., Jr. (interview). USSR. 1970's. *410*
—. Congress. 1970's. *600*
—. Congress. Executive Branch. Foreign Policy. 1972-78. *531*
—. Congress. Foreign policy. Panama. Politics. Treaties. 1977-78. *512*
—. Defense Policy. Manpower. Military. USSR. 1973-74. *384*
—. Defense Policy. Multiple delivery system. Naval Strategy (Blue Water Strategy). 1974. *683*
—. Defense Policy. Nuclear arms. Technology. 1975. *231*
—. Defense Policy. Nuclear arms. USSR. 1972-79. *582*
—. Defense policy. Nuclear War. USSR. 1969-79. *293*
—. Defense Policy. USSR. Weapons. 1972-78. *643*
—. Detente. USA. USSR. 1950's-70's. *679*
—. Detente. USSR. 1973-77. *677*
—. Deterrence. Nuclear Arms. USA. USSR. 1973. *623*
—. Deterrence. Nuclear arms. USSR. 1969-78. *496*
—. Diplomacy. Personal narratives. Rowny, Edward L. 1973-79. *638*
—. Diplomacy. USSR. 1968-76. *537*
—. Disarmament. National Security. USA. USSR. Verification methods. 1972-76. *593*
—. Disarmament. Nuclear arms. 1974-76. *513*
—. Europe, Western. Japan. Mutual Balanced Force Reductions. USA. USSR. 1974. *695*
—. Federal Government. Military Strategy. Nuclear Arms. Public Opinion. 1945-78. *498*
—. Foreign Policy. Kissinger, Henry A. Newhouse, John (review article). Nuclear arms. 1973. *461*
—. Foreign Policy. Middle East. USSR. 1977. *673*
—. Foreign Policy. National Security. Nuclear Arms. USSR. 1950's-70's. *565*
—. Foreign Policy. Nuclear Arms. 1969-72. *479*
—. Foreign Relations. Kissinger, Henry A. Nixon, Richard M. Secret agreements. USA. USSR. 1972-74. *539*
—. Foreign Relations. Military Strategy. Nuclear Arms. USSR. 1972-77. *656*
—. Foreign Relations. National security. Nixon Doctrine. 1970-71. *632*
—. Foreign Relations. Nuclear arms. USA. USSR. 1972-75. *602*
—. ICBM's. Military strategy. Nuclear arms. Schlesinger, James R. 1974. *327*
—. Intelligence Service. Nuclear arms. USSR. 1970-79. *642*
—. International Law. Strategic vicinity (concept). USSR. 1972-74. *523*
—. Joint Chiefs of Staff. Military Strategy. Nuclear Arms. Single Integrated Operations Plan. Target planning. 1972-74. *383*
—. Military Strategy. Nuclear Arms. 1970's. *551*

—. Military Strategy. Nuclear arms. USSR. 1969-78. *567*
—. Military strategy. Nuclear Arms. USSR. 1970-81. *204*
—. National security. Nuclear Arms. USSR. 1972-78. *696*
—. Nuclear arms. 1974. *682*
—. Nuclear arms. 1978. *577*
—. Nuclear Arms. Population vulnerability. 1974. *502*
—. Nuclear Arms. USA. USSR. 1972. *549*
—. Nuclear Arms. War. 1805-1972. *597*
—. Nuclear war, likelihood of. 1974. *457*
—. Reagan, Ronald (administration). USSR. 1981. *493*
—. ULMS. USSR. 1960's-72. *168*
—. USA. USSR. 1969-75. *478*
—. USA. USSR. 1972-75. *568*
—. USA. USSR. 1972. *471*
—. USSR. 1963-79. *671*
—. USSR. 1969-75. *550*
—. USSR. 1970's. *576*
—. USSR. 1970's. *647*
—. USSR. 1972-79. *658*
—. USSR. 1973-78. *491*
—. USSR. Verification. 1975-80. *509*
—. USSR. Vladivostok Accord. 1960-75. *477*
Strategic Arms Limitation Talks (II). Air Force Association (symposium). Military Strategy. Weapons. 1976. *670*
—. Arms control. USSR. 1970's. *475*
—. Arms race. 1816-1965. *426*
—. Balance of Power. Cruise missiles. Defense Department. Military Capability. ca 1950-79. *125*
—. Balance of Power. Europe. USSR. 1972-78. *542*
—. Carter, Jimmy. USSR. 1972-77. *666*
—. Carter, Jimmy (administration). Defense Policy. Diplomacy. Politics. USSR. 1978. *487*
—. Carter, Jimmy (administration). Foreign policy. USSR. 1970. *464*
—. Carter, Jimmy (administration). USSR. 1977. *470*
—. China. Defense Policy. USSR. 1968-79. *534*
—. Defense Policy. Foreign policy. Military Strategy. USSR. 1960's-70's. *521*
—. Defense Policy. Nixon, Richard M. Nuclear arms. 1973. *665*
—. Detente. Military strategy. Missile development. 1974-75. *548*
—. Diplomacy. Nuclear arms. USSR. 1972-76. *639*
—. Diplomacy. Talbott, Strobe (review article). USSR. 1972-79. *629*
—. Europe. USSR. 1979. *511*
—. Europe, Western. USSR. 1970's. *500*
—. Foreign Policy. Military Capability. Nitze, Paul. Senate. 1945-79. *668*
—. Foreign Policy. National Security. USSR. 1974-78. *592*
—. Foreign relations. Military strategy. 1972. *583*
—. Garn, Jake. National Security. Political Speeches. Senate. 1979. *533*
—. Military. USSR. 1973. *566*
—. Military Capability. Treaties. USSR. 1950's-70's. *613*
—. Military Intelligence. Telemetry (encryption). Verification. 1977-79. *663*
—. MIRV. USA. USSR. 1970's. *645*
—. Missiles. Naval Strategy. 1979. *541*
—. Mutual Assured Destruction. National security. Nonintervention. 1980. *625*
—. National Security. Politics. Weapons. 1977-78. *612*
—. Nuclear arms, offensive. USA. USSR. 1970. *490*
—. Political Surveys. Presidents. USSR. 1979. *608*
—. Senate. 1970's. *472*
—. USA. USSR. Vladivostok Accord (1974). 1950's-74. *618*
—. USSR. 1964-79. *669*
—. USSR. 1970's. *494*
—. USSR. Vladivostok Accord. 1963-75. *672*
—. USSR. Vladivostok Accord (1974). 1974. *680*
Strategic Arms Limitation Talks (Standing Consultative Commission). USA. USSR. 1970's. *469*
Strategic Arms Limitation Treaty. 1972-80. *609*
—. ABM. Defense Policy. USSR. 1982. *610*
—. USSR. 1945-81. *560*
Strategic Arms Limitation Treaty (1972). Arms race, qualitative. International security. Military innovation. Nuclear test ban. 1972-75. *220*
Strategic arms proposals. Flexible response options. Nuclear Arms. Schlesinger, James R. 1975. *366*
Strategic arms race. Disarmament. Nuclear Arms. USA. USSR. 1960-74. *446*
Strategic balance. Arms control (myths of). International Security. USA. USSR. 1975. *286*
—. Foreign policy. 1974. *266*
Strategic materials. Corporations. Politics. Shortages. Stockpiling decisions. 1946-70's. *103*
Strategic power. China. Coffey, Joseph I. Military strength. USA. USSR. ca 1965-73. *203*
Strategic response. Defense Policy. Economic policy. 1974. *313*
Strategic stability. Deterrence. USA. USSR. 1975. *348*
Strategic studies. Deterrence (military). Disarmament. Nuclear Arms. 1973. *364*
—. Nuclear Energy. Public Policy. 1490-1977. *265*
Strategic vicinity (concept). International Law. Strategic Arms Limitation Talks. USSR. 1972-74. *523*
Strategic weapons limitation (review article). National Security. 1958-74. *575*
Strategy. Deterrence. Military policy. -1972. *291*
Strauss, Lewis. Atomic Energy Commission. Executive Order 10450. Loyalty. Oppenheimer, J. Robert. 1946-54. *117*
—. Defense Policy. Dulles, John Foster. Eisenhower, Dwight D. Goodpaster, Andrew. Nuclear test ban issue. USSR. 1958. *240*
Submarine Warfare. Defense Department. Military Finance. Naval Strategy. Nuclear War. Tidal waves. 1950's-80. *131*
—. *George Washington* (vessel). Navies (Special Projects Office). Nuclear Arms. Polaris (missile). Whitmore, William F. (account). 1955-60. *186*
Submarines. Naval strategy. USA. USSR. 1970's. *279*
—. Navies. Nuclear Arms. 1975. *111*
—. Nuclear power. USSR. 1958-80. *154*
—. Nuclear Vessels. *Ohio* (vessel). 1981. *91*
Subsidies. Economic conditions. Federal government. Nuclear power plants. 1954-78. *718*
Supreme Court. Cold War. Espionage. Nuclear Arms. Rosenberg case. 1953. *152*
Surrender. B-29 (aircraft). Japan. Lemay, Curtis. Nuclear War. World War II. 1945. *67*
Surrender offer. Japan. Nuclear War. Politics. Truman, Harry S. (administration). World War II. 1945. *12*

Survivors. Disasters. Psychology. War. 1941-72. *34*
Sweden. Balance of Power. Defense policy. USSR (Kola Peninsula). 1945-74. *322*
Szilard, Leo. Anderson, Herbert L. (reminiscences). Chicago, University of. Columbia University. Fermi, Enrico. Nuclear Science and Technology. 1933-45. *6*
—. Byrnes, James F. Federal Government. Germany. Memoirs. Nuclear Science and Technology. 1898-1945. *59*
—. Einstein, Albert. Hungarian Americans. Nuclear Arms (program). USA. World War II. 1941-74. *26*

T

Talbott, Strobe (review article). Diplomacy. Strategic Arms Limitation Talks (II). USSR. 1972-79. *629*
Target planning. Joint Chiefs of Staff. Military Strategy. Nuclear Arms. Single Integrated Operations Plan. Strategic Arms Limitation Talks. 1972-74. *383*
Targeting. Military Strategy. Nuclear War. 1980. *74*
Taxation. Capacity, idle. Nuclear power plants. Public utilities. 1973-74. *715*
—. Government. Insurance. Liability. Public Opinion. Risk management. 1948-78. *730*
Technical cooperation. Science. USSR. 1972. *790*
Technical innovation. Arms control. USA. USSR. 1945-73. *540*
Technological innovations. Deterrence. Military Strategy. Nuclear Arms. USSR. 1969-78. *338*
Technologies. Arms race. Foreign relations. USA. USSR. 1973. *343*
Technology. Airplanes, Military. B-52 bomber. Vietnam. 1952-75. *124*
—. Arms control. Balance of Power. USSR. 1960-80. *530*
—. Arms control. Balance of Power. Weapons. 1970's. *135*
—. Arms race. Management. Military. Scientific Experiments and Research. 1945-75. *190*
—. Arms race. Nuclear Arms. Values, human. 1973. *528*
—. Attitudes. Economic development. Military Strategy. Nuclear arms. War. 1914-68. *351*
—. Business. Energy. Government. Public Policy. 1978. *731*
—. Communist Countries. Exports. National security. 1949-78. *242*
—. Defense Policy. ICBM's. Land-mobile system, proposed. 1950's-74. *80*
—. Defense Policy. Nuclear arms. Strategic Arms Limitation Talks. 1975. *231*
—. Defense Policy. USSR. Weapons. 1978. *283*
—. Energy crisis. Industry. Nuclear power. ca 1940-80. *745*
—. Exports. Foreign Policy. Nuclear Arms (control). 1976. *435*
—. Foreign Policy. Nuclear proliferation. 1946-80. *684*
—. Foreign policy. Nuclear reactors. 1953-74. *290*
—. Foreign Relations. International Trade. Nuclear Energy. Nuclear Nonproliferation Act (US, 1978). 1978-79. *614*
—. ICBM's. Military strategy. Nuclear arms limitation. Vladivostok Accord. 1970's. *611*
—. Military Capability. NATO. Nuclear arms. USA. USSR. 1970's. *98*
—. Military Strategy. Weapons. 1970's. *409*
—. Navies. *Norton Sound* (vessel). Research. Weapons. 1943-79. *155*
—. Nuclear Energy. 1946-77. *805*
—. Nuclear power industry. Power Resources. 1954-73. *716*
—. Power Resources (alternatives). 1900's-1985. *777*
Technology assessment. Atomic Energy Commission. Environment. Nuclear power. Public participation. 1946-73. *742*
Technology, military. Arms control. 1970's. *646*
Telemetry (encryption). Military Intelligence. Strategic Arms Limitation Talks (II). Verification. 1977-79. *663*
Television. Nuclear accidents. Reporters and Reporting. Three Mile Island power plant. 1979. *773*
—. Nuclear energy. *Plutonium: An Element of Risk* (documentary). 1970's. *743*
Tennessee. Army Corps of Engineers. Journalism. Nuclear energy. Oak Ridge *Journal*. 1943-49. *2*
Terrorism. Civil Rights. Martial Law. Nuclear Power Plants. Plutonium. 1970's. *728*
—. ICBM's. Nuclear Science and Technology. 1960's-70's. *704*
—. Nuclear Arms. ca 1972-74. *749*
—. Nuclear power plants. 1974. *788*
TG-2 (aircraft). Drones. Fahrney, Delmar S. (account). Missiles. Navy Bureau of Aeronautics. N2C-2 (aircraft). Radio. Research and development. World War II. 1936-45. *107*
Theater Nuclear Forces. Disarmament. NATO. Nuclear arms. 1977-79. *278*
—. NATO. Nuclear Arms. 1950's. *395*
Theater Production and Direction. Little Theatre Group. Manhattan Project. New Mexico (Los Alamos). Nuclear Arms. World War II. 1943-46. *3*
Theft. Public opinion. Public policy. Radioactive waste. 1970's. *821*
Three Mile Island power plant *See also* Pennsylvania (Three Mile Island).
—. Carter, Jimmy (administration). Nuclear accidents. Pennsylvania. President's Commission on the Accident at Three Mile Island. 1979. *761*
—. Housing. Nuclear accidents. Pennsylvania. Prices. 1977-79. *771*
—. Licensing. Nuclear accidents. Reform. 1977-79. *802*
—. Mental health. Mothers. Nuclear accidents. Pennsylvania. 1979. *708*
—. Nuclear accidents. Pennsylvania. 1977-81. *776*
—. Nuclear accidents. Pennsylvania. Political Protest. 1979-80. *810*
—. Nuclear accidents. Pennsylvania. Politics. Research. Social Problems. 1969-80. *823*
—. Nuclear Accidents. Pennsylvania. Public Relations. Reporters and Reporting. 1976-79. *724*
—. Nuclear accidents. Reporters and Reporting. Television. 1979. *773*
—. Nuclear Accidents. Rhetoric. 1979. *721*
Tidal waves. Defense Department. Military Finance. Naval Strategy. Nuclear War. Submarine Warfare. 1950's-80. *131*
Tlatelolco, Treaty of. Defense Policy. Foreign Relations. Latin America. Nuclear arms. 1960's-70's. *637*
Tlatelolco, Treaty of (protocol). Latin America. Nuclear arms. 1973-77. *520*
Tomahawk (missile). Arms race. Carter, Jimmy (administration). Cruise missiles. World War II. 1977-80. *160*

Trade. India. Nuclear Power. Treaties. Uranium. 1963-80. *774*
Transportation. Government Regulation. Nuclear waste. Safety. South. ca 1960-79. *781*
Treaties. Arms control. Disarmament. USSR. 1977. *570*
—. Arms control. Executive Branch. Public opinion. USSR. 1972-82. *517*
—. Congress. Foreign policy. Panama. Politics. Strategic Arms Limitation Talks. 1977-78. *512*
—. Energy. Nuclear Science and Technology. 1940's-70's. *476*
—. India. Nuclear Power. Trade. Uranium. 1963-80. *774*
—. Military Capability. Strategic Arms Limitation Talks (II). USSR. 1950's-70's. *613*
—. Nuclear arms. Space. USSR. 1962-80. *536*
—. Nuclear testing. USSR. 1963-73. *631*
Trials. Atomic Energy Commission. Espionage. Federal Bureau of Investigation. Greenglass, David. Nuclear Arms. Rosenberg case. 1942-74. *142*
—. Atomic Energy Commission. Greenglass, David. Nuclear Science and Technology. Rosenberg case. Secrecy. 1945-51. *75*
Trident (missile). Arms race. Missiles. Navies. Nuclear submarines. 1960-80. *158*
Trident submarine. B-1 (aircraft). Congress. Defense budget cuts. Pressure groups. 1970's. *78*
Tritium. American Atomics Corp. Arizona (Tucson). Manufactures. Public Health. Radiation. Working Conditions. 1979. *709*
Truman, Harry S. Churchill, Winston. Communism. Diaries. Nuclear Arms. Potsdam Conference. Stalin, Joseph. 1945. *37*
—. Diplomacy. Foreign Policy. Nuclear War. Roosevelt, Franklin D. 1945-51. *14*
—. Domestic policy. Foreign policy. Historiography. Liberalism. 1945-52. *275*
—. Executive Power. Foreign policy. 1945-53. *43*
—. Hydrogen bomb. Joint Chiefs of Staff. Military Strategy. Nuclear Arms. USSR. 1945-50. *161*
Truman, Harry S. (administration). Arms race. Cold War. Economic Conditions. International Trade. 1948-49. *347*
—. Atomic bomb. Franck, James. Nuclear arms race. Political Attitudes. World War II. 1944-45. *62*
—. Defense Policy. Documents. 1946-75. *360*
—. Defense Policy. Espionage. Groves, Leslie R. Nuclear arms. Secrecy. USSR. 1945-50. *122*
—. Japan. Nuclear War. Politics. Surrender offer. World War II. 1945. *12*
Trusts, industrial. Canada. Foreign policy. Uranium industry. USA. 1970's. *798*

U

ULMS. Strategic Arms Limitation Talks. USSR. 1960's-72. *168*
UN. Arms trade. 1945-77. *595*
—. Baruch Plan. Foreign Relations. Nuclear Energy. Nuclear proliferation. 1945-77. *712*
—. Brown, Harrison (account). Gromyko, Andrei. Nuclear arms. Scientists. USSR. 1947. *482*
—. Diplomacy. Disarmament. Nuclear nonproliferation. 1945-77. *467*
—. Nuclear arms. 1961-78. *519*
Unemployment. Airplanes, Military. B-1 (aircraft). Lobbyists. Rockwell International Corp. 1975. *149*
United States v. *The Progressive* (US, 1979). Freedom of the press. Nuclear arms. Press. *Progressive* (periodical). 1979. *800*
Uranium. Arizona. Grand Canyon. Mines. 1540-1969. *156*
—. Central Intelligence Agency. Israel. Nuclear Arms. Nuclear Materials and Equipment Corporation. Pennsylvania (Apollo). 1960's. *251*
—. Foreign policy. South Africa. ca 1961-79. *428*
—. India. Nuclear Power. Trade. Treaties. 1963-80. *774*
Uranium cartel (possible). Foreign policy. International Trade. Nuclear power industry. USA. 1970's. *815*
Uranium industry. Canada. Foreign policy. Trusts, industrial. USA. 1970's. *798*
Uranium Mill Tailings Radiation Control Act (US, 1978). Interest groups. Nuclear Regulatory Commission. Radioactive waste. 1978. *801*
Uranium mining. Federal Policy. Indian-White Relations. Navajo Indian Reservation. Southwest. 1946-79. *82*
Uranium rushes. Cooper, Joe. Pick, Vernon. Prospecting. Steen, Charles. Western States. 1948-57. *144*
USA (role of). Military Strategy. NATO. Warsaw Pact. 1960's-75. *247*
USSR. ABM. Defense Policy. Strategic Arms Limitation Treaty. 1982. *610*
—. Action-reaction theory. Arms race. International Relations (discipline). Models. 1970's. *331*
—. Aeronautics, Military. Bombers, long-range. Strategic Arms Limitation Talks. 1950-77. *651*
—. Afghanistan. Arms race. Cold War. Foreign Relations. 1960-80. *355*
—. Allies. Arms control. Balance of power. Europe. 1979. *598*
—. American Aeronautics and Astronautics Associations. ICBM's. 1970's. *406*
—. Antisatellite weapons. Armaments. Space. 1960-80. *145*
—. Armaments. Defense Policy. 1960's-70's. *396*
—. Arms control. Balance of Power. Technology. 1960-80. *530*
—. Arms control. Brezhnev, Leonid. Ford, Gerald R. Foreign Relations. Kissinger, Henry A. Vladivostok Accord. 1974. *674*
—. Arms control. Competition. Foreign Policy. Institutions. 1970-80. *585*
—. Arms control. Defense policy. Foreign Relations. 1960-80. *584*
—. Arms control. Detente. Foreign relations. USA. 1961-75. *574*
—. Arms control. Detente. Mutual Balanced Force Reductions. USA. 1974. *634*
—. Arms control. Disarmament. Foreign Policy. Military strategy. 1945-70's. *214*
—. Arms control. Disarmament. Foreign Relations. USA. 1970's. *664*
—. Arms control. Disarmament. Seabed talks. 1967-70. *633*
—. Arms control. Disarmament. Treaties. 1977. *570*
—. Arms control. Disarmament. USA. 1974. *649*
—. Arms control. Domestic Policy. USA. Weapons. 1963-75. *462*
—. Arms control. Executive Branch. Public opinion. Treaties. 1972-82. *517*
—. Arms control. Foreign Policy. 1979-82. *622*
—. Arms control. Military Intelligence. 1960-82. *518*

—. Arms control. Political change. USA. 1945-73. *603*
—. Arms control. Strategic Arms Limitation Talks (II). 1970's. *475*
—. Arms control. Technical innovation. USA. 1945-73. *540*
—. Arms control. USA. World order. 1950's-75. *483*
—. Arms control agreements. China. Detente. Nuclear strategy. USA. 1970's. *401*
—. Arms control (myths of). International Security. Strategic balance. USA. 1975. *286*
—. Arms race. 1960's-70's. *349*
—. Arms race. Balance of Power. Deterrence. Foreign Relations. 1970's. *312*
—. Arms race. Beam energy. Grechko, Andrei. ICBM's. 1960's-70's. *414*
—. Arms race. Detente. Nuclear strategy. USA. 1972-74. *319*
—. Arms race. Diplomacy. Negotiations. Nuclear test ban. Richardson models. 1962-63. *558*
—. Arms race. Disarmament. USA. 1945-70's. *277*
—. Arms race. Foreign Relations. Great Powers. 1945-78. *660*
—. Arms race. Foreign relations. Technologies. USA. 1973. *343*
—. Arms race. International relations theory. Models. 1948-70. *289*
—. Arms race. Military Capability. Nuclear Arms. 1960-80. *404*
—. Arms race. Military potential. USA. 1940-74. *445*
—. Arms race model. USA. 1945-73. *315*
—. Attitudes. Democracy. Europe. Interest Groups. Japan. Nuclear Science and Technology. 1970's. *739*
—. Attitudes. Economics. Foreign relations. Nuclear arms limitation. USA. 1972-73. *465*
—. Austria. Foreign Relations. Nuclear Test Ban Treaty. 1955. 1963. *581*
—. Balance of power. China. Detente. Foreign policy. 1945-79. *218*
—. Balance of Power. Communism. Foreign Policy. 1965-78. *336*
—. Balance of power. Cuban Missile Crisis. Kennedy, John F. Military Strategy. USA. 1962. *236*
—. Balance of Power. Defense Policy. Nuclear Arms. Schlesinger, James R. 1960's-70's. *380*
—. Balance of Power. Defense Policy. Nuclear Arms. USA. 1974-75. *241*
—. Balance of power. Developing Nations. Foreign policy. 1961-79. *556*
—. Balance of Power. Europe. Strategic Arms Limitation Talks (II). 1972-78. *542*
—. Balance of Power. Foreign Relations. 1960-69. *363*
—. Balance of Power. France. Great Britain. Newspapers. 1948-73. *262*
—. Balance of Power. ICBM's. 1972-75. *453*
—. Balance of Power. Metaphysics. Nuclear Arms. 1970's. *222*
—. Balance of Power. Meyer, John C. (interview). Military Strategy. 1970's. *412*
—. Balance of Power. Military Capability. 1980. *238*
—. Balance of Power. Military Ground Forces. Nuclear Arms. 1969-77. *193*
—. Balance of power. Strategic Arms Limitation Talks. USA. 1972. *554*
—. Baruch, Bernard. Cold War. 1946. *110*
—. Brown, George S. (views). Joint Chiefs of Staff. Military Strategy. Nuclear Arms. 1970's. *408*
—. Brown, Harrison (account). Gromyko, Andrei. Nuclear arms. Scientists. UN. 1947. *482*
—. Canada. Foreign Policy. Great Britain. Nuclear Nonproliferation. 1943-76. *545*
—. Carter, Jimmy. Strategic Arms Limitation Talks (II). 1972-77. *666*
—. Carter, Jimmy (administration). Defense Policy. Diplomacy. Politics. Strategic Arms Limitation Talks (II). 1978. *487*
—. Carter, Jimmy (administration). Foreign policy. Strategic Arms Limitation Talks (II). 1970. *464*
—. Carter, Jimmy (administration). Strategic Arms Limitation Talks (II). 1977. *470*
—. Carter, Jimmy (views). Foreign policy. 1977. *499*
—. China. Coffey, Joseph I. Military strength. Strategic power. USA. ca 1965-73. *203*
—. China. Defense Policy. Strategic Arms Limitation Talks (II). 1968-79. *534*
—. China. Nuclear stalemate, function of. USA. World order. 1970's. *201*
—. Civil defense. Defense Policy. Military. Nuclear War. Public Opinion. 1948-78. *424*
—. Civil defense. Military Strategy. Nuclear War. 1950's-77. *392*
—. Civil defense. Nuclear War. 1950-78. *133*
—. Clements, William P., Jr. (interview). Strategic Arms Limitation Talks. 1970's. *410*
—. Cold War. 1981. *346*
—. Cold War. Communism. Diplomacy. Middle East. Morgenthau, Hans J. Nuclear war. 1945-76. *399*
—. Cold War. Decisionmaking. Disarmament. Eisenhower, Dwight D. Nuclear arms. 1953-61. *657*
—. Cold War. Detente. Foreign Relations. Great Powers. 1945-75. *252*
—. Cold War. Diplomacy. Foreign policy. Nuclear Arms. USA. 1942-46. *13*
—. Cold War (origins). Marshall Plan. Nuclear Arms. Second Front. World War II. 1940's. *429*
—. Command structure. Defense policy. Nuclear Arms. Reagan, Ronald (administration). 1980-81. *393*
—. Communications, Military. Defense Policy. 1948-82. *174*
—. Communications, Military. Defense Policy. Nuclear weapons. 1940-81. *99*
—. Conflict theory. Deterrence. Military Strategy. USA. 1945-75. *258*
—. Decisionmaking. Divine, Robert A. (review article). Foreign relations. Nuclear arms. 1954-60. 1978. *195*
—. Defense. ICBM's. Military Capability. Nuclear War. Research and development. 1970's. *413*
—. Defense budget. Detente. USA. 1959-74. *248*
—. Defense policy. Detente. Deterrence. USA. 1975. *379*
—. Defense Policy. Disarmament. Foreign policy. Nuclear arms. Panel of Consultants on Disarmament (report). 1953. *486*
—. Defense Policy. Dulles, John Foster. Eisenhower, Dwight D. Goodpaster, Andrew. Nuclear test ban issue. Strauss, Lewis. 1958. *240*
—. Defense Policy. Espionage. Groves, Leslie R. Nuclear arms. Secrecy. Truman, Harry S. (administration). 1945-50. *122*
—. Defense Policy. Europe, Western. Mutual balanced force reductions. 1970's. *689*
—. Defense Policy. Foreign policy. Military Strategy. Strategic Arms Limitation Talks (II). 1960's-70's. *521*
—. Defense Policy. Manpower. Military. Strategic Arms Limitation Talks. 1973-74. *384*

—. Defense Policy. Military Capability. 1977-80. *374*
—. Defense Policy. Military Capability. Nuclear Arms. 1970's. *297*
—. Defense Policy. Military Strategy. ca 1960-80. *210*
—. Defense policy. Military Strategy. Mutual assured destruction. Nuclear War. USA. 1960-76. *196*
—. Defense Policy. Military strategy. Stability, strategic. 1960-80. *271*
—. Defense Policy. Nuclear arms. Strategic Arms Limitation Talks. 1972-79. *582*
—. Defense policy. Nuclear War. Strategic Arms Limitation Talks. 1969-79. *293*
—. Defense Policy. Nuclear War (risk of). USA. 1945-74. *285*
—. Defense Policy. Strategic Arms Limitation Talks. Weapons. 1972-78. *643*
—. Defense Policy. Technology. Weapons. 1978. *283*
—. Defense Policy. USA. 1974. *381*
—. Defense spending. ICBM's. Military Capability. Research and development. 1970's. *182*
—. Defense spending. Intelligence Service. Military capability. 1960's-76. *434*
—. Defense spending. Military Strategy. National security. NATO. 1980. *359*
—. Defense spending. Missile capability. Nuclear arms race. USA. 1970's. *433*
—. Detente. 1960's-70's. *676*
—. Detente. Diplomacy. Europe. International Security. 1970-74. *510*
—. Detente. Disarmament. Nuclear Arms. USA. 1970's. *573*
—. Detente. Disarmament. USA. 1976-77. *693*
—. Detente. Economic relations. Nixon, Richard M. (administration). Strategic arms limitation. 1969-73. *648*
—. Detente. Europe. Military Strategy. 1950's-70's. *444*
—. Detente. Europe. USA. 1879-1906. 1945-75. *372*
—. Detente. Europe, Western. European Security Conference. Propaganda. 1960's-73. *463*
—. Detente. Foreign Policy. Human rights. 1973-79. *552*
—. Detente. Strategic Arms Limitation Talks. 1973-77. *677*
—. Detente. Strategic Arms Limitation Talks. USA. 1950's-70's. *679*
—. Detente. USA. 20c. *624*
—. Deterrence. 1963-80. *288*
—. Deterrence. ICBM's. 1960's-70's. *416*
—. Deterrence. Military Capability. Nuclear arms. 1945-50. *337*
—. Deterrence. Military strategy. Nuclear arms. 1950-80. *318*
—. Deterrence. Military Strategy. Nuclear Arms. Technological innovations. 1969-78. *338*
—. Deterrence. Nuclear Arms. 1952-80. *259*
—. Deterrence. Nuclear arms. Strategic Arms Limitation Talks. 1969-78. *496*
—. Deterrence. Nuclear Arms. Strategic Arms Limitation Talks. USA. 1973. *623*
—. Deterrence. Nuclear stockpiling. USA. 1960's-70's. *440*
—. Deterrence. Strategic stability. USA. 1975. *348*
—. Developing nations. Foreign policy. 1970-79. *282*
—. Diplomacy. Nuclear arms. Strategic Arms Limitation Talks (II). 1972-76. *639*
—. Diplomacy. Strategic Arms Limitation Talks. 1968-76. *537*
—. Diplomacy. Strategic Arms Limitation Talks (II). Talbott, Strobe (review article). 1972-79. *629*
—. Diplomacy (review article). 1933-80. *532*
—. Disarmament. Europe. Mutual Balanced Force Reductions. 1973. *586*
—. Disarmament. Foreign policy. Nuclear Arms. USA. 1970's. *300*
—. Disarmament. Foreign Relations. Military. Political power. 1945-82. *333*
—. Disarmament. Foreign Relations. USA. 1945-70's. *561*
—. Disarmament. Great Britain. Nuclear Test Ban Treaty (1963). 1954-70's. *675*
—. Disarmament. Military (force reductions). NATO. 1955-73. *667*
—. Disarmament. National Security. Strategic Arms Limitation Talks. USA. Verification methods. 1972-76. *593*
—. Disarmament. NATO. 1932-82. *626*
—. Disarmament. Nuclear Arms. 1943-78. *294*
—. Disarmament. Nuclear Arms. Strategic arms race. USA. 1960-74. *446*
—. Disarmament. Nuclear Arms. USA. 1940's-74. *563*
—. Disarmament. Space. 1957-81. *659*
—. Documents. Military Strategy. Nuclear Arms. 1954-55. *373*
—. Dropshot (plan). Military Strategy. 1949-57. *87*
—. Europe. Foreign relations. Military strategy. Nuclear arms. USA. 1950's-70's. *436*
—. Europe. Strategic Arms Limitation Talks (II). 1979. *511*
—. Europe, Western. Japan. Mutual Balanced Force Reductions. Strategic Arms Limitation Talks. USA. 1974. *695*
—. Europe, Western. Military Strategy. NATO. Nuclear Arms. 1950-78. *85*
—. Europe, Western. Military strategy. NATO. Nuclear arms, tactical. USA. 1970's. *256*
—. Europe, Western. NATO. Nuclear arms. 1979. *402*
—. Europe, Western. Strategic Arms Limitation Talks (II). 1970's. *500*
—. Foreign Policy. 1960's-75. *421*
—. Foreign policy. Hydrogen bomb testing. Nuclear Arms. York, Herbert (hypothesis). 1953-55. 1970's. *84*
—. Foreign Policy. Middle East. Strategic Arms Limitation Talks. 1977. *673*
—. Foreign Policy. Military Capability. 1952-82. *329*
—. Foreign Policy. National Security. Nuclear Arms. Strategic Arms Limitation Talks. 1950's-70's. *565*
—. Foreign Policy. National Security. Strategic Arms Limitation Talks (II). 1974-78. *592*
—. Foreign Policy. Nuclear Arms. October War. 1973. *213*
—. Foreign Policy. Nuclear Arms. USA. 1969-74. *438*
—. Foreign policy. USA. 1945-75. *378*
—. Foreign Relations. 1933-78. *628*
—. Foreign Relations. 1958-73. *591*
—. Foreign Relations. 1975-76. *200*
—. Foreign Relations. Kissinger, Henry A. Nixon, Richard M. Secret agreements. Strategic Arms Limitation Talks. USA. 1972-74. *539*
—. Foreign Relations. Military Strategy. Nuclear Arms. Strategic Arms Limitation Talks. 1972-77. *656*
—. Foreign Relations. NATO. Nuclear Arms. 1970's-81. *455*
—. Foreign Relations. Nuclear arms. Strategic Arms Limitation Talks. USA. 1972-75. *602*
—. Foreign relations. Nuclear arms limitation. USA. 1959-75. *527*
—. Fusion, Thermonuclear. Lasers. Nuclear Science and Technology. Scientific Experiments and Research. USSR. 1971-75. *701*

—. Fusion, Thermonuclear. Lasers. Nuclear Science and Technology. Scientific Experiments and Research. USSR. 1971-75. *701*
—. Hydrogen bomb. Joint Chiefs of Staff. Military Strategy. Nuclear Arms. Truman, Harry S. 1945-50. *161*
—. ICBM's (survivability). Nuclear War. 1960's-70's. *332*
—. Intelligence Service. Nuclear arms. Strategic Arms Limitation Talks. 1970-79. *642*
—. International Law. Strategic Arms Limitation Talks. Strategic vicinity (concept). 1972-74. *523*
—. International Security. 1980. *681*
—. Japan. Leftists. Nuclear War. World War II. 1945. *17*
—. Japan (Hiroshima, Nagasaki). Nuclear War. USA. World War II. 1942-47. *15*
—. Military. Missiles. 1955-80. *170*
—. Military. Strategic Arms Limitation Talks (II). 1973. *566*
—. Military Capability. 1973-74. *451*
—. Military Capability. 1974-75. *448*
—. Military Capability. 1979. *442*
—. Military Capability. NATO. Nuclear arms. Technology. USA. 1970's. *98*
—. Military Capability. Nuclear arms. USA. 1946-72. *89*
—. Military Capability. Strategic Arms Limitation Talks (II). Treaties. 1950's-70's. *613*
—. Military strategy. 1950's-80. *308*
—. Military Strategy. Missiles (vulnerability). USA. 1975. *394*
—. Military Strategy. National security. Nuclear Arms. USA. 1954-74. *309*
—. Military Strategy. Nuclear arms. 1945-78. *219*
—. Military Strategy. Nuclear Arms. 1950's-79. *296*
—. Military Strategy. Nuclear Arms. 1964-81. *377*
—. Military strategy. Nuclear arms. Schlesinger, James R. USA. 1970's. *229*
—. Military Strategy. Nuclear arms. Strategic Arms Limitation Talks. 1969-78. *567*
—. Military strategy. Nuclear Arms. Strategic Arms Limitation Talks. 1970-81. *204*
—. Military Strategy. Nuclear War. 1977. *356*
—. Military Strategy. Planning. 1945-79. *339*
—. MIRV. Strategic Arms Limitation Talks (II). USA. 1970's. *645*
—. National security. Nuclear Arms. Strategic Arms Limitation Talks. 1972-78. *696*
—. Naval construction. USA. Western nations. 1965-76. *143*
—. Naval strategy. Nuclear arms. 1950's-80. *330*
—. Naval strategy. Submarines. USA. 1970's. *279*
—. Navies. Nuclear weapons. USA. 1945-74. *264*
—. Nuclear Arms. Peace. 1948-80. *423*
—. Nuclear arms. Space. Treaties. 1962-80. *536*
—. Nuclear Arms. Strategic Arms Limitation Talks. USA. 1972. *549*
—. Nuclear Arms. USA. 1965-73. *215*
—. Nuclear arms, offensive. Strategic Arms Limitation Talks (II). USA. 1970. *490*
—. Nuclear power. Submarines. 1958-80. *154*
—. Nuclear testing. Treaties. 1963-73. *631*
—. Political Surveys. Presidents. Strategic Arms Limitation Talks (II). 1979. *608*
—. Reagan, Ronald (administration). Strategic Arms Limitation Talks. 1981. *493*
—. Science. Technical cooperation. 1972. *790*
—. Strategic Arms Limitation Talks. 1963-79. *671*
—. Strategic Arms Limitation Talks. 1969-75. *550*
—. Strategic Arms Limitation Talks. 1970's. *576*
—. Strategic Arms Limitation Talks. 1970's. *647*
—. Strategic Arms Limitation Talks. 1972-79. *658*
—. Strategic Arms Limitation Talks. 1973-78. *491*
—. Strategic Arms Limitation Talks. ULMS. 1960's-72. *168*
—. Strategic Arms Limitation Talks. USA. 1969-75. *478*
—. Strategic Arms Limitation Talks. USA. 1972-75. *568*
—. Strategic Arms Limitation Talks. USA. 1972. *471*
—. Strategic Arms Limitation Talks. Verification. 1975-80. *509*
—. Strategic Arms Limitation Talks. Vladivostok Accord. 1960-75. *477*
—. Strategic Arms Limitation Talks (II). 1964-79. *669*
—. Strategic Arms Limitation Talks (II). 1970's. *494*
—. Strategic Arms Limitation Talks (II). USA. Vladivostok Accord (1974). 1950's-74. *618*
—. Strategic Arms Limitation Talks (II). Vladivostok Accord. 1963-75. *672*
—. Strategic Arms Limitation Talks (II). Vladivostok Accord (1974). 1974. *680*
—. Strategic Arms Limitation Talks (Standing Consultative Commission). USA. 1970's. *469*
—. Strategic Arms Limitation Treaty. 1945-81. *560*
USSR (Kola Peninsula). Balance of Power. Defense policy. Sweden. 1945-74. *322*

V

Value systems. Education. Nuclear Science and Technology. 1974. *109*
Values, human. Arms race. Nuclear Arms. Technology. 1973. *528*
Verification. Military Intelligence. Strategic Arms Limitation Talks (II). Telemetry (encryption). 1977-79. *663*
—. Strategic Arms Limitation Talks. USSR. 1975-80. *509*
Verification methods. Disarmament. National Security. Strategic Arms Limitation Talks. USA. USSR. 1972-76. *593*
Vietnam. Airplanes, Military. B-52 bomber. Technology. 1952-75. *124*
Vietnam War. Attitudes. Institute of Defense Analysis (Jason group). Scientific Experiments and Research. 1966. *223*
—. Foreign Policy. Graduated response (policy). Military strategy. 1950-75. *230*
Vietnam War (ramifications). Military Strategy (studies). 1945-73. *232*
Violence. Developing Nations. Heilbroner, Robert. Nuclear Arms. Peace, prospects for. 1976. *535*
Vladivostok Accord. ABM Treaty. Arms control. 1974. *654*
—. Arms control. Brezhnev, Leonid. Ford, Gerald R. Foreign Relations. Kissinger, Henry A. USSR. 1974. *674*
—. Brezhnev, Leonid. Detente. Ford, Gerald R. 1974. *617*
—. ICBM's. Military strategy. Nuclear arms limitation. Technology. 1970's. *611*
—. Strategic Arms Limitation Talks. USSR. 1960-75. *477*

—. Strategic Arms Limitation Talks (II). USSR. 1963-75. *672*
Vladivostok Accord (1974). Strategic Arms Limitation Talks (II). USA. USSR. 1950's-74. *618*
—. Strategic Arms Limitation Talks (II). USSR. 1974. *680*
Voting and Voting Behavior. Attitudes. Decisionmaking. Nuclear power plants. State Government. Western States. 1963-78. *703*

W

War. Arms race. 1816-1965. *425*
—. Attitudes. Economic development. Military Strategy. Nuclear arms. Technology. 1914-68. *351*
—. Defensive war argument. Foreign Policy. Ideology. 1941-73. *398*
—. Disasters. Psychology. Survivors. 1941-72. *34*
—. Education. Public Opinion. 1944-63. *146*
—. Nuclear Arms. Strategic Arms Limitation Talks. 1805-1972. *597*
War Department. Military Strategy. Nuclear energy. Publishers and Publishing. Smyth Report, 1945. 1946-73. *172*
War Powers Resolution (1973). Congress. Federation of American Scientists. Nuclear Arms. Presidency. 1975-76. *578*
War, principles of. Clausewitz, Karl von (theories). Military Strategy. Nuclear War, limited. 1970's. *342*
War, risks of. Deterrence logic. Disarmament. World War II (antecedents). 1933-62. *307*
Wars, significance of. Armaments. Disarmament. Nuclear proliferation. 1945-75. *205*
Warsaw Pact. Arms control. Mutual Balanced Force Reductions. NATO. 1973-74. *635*
—. Balance of Power. Military. NATO. 1962-75. *449*
—. Defense Policy. Deterrence. Europe, Western. NATO. Nuclear arms. 1970's. *326*
—. Deterrence. NATO. Neutron bomb. Nuclear Arms. 1954-78. *83*
—. Military Capability. NATO. Nuclear Arms. 1970. *226*
—. Military Strategy. NATO. USA (role of). 1960's-75. *247*
Warsaw Pact (negotiations). Military Strategy. Mutual Balanced Force Reductions. NATO. 1973-74. *606*
Washington (Hanford). Accountability. Nuclear waste. Safety. 1970's. *699*
WASH-1400 (report). Nuclear Power Plants. Rasmussen, Norman (interview). Safety. 1970-79. *779*
Waste, toxic. New York (Love Canal). 1894-1980. *819*
Wattenberg, Albert (reminiscences). Chicago, University of. Illinois. Nuclear chain reaction. World War II. 1942. *64*
Weapons. Air Force Association (symposium). Military Strategy. Strategic Arms Limitation Talks (II). 1976. *670*
—. Air Forces. Military Capability. 1970's. *183*
—. Annexation. Intergovernmental Relations. New Mexico (Albuquerque). Nuclear Science and Technology. 1940-78. *157*
—. Arms control. Balance of Power. Technology. 1970's. *135*
—. Arms control. Domestic Policy. USA. USSR. 1963-75. *462*
—. Attitudes. Cruise missiles. Defense Department. Politics and the Military. 1964-76. *140*
—. California (Livermore, Los Altos). Laboratories. Nuclear Science and Technology. 1942-77. *101*
—. Defense Policy. Strategic Arms Limitation Talks. USSR. 1972-78. *643*
—. Defense Policy. Technology. USSR. 1978. *283*
—. Defense spending. Disarmament. 1978. *206*
—. Military. National security. Research and development. 1978. *123*
—. Military Strategy. Sea-launched ballistic missiles. 1969-82. *187*
—. Military Strategy. Technology. 1970's. *409*
—. National Security. Politics. Strategic Arms Limitation Talks (II). 1977-78. *612*
—. Naval Vessels. 1930-76. *178*
—. Navies. *Norton Sound* (vessel). Research. Technology. 1943-79. *155*
Weapons acquisition. Defense Policy. Political Reform. 1945-75. *173*
—. Federal Policy. Politics. 1950's-70's. *115*
Weapons, conventional. Military strategy. NATO. Nuclear arms, tactical. USA. 1970's. *225*
Weapons, strategic. Defense Policy. Europe. 1973. *447*
Weiner, Charles. Letters. Nuclear Science and Technology. Oppenheimer, J. Robert (reminiscences). Physics. Smith, Alice Kimball. ca 1900-45. 1980. *47*
Wendover Field. Air Warfare. Atomic bomb. *Enola Gay* (aircraft). World War II (transportation). 1944-45. *60*
Western nations. Naval construction. USA. USSR. 1965-76. *143*
Western States. Attitudes. Decisionmaking. Nuclear power plants. State Government. Voting and Voting Behavior. 1963-78. *703*
—. Cooper, Joe. Pick, Vernon. Prospecting. Steen, Charles. Uranium rushes. 1948-57. *144*
—. Nuclear power plants. Ohio. Referendum. 1976. *812*
Westinghouse. Contracts. Economic Regulations. General Electric Co. Nuclear power plants. 1955-78. *711*
Whitmore, William F. (account). *George Washington* (vessel). Navies (Special Projects Office). Nuclear Arms. Polaris (missile). Submarine Warfare. 1955-60. *186*
Wilson, Robert R. (reminiscences). Cyclotrons. Manhattan Project. New Mexico (Los Alamos). Nuclear Science and Technology. 1942-43. *66*
Working Conditions. American Atomics Corp. Arizona (Tucson). Manufactures. Public Health. Radiation. Tritium. 1979. *709*
World government. Arms control. International Security. 1975. *524*
World order. Arms control. USA. USSR. 1950's-75. *483*
—. China. Nuclear stalemate, function of. USA. USSR. 1970's. *201*
World War I. Cannon, Clarence. House of Representatives. Mexico. Military Service. Spanish-American War. World War II. 1898-1945. *35*
World War II. Allison, Graham *(Essence of Decision)*. Committees. Decisionmaking. Foreign policy. Nuclear War. 1944-45. *54*
—. Alperovitz, Gar. Diplomacy. Historiography. Nuclear Arms. 1943-46. 1965-73. *323*
—. Alperovitz, Gar (review article). Decisionmaking. Historiography. Japan (Hiroshima). Nuclear War. 1945-65. *52*
—. Anderson, Herbert L. (reminiscences). Chicago, University of. Illinois. Nuclear chain reaction. 1942. *5*

—. Arms race. Carter, Jimmy (administration). Cruise missiles. Tomahawk (missile). 1977-80. *160*
—. Atomic bomb. Franck, James. Nuclear arms race. Political Attitudes. Truman, Harry S. (administration). 1944-45. *62*
—. Atomic bomb. Franck, James. Policymaking. Scientists. Stimson, Henry L. 1945. *58*
—. Atomic bomb. New Mexico (Los Alamos). 1943-46. *41*
—. Atomic bomb development. Diplomacy. Nuclear Science and Technology. 1942-45. *46*
—. Atomic bombs. Japan. Public opinion. USA. 1945-49. *68*
—. Bainbridge, Kenneth T. (reminiscences). Manhattan Project. New Mexico (Los Alamos). Nuclear Arms. 1945. *8*
—. Battles. Deterrence. Morale. Nuclear War. 1944-81. *130*
—. Bohr, Niels. Nuclear energy. 1943. *51*
—. Bombing. Japan. 1937-45. *18*
—. B-29 (aircraft). Japan. Lemay, Curtis. Nuclear War. Surrender. 1945. *67*
—. Cannon, Clarence. House of Representatives. Mexico. Military Service. Spanish-American War. World War I. 1898-1945. *35*
—. Censorship, voluntary. National security. Press. 1941-80. *515*
—. Chicago, University of. Illinois. Nuclear chain reaction. Wattenberg, Albert (reminiscences). 1942. *64*
—. Churchill, Winston. Diplomacy. Great Britain. Military Strategy. Roosevelt, Franklin D. 1942-44. *61*
—. Cold War. Japan (Hiroshima). Nuclear War. 1945. *31*
—. Cold War (origins). Marshall Plan. Nuclear Arms. Second Front. USSR. 1940's. *429*
—. Decisionmaking. Japan (Hiroshima, Nagasaki). Military Strategy. Nuclear War. 1944-45. *23*
—. Drones. Fahrney, Delmar S. (account). Missiles. Navy Bureau of Aeronautics. N2C-2 (aircraft). Radio. Research and development. TG-2 (aircraft). 1936-45. *107*
—. Einstein, Albert. Hungarian Americans. Nuclear Arms (program). Szilard, Leo. USA. 1941-74. *26*
—. Foreign Relations. Great Britain. Nuclear Arms. Roosevelt, Franklin D. USA. 1940-45. *16*
—. France. Great Britain. Nuclear arms testing. Secrecy. 1939-45. *25*
—. Frisch, Otto R. (account). Nuclear Physics. Nuclear War. 1945. *24*
—. History Teaching. Japan. Nuclear War. Simulation and Games. 1945. 1978. *19*
—. Japan. Leftists. Nuclear War. USSR. 1945. *17*
—. Japan. Military Strategy. Nuclear energy. Smyth Report, 1945. USA. 1940-45. *56*
—. Japan. Military strategy. Nuclear War. USA. 1945. *29*
—. Japan. Nuclear War. Politics. Surrender offer. Truman, Harry S. (administration). 1945. *12*
—. Japan (Hiroshima, Nagasaki). Military Strategy. Nuclear War. 1945. *20*
—. Japan (Hiroshima, Nagasaki). Nuclear War. 1945. *40*
—. Japan (Hiroshima, Nagasaki). Nuclear War. USA. USSR. 1942-47. *15*
—. Japan (Kyoto). Nuclear War. 1945. *45*
—. Little Theatre Group. Manhattan Project. New Mexico (Los Alamos). Nuclear Arms. Theater Production and Direction. 1943-46. *3*
—. Manley, John H. (reminiscences). New Mexico (Los Alamos). Nuclear Science and Technology. 1942-45. *36*
—. McDaniel, Boyce (reminiscences). New Mexico (Los Alamos). Nuclear Arms. Plutonium. 1940-45. *39*
World War II (aerial operations). Baldwin, Paul H. (account). Fighter squadrons, night. Nuclear War. Philippines. 1942-45. *9*
World War II (antecedents). Deterrence logic. Disarmament. War, risks of. 1933-62. *307*
World War II (Pacific Theater). Decisionmaking. Japan (Hiroshima, attack on). Nuclear War. 1944-45. *50*
World War II (review article). Alperovitz, Gar. Foreign Relations. Nuclear War. 1945. *27*
World War II (strategy). Japan (Hiroshima, Nagasaki). Nuclear War. USA. 1945. *11*
World War II (transportation). Air Warfare. Atomic bomb. *Enola Gay* (aircraft). Wendover Field. 1944-45. *60*

Y

York, Herbert. Lawrence Livermore laboratory. Nuclear arms. ca 1949-53. *191*
York, Herbert (hypothesis). Foreign policy. Hydrogen bomb testing. Nuclear Arms. USSR. 1953-55. 1970's. *84*
York, Herbert (review article). Decisionmaking. Hydrogen bomb. Nuclear Arms policy. 1949-75. *192*
Youth. Elites. National Security. Political Attitudes. 1945-75. *431*

Z

Zionism. Einstein, Albert. Germany. Nuclear arms. Pacifism. 1939-55. *21*

AUTHOR INDEX

A

Aaron, David 446
Abbotts, John 697
Abelson, Philip H. 1
Ackley, Richard T. 193 457
Adams, Gordon 194
Adamson, June 2
A.G 698
Alexander, Charles C. 195
Alfven, Hannes 72
Alger, Chadwick F. 458
Ali, Mehrunnisa 459
Alkhimenko, A. 73
Allen, H. C. 460
Allen, Scott 196
Allison, Graham 461 462
Almaráz, Felix Díaz, Jr. 3
Alsop, Joseph 445
Alsterdal, Alvar 463
Altherr, Marco 197
Althoff, Phillip 706
Alvarez, Luis W. 4
Ambri, Mariano 74 464
Anders, Roger M. 75
Andersen, A. E. 198
Anderson, Herbert L. 5 6
Anthony, Robert 699
Appleby, Charles A. 199
Arbab, John 76
Arbatov, G. A. 200 465
Arkhipov, M. 77
Aspaturian, Vernon D. 201
Aspin, Les 78
Austin, Nancy 700

B

Badash, Lawrence 7
Bailey, Martin J. 202
Bainbridge, Kenneth T. 8
Baker, Jeffrey J. W. 466
Baldwin, Paul H. 9
Ball, Desmond J. 79 80
Ballard, William T. 203
Bamière, Christine 701
Bargman, Abraham 467
Barkan, Steven E. 702
Barlow, William J. 204
Barnaby, Frank 205 206
Barnett, Lincoln 10
Barnett, Roger W. 419
Barry, Hamlet J. 81
Barry, Tom 82
Barton, John H. 468
Bates, E. A., Jr. 469
Beard, Robin 470
Beavers, Roy L., Jr. 471
Behr, Robert M. 313
Behuncik, John G. 83
Bellamy, Ian 84
Benard, Cheryl 745
Benedict, Robert 703
Bennett, Roy 472
Benoit, Emile 473
Beres, Louis René 207 474
Bernstein, Barton J. 11 12 13 14 15 16 208 209
Bertram, Christoph 475
Beukel, Erik 210
Bhatia, Shyam 476
Black, Edwin F. 85 211
Blair, Bruce G. 704
Blechman, Barry M. 212 213
Bloomfield, Lincoln P. 214
Blosser, Henry G. 86
Boitsov, M. 87
Boller, Paul F., Jr. 17
Bone, Hugh 703
Bonetti, Alberto 705
Bonnemaison, Jacques 477
Booth, Kenneth 478
Borowski, Harry R. 88
Borst, Gert 89 215
Boskma, Peter 171
Boulding, Elise 458
Bowman, Richard C. 216
Božić, Nemanja 217 479
Brach, Hans Günter 218
Brady, David W. 706
Brauzzi, Alfredo 90 91
Brenner, Michael 480
Brewer, Garry D. 704
Brim, Raymond E. 92
Brodie, Bernard 219 481
Bromberg, F. 707
Bromet, Evelyn 708
Bromwich, David 822
Brooks, Bill 791
Brooks, Harvey 220
Brown, Harrison 482
Brown, William D. 93
Bruce, Geoffrey F. 221
Brucer, Marshall 709
Bruer, Patrick J. 806
Bull, Hedley 483 484 485
Bundy, McGeorge 94 486 487
Bupp, Irvin C. 710
Burke, Gerard K. 222
Burlingham, Bo 793
Burness, H. Stuart 711
Burns, Richard Dean 488 489
Burshop, E. H. S. 223
Burt, Richard 490 491
Buteux, Paul 224
Byers, R. B. 492

C

Caldwell, Lawrence T. 493 494
Camilleri, J. A. 712
Campbell, John 495
Canby, Steven L. 225 226
Caracciolo, Roberto 496
Carlin, Robert J. 227
Carnesale, Albert 513
Carreras Matas, Narciso 228
Carter, Barry 173 229 497 498
Carter, Jimmy 499
Cavers, David F. 95
Chalfont, Lord 500
Chandra, Romesh 693
Chayes, Abram 501
Chester, Conrad V. 502
Chistiakov, I. 96
Christopher, Robert C. 503
Clark, Wesley K. 230
Clarke, Duncan L. 274 504 505 506 507
Clemens, Walter C., Jr. 508
Cochran, Thomas B. 797
Cockburn, Alexander 97
Cockburn, Andrew 97
Cohen, Bernard L. 713 714
Cohen, S. T. 85 98 231
Cohen, Stuart A. 509
Coleman, Edwin 80
Coles, Harry L. 232
Collins, John M. 233
Combs, James E. 773
Comey, David Dinsmore 715
Commager, Henry Steele 234
Condon, Patricia 92
Connolly, Peter 822
Connor, James E. 716
Cook, Earl 717
Coulam, Robert F. 99
Cowhey, Peter F. 235
Crandell, Michael 779
Critchley, Julian 510
Crosby, Ralph D., Jr. 236

D

Daneke, Gregory A. 718
Daniels, Gordon 18
Davis, Lynn Etheridge 100
Day, Samuel H., Jr. 101
DeForth, Peter W. 237
Delbourg, Denis 511
DelSesto, Steven L. 719
Denis, Jacques 238
DePuy, William E. 239
Destler, I. M. 512
DeVolpi, A. 720
Dimsdale, Joel E. 34
Dine, Thomas A. 102
Divine, Robert A. 240
Dobney, Frederick J. 103
Donato, Alberto 241
Dornan, Robert K. 242
Doty, Paul 513 514
Dumas, Lloyd J. 243
Dunn, Leslie 708
Dunn, Lewis A. 244

E

Easterbrook, Gregg 104
Eayrs, James 105
Eberhard, Wallace B. 515
Eckhardt, William 516
Eggleston, Noel C. 19
Einhorn, Robert J. 517
Ekebjär, Göran 20
Ellsberg, Daniel 245
Emel'ianov, V. 246
Enthoven, Alain C. 247
Epstein, Edward Jay 518
Epstein, William 248 519
Espiell, Hector Gros 520
Etzold, Thomas H. 249 256 521
Evgen'ev, G. 522

F

Fahl, Gundolf 523
Fahrney, D. S. 106
Fahrney, Delmar S. 107
Falk, Richard A. 524 525
Farrell, Thomas B. 721
Fascell, Dante B. 526
Faulkner, Peter 722
Feiveson, H. A. 250
Feld, Bernard T. 21 527 528 529
Fialka, John J. 251
Finan, J. S. 530

Fitch, Val L. 22
Flanagan, Stephen J. 531
Flood, Michael 723
Flores Pinel, Fernando 252
Floyd, David 532
Fox, Daniel J. 312
Freedman, Lawrence 23 253 254
Friedberg, Aaron L. 255
Friedman, Sharon M. 724
Frisch, Otto R. 24
Furems, M. 108

G

Gaboury, L. R. 256
Gall, Norman 257
García Ferrando, Manuel 725
Garigue, Philippe 258
Garn, Jake 533
Garner, William V. 534
Garrett, Stephen A. 535
Garthoff, Raymond L. 259 536 537 538
Garwin, Thomas M. 394
Gelb, Leslie H. 539
Gelber, Harry G. 540
Gelbond, Florence 109
Geneste, Marc 260
George, James L. 541
Gerber, Larry G. 110
Gergorin, Jean-Louis 542
Geyer, Alan 543
Ghêbali, Victor-Yves 544
Gilinsky, Victor 726 824
Giorgerini, Giorgio 111
Gnevushev, N. 261
Goldhamer, Herbert 262
Goldschmidt, Bertrand 545
Gontaev, A. 263
Goodnight, G. Thomas 721
Gordon, Suzanne 727
Gorshkov, S. G. 264
Gowing, Margaret 25
Gravel, Mike 728
Gray, Colin S. 112 113 114 265 266 267 268 269 270 271 272 273 546 547 548 549 550 551
Gray, Robert C. 115
Graybar, Lloyd J. 116
Greb, G. Allen 190
Green, Harold P. 117 729
Green, William C. 552
Greene, Mark R. 730
Greenwood, John T. 118
Grieco, Joseph M. 274 553
Griffith, Robert 275
Griffith, William E. 276
Gurney, Ramsdell, Jr. 277

H

Haakonsen, Per 278
Hagan, Kenneth J. 279
Halász, Nicholas 26
Halász, Robert 26
Hall, Gus 554
Hall, R. Cargill 119
Halperin, Morton 280 445
Hamburg, Roger 281
Hamilton, John A. 555
Hamilton, Mary A. 731
Hammond, Thomas T. 27
Hansen, Chuck 120
Harnik, Peter 732
Harrington, Anne 121
Harrington, Michael 282 822
Hart, Douglas M. 213
Hartley, Anthony 556
Head, Richard G. 283
Held, Margery 786
Helms, Robert F., II 284
Herken, Gregg 122
Herzfeld, Charles M. 123
Heurlin, Bertel 285
Hewlett, Richard G. 28 733
Hill, Gladwin 734
Hoeber, Amoretta M. 286 288
Hoeber, Francis P. 287 288
Hoffman, Walter 557
Holder, William G. 124
Holdren, John P. 735
Hollist, W. Ladd 289
Holst, Johan Jorgen 446
Hopmann, P. Terrence 558
Horder, Mervyn 29
Howard, Michael 559
Howe, Irving 822
Huff, Rodney L. 290
Hughes, Peter C. 588
Huisken, Ron 125
Huntington, Samuel P. 291
Hyland, William G. 560

I

Ichord, Robert F., Jr. 736
Ignatieff, George 561
Iklé, Fred Charles 126 292 444 562 563 564 565
Inglis, Alex I. 42
Ingram, Timothy H. 737
Ioisysh, A. I. 738

J

Jackson, Henry M. 566
Jackson, William D. 293 294 567
Jacoby, Henry D. 614
Jain, J. P. 568
Jaipal, Rikhi 456
Janke, Peter 739
Jeffries, Vincent 295
Jenson, John W. 296
Jervis, Robert 297
Jezer, Marty 740
Johnson, Giff 127
Johnson, M. Glen 298
Johnston, Whittle 569
Jones, David C. 128
Jones, Kenneth Macdonald 299
Jones, Thomas P., Jr. 313
Juda, Lawrence 570

K

Kade, Gerhard 693
Kaiser, Karl 571 572
Kamen, Martin D. 30
Kaplan, Barbara G. 786
Kaplan, Fred 300
Kaplan, Stephen S. 212
Kapur, Ashok 301
Karenin, A. 573
Karp, Walter 129
Kaspi, André 31
Katz, Neil H. 741
Keating, William Thomas 742
Keegan, John 32 130
Kelleher, Catherine McArdle 302
Keller, Bill 131
Keller, Edward B. 743
Kemeny, John G. 744
Kennedy, Edward M. 574
Kevles, Daniel J. 132
Khalilzad, Zalmay 745
Kincade, William H. 133 575
Kinnard, Douglas 303
Kipp, Jacob W. 279
Kirgis, Frederic L., Jr. 304
Kiselyak, Charles 576
Kistiakowsky, G. B. 577 578
Klare, Michael T. 305
Klein, Jean 306
Klein, Jeffrey S. 746
Knight, Jonathan 307
Kohl, Wilfrid L. 579
Kokoshin, A. 397
Kolowicz, Roman 308
Kopkind, Andrew 747
Kopp, Carolyn 134
Korb, Lawrence J. 309 580
Kortunov, Sergei Vadimovich 317
Kowarski, L. 748
Kreith, Frank 188
Krell, Gert 135 310
Krepon, Michael 136
Krieger, David M. 749
Kriesberg, Louis 581
Krohn, Richard W. 582
Kruger, Lewis 786
Kruzel, Joseph 583 584
Kublig, Bernd W. 311
Kugler, Jacek 312
Kupperman, Robert H. 313
Kurenkov, Iu 707
Kuter, Laurence S. 137
Kuz'min, I. 138

L

La Tournelle, Guy de 314
Laitin, David D. 235
Lamb, John 585
Lambelet, John C. 315
Lanouette, William J. 750 824
Lapp, Ralph H. 751
Lauren, Paul Gordon 316 391
Leavel, Willard 703
Lebedev, Nikolai Ivanovich 317
Lee, Kai N. 752
Legault, Albert 586 587
Lehman, Christopher M. 588
Lehman, John F., Jr. 589
Leitenberg, Milton 139
Lellouche, Pierre 590 753
Leopold, Richard W. 33 754
Lester, Richard 755
Levine, Henry D. 140
Lifton, Robert Jay 34
Light, Alfred R. 756 757 758
Lilley, Stephen Ray 35
List, David C. 741
Listerud, Gunnar 591
Lodal, Jan M. 318 592 593
Long, Clarence D. 456
Long, F. A. 594
Lovins, Amory B. 759
Lovins, L. Hunter 759
Luck, Edward C. 595
Lugato, Giuseppe 319
Luttwak, Edward N. 320 596
Lynn, Laurence E., Jr. 321
Lyons, W. C. 98
Lyth, Einar 322

M

Mabbutt, Fred R. 760
Macdonald, Hugh 141
Mack, Charles 597
Maddox, Robert James 323
Makins, Christopher J. 598
Maland, Charles 599
Mandelbaum, Michael 324 600
Mandell, Brian 585
Manley, J. H. 36
Mann, Robert 814
Marawitz, Wayne L. 325
Mark, Eduard 37
Markov, M. A. 38
Markowitz, Gerald E. 142
Marrett, Cora Bagley 761
Marshall, Charles Burton 601
Marston, Philip M. 815
Martin, J. J. 326
Martin, Laurence 327
Marx, Leo 762
Mayer, Laurel A. 328
Maynes, Charles William 329
Mazuzan, George T. 763
MccGwire, Michael 143
McCormick, Gordon H. 330
McCracken, Samuel 764 765
McDaniel, Boyce 39
McDonald, W. Wesley 331
McGlinchey, Joseph J. 332
McNeill, William H. 333
Meeropol, Michael 142
Melby, Svein 334
Menchén Benítez, Pedro 602
Mets, David R. 603
Meyer, Larry L. 144
Meyer, Stephen M. 145 604
Migolatyev, A. 605
Mikulín, Antonín 335
Miller, Kenneth G. 336
Miller, Mark E. 330
Millett, Stephen M. 337
Minter, Charles S. 606
Mitchell, Robert Cameron 766
Molineu, Harold 338 607
Monin, M. 339
Montgomery, W. David 711
Moore, David W. 608
Moore, William C. 340
Morgenthau, Hans J. 341
Morris, Frederic A. 462
Morton, Louis 40
Moylan, Maurice P. 767
Mueller, John E. 146
Mulkin, Barb 41
Munro, John A. 42
Muravchik, Joshua 609
Murphy, Dervla 768
Murry, William V. 342
Myrdal, Alva 343 446

N

Nacht, Michael 344 345 446 513 610 611
Nash, Henry T. 147
Naumov, P. 612
Navrozov, Lev 613
Neal, Fred Warner 346
Neff, Thomas L. 614
Nelkin, Dorothy 769 770 823
Nelson, Jon P. 771
Neubroch, H. 615
Neumann, Gerd 347
Newcomb, Richard 772
Newcombe, Alan G. 616
Nikolayev, Y. 617
Nimmo, Dan 773
Nitze, Paul H. 348 445 446 618
Noorani, A G. 774
Norton, Augustus R. 349
Novikov, N. 160 170
Nunn, Jack H. 148
Nye, Joseph S. 619 620
Nye, Joseph S., Jr. 621 622
Nyström, Sune 350 351

O

Obrador Serra, Francisco 352
O'Connor, Raymond G. 43
Ognibene, Peter J. 149
Oppenheimer, J. Robert 44
Organski, A. F. K. 312
Orr, David W. 822
Osgood, Robert E. 353
Otis, Cary 45
Outrey, Georges 354
Overton, Jim 775

P

Paarlberg, Rob 150
Paine, Christopher 151
Panofsky, Wolfgang K. H. 623
Parrish, Michael E. 152
Pastusiak, Longin 624
Paul, Ron 625
Payne, Keith 272
Peck, Earl G. 153
Perrow, Charles 776
Perry, Harry 777
Petrovski, V. F. 626 627 628
Phillips, Dennis 355
Pierre, Andrew J. 629
Pipes, Richard 356
Pogodin, V. 630
Polmar, Norman 154 155
Polyanov, N. 357
Pomerance, Jo 631
Porro, Jeffrey D. 358
Posen, Barry R. 359
Poss, John R. 156
Postbrief, Sam 360
Power, Paul F. 361
Preda, Eugen 46
Preston, Adrian 362
Primack, Joel 778
Prince, Howard T., II 632

Q

Quaroni, Pietro 363
Quester, George H. 364 365 454
Quirk, James P. 711

R

Rabinowitz, Howard N. 157
Rakitin, E. 158
Ramberg, Bennett 633
Ramsey, Bill 159
Ranger, Robin 634 635 661
Rasmussen, Norman 779
Rathjens, G. W. 366 636 655
Ravenal, Earl C. 367
Record, Jeffrey 368 369
Redick, John R. 637
Reed, Adam V. 780
Reed, Robert H. 370
Reingold, Nathan 47
Reynolds, William 781 782
Rhodes, Richard 48 783
Rhodes, Suzanne 784
Rice, Ross 703
Richardson, Robert A. 785
Rochlin, Gene I. 786
Rodionov, B. 160
Roschlau, Wolfgang 371
Rosecrance, Richard 372
Rosenberg, David Alan 161 162 373
Ross, Leonard 759
Rowny, Edward L. 638
Ruehl, Lothar 444 639
Rumsfeld, Donald H. 374
Rushford, Greg 163
Russett, Bruce M. 375
Rycroft, Robert W. 787

S

Sakamoto, Yoshikazu 640
Salaff, Stephen 49 164
Salomon, Michael D. 376
Santoni, Alberto 50
Schelling, Thomas C. 641
Schilling, Warner R. 100 377
Schlafly, Phyllis 642 643
Schleimer, Joseph D. 788
Schlesinger, James R. 378 379
Schneider, Barry 165
Schneider, Mark B. 380
Schneider, Steven A. 789
Schorr, Brian L. 644
Schroeder, Pat 381
Schulze, Ludwig 382
Schütze, Walter 166
Schwartz, Charles 167 383
Schwellen, Joachim 168
Scott, William F. 384
Scoville, Herbert, Jr. 385 645
Seelig, Jakob W. 332
Semenishchev, Iu. P. 51
Seryogin, I. 790
Shapely, Deborah 646
Shapiro, Edward S. 52
Sharp, Jane M. O. 386 647
Sherwin, Martin Jay 53
Shields, Mitchell J. 791
Shrivastava, B. K. 648
Shulman, Marshall D. 649 650
Sigal, Leon V. 54
Sills, David L. 823
Simon, Anne L. 651
Simpson, John A. 792
Singh, L. P. 387
Skinner, Scott 793
Skogan, John Kristen 388
Skogmar, Gunnar 794
Slay, Alton D. 169
Smart, Ian 389 652
Smirnov, A. 170
Smit, Wim A. 171
Smith, A. Robert 653
Smith, Alice Kimball 55
Smith, Datus C., Jr. 172
Smith, Gaddis 390
Smith, Gerald C. 654
Smith, Gerard C. 655 656
Smith, Theresa C. 558
Smoke, Richard 391
Smyth, H. D. 56
Snelders, H. A. M. 57
Soapes, Thomas F. 657
Soutou, Georges-Henri 658
Spector, Judith A. 795
Speth, J. Gustave 796 797

Stakh, G. 659
Stashevski, S. 659
Staudenmaier, William O. 392
Steinbruner, John D. 173 393 394
Steiner, Arthur 58
Steiner, Barry H. 660
Stevenson, Adlai E., III 454
Stewart, Larry R. 798
Stockman, David A. 799
Stone, Jeremy J. 445 578
Struck, Myra 395
Stubbs, Richard 661
Stupak, Ronald J. 328
Sussman, Leonard R. 174
Sviatov, G. I. 396 397
Swain, Bruce M. 800
Sweet, William 662 801
Sylves, Richard T. 802
Szilard, Leo 59

T

Talbott, Strobe 663
Tamplin, Arthur R. 797
Tell, Geoffrey 175
Temples, James R. 803
Terzibaschitsch, Stefan 176 177 178
Thee, Marek 664
Thomas, James A. 398
Thompson, Kenneth W. 399
Tiezzi, Enzo 804
Toinet, Marie-France 665 666
Tomilin, Y. 667
Tonelson, Alan 668
Toner, James H. 400
Torkelson, Richard 819
Tow, William T. 401
Treverton, Gregory F. 402
Trofimenko, G. A. 403
Trofimenko, Henry A. 404
Tucker, Robert W. 405
Tumkovskii, R. G. 669

U

Ulsamer, Edgar 179 180 181 182 183 406 407 408 409 410 411 412 413 414 415 416 417 670 805
Urban, Josef 418 671

V

Vallaux, François 672
Van Cleave, William R. 419
VanEvera, Stephen W. 359
Vardamis, Alex A. 420
Vernant, Jacques 421 422 673 674
Vidich, Arthur J. 423
Vig, Norman J. 806
Viktorov, V. 675
Villa, Brian Loring 61 62
VonHippel, Frank 778
vonHippel, Frank 807
Vorontsov, G. A. 676
Vukadinović, Radovan 677

W

Walker, George K. 696
Walker, J. Samuel 808 809
Wallace, J. F. 424
Wallace, Michael D. 425 426
Walsh, Edward J. 810
Walter, Franz 89
Walters, Ronald W. 427 428
Warnecke, Steven J. 678
Warnke, Paul C. 446 696
Warren, Shields 63
Wattenberg, Albert 64
Webb, B. H. 429
Weinberg, Alvin M. 811
Weiner, Charles 55 65
Weisgall, Jonathan M. 184
Welch, Richard E., Jr. 679
Wells, Samuel F., Jr. 430
Wenner, Lettie McSpadden 812
Wenner, Manfred W. 812
Werner, Roy A. 431 680
Werrell, Kenneth P. 185
Wessell, Nils H. 681
West, Ronald E. 188
Westervelt, Donald R. 682
Wheeler, Keith 813
Whitmore, William F. 186
Wigner, Eugene P. 502
Wilkes, Owen 814
Willhelm, Sidney M. 432
Willrich, Mason 815 816
Wilson, Carroll L. 817
Wilson, Robert R. 66
Winne, Clinton H., Jr. 683
Winters, Francis X. 818
Wit, Joel S. 187
Wohlstetter, Albert 433 434 435 436 445 446
Wolfe, Bertram 684
Wolfe, Gary K. 437
Wolk, Herman S. 67
Wood, Robert S. 438
Woolsey, R. James 685
Worthley, John A. 819
Wrenn, Catherine 188
Wright, Frank 189
Wyle, Frederick 686

Y

Yanarella, Ernest 820
Yankelovich, Daniel 687
Yavenditti, Michael J. 68 69 70
Yochelson, John 688 689
York, Herbert F. 71 190 191 439 440
Young, Wayland 690

Z

Zacharias, Jerrold R. 192
Zhigalov, I. I. 441
Zinberg, Dorothy 821
Zumwalt, Elmo R. 442

LIST OF ABBREVIATIONS

A.	Author-prepared Abstract
Acad.	Academy, Academie, Academia
Agric.	Agriculture, Agricultural
AIA	Abstracts in Anthropology
Akad.	Akademie
Am.	America, American
Ann.	Annals, Annales, Annual, Annali
Anthrop.	Anthropology, Anthropological
Arch.	Archives
Archaeol.	Archaeology, Archaeological
Art.	Article
Assoc.	Association, Associate
Biblio.	Bibliography, Bibliographical
Biog.	Biography, Biographical
Bol.	Boletim, Boletin
Bull.	Bulletin
c.	century (in index)
ca.	circa
Can.	Canada, Canadian, Canadien
Cent.	Century
Coll.	College
Com.	Committee
Comm.	Commission
Comp.	Compiler
DAI	Dissertation Abstracts International
Dept.	Department
Dir.	Director, Direktor
Econ.	Economy, Econom-.
Ed.	Editor, Edition
Educ.	Education, Educational
Geneal.	Genealogy, Genealogical, Genealogique
Grad.	Graduate
Hist.	History, Hist-.
IHE	Indice Historico Espanol
Illus.	Illustrated, Illustration
Inst.	Institute, Institut-.
Int.	International, Internacional, Internationaal, Internationaux, Internazionale
J.	Journal, Journal-prepared Abstract
Lib.	Library, Libraries
Mag.	Magazine
Mus.	Museum, Musee, Museo
Nac.	Nacional
Natl.	National, Nationale
Naz.	Nazionale
Phil.	Philosophy, Philosophical
Photo.	Photograph
Pol.	Politics, Political, Politique, Politico
Pr.	Press
Pres.	President
Pro.	Proceedings
Publ.	Publishing, Publication
Q.	Quarterly
Rev.	Review, Revue, Revista, Revised
Riv.	Rivista
Res.	Research
RSA	Romanian Scientific Abstracts
S.	Staff-prepared Abstract
Sci.	Science, Scientific
Secy.	Secretary
Soc.	Society, Societe, Sociedad, Societa
Sociol.	Sociology, Sociological
Tr.	Transactions
Transl.	Translator, Translation
U.	University, Universi-.
US	United States
Vol.	Volume
Y.	Yearbook

APR 3 0 1984
APR 2 7 1987